Energy and Entropy

Energy and Entropy

A Dynamic Duo

Harvey S. Leff

CRC Press
Taylor & Francis Group
Boca Raton London New York

CRC Press is an imprint of the
Taylor & Francis Group, an **informa** business

First edition published 2021
by CRC Press
6000 Broken Sound Parkway NW, Suite 300, Boca Raton, FL 33487-2742

and by CRC Press
2 Park Square, Milton Park, Abingdon, Oxon, OX14 4RN

© 2021 Taylor & Francis Group, LLC

CRC Press is an imprint of Taylor & Francis Group, LLC

Library of Congress Cataloging-in-Publication Data
Names: Leff, Harvey S., 1937- author.
Title: Energy and entropy : a dynamic duo / Harvey S. Leff.
Description: Boca Raton : CRC Press, 2020. | Includes bibliographical
    references and indexes.
Identifiers: LCCN 2020015149 | ISBN 9780367349066 (paperback) |
    ISBN 9780367351410 (hardback) | ISBN 9780429330018 (ebook)
Subjects: LCSH: Thermodynamics. | Entropy. | Force and energy.
Classification: LCC QC311 .L347 2020 | DDC 536/.7--dc23
LC record available at https://lccn.loc.gov/2020015149

ISBN: 9780367351410 (hbk)
ISBN: 9780367349066 (pbk)
ISBN: 9780429330018 (ebk)

Typeset in SFRM1000
by Nova Techset Private Limited, Bengaluru & Chennai, India

To my wife, Ellen
who has endured my obsession to understand
thermodynamics and entropy for more than sixty years.
Her love and compassion have kept me going.

Contents

CHAPTER 6 ▪ Numerical Entropy   179

CHAPTER 7 ▪ Language & Philosophy of Thermodynamics   195

CHAPTER 8 ■ Working, Heating, Cooling 213

Preface

The primary goal of this book is to illuminate energy and entropy, the two pillars of thermodynamics, and to clarify their essential linkage. The science of thermodynamics is generally considered to be the study of energy storage within, and transfers between, macroscopic objects. The existence of *stored* internal energy distinguishes thermodynamics from classical mechanics, which focuses on imagined point particles and rigid bodies. Closely linked with internal energy is entropy. Cooling an object decreases its entropy as its internal energy decreases. When heated, its entropy increases as its internal energy increases. These experiments show that thermodynamic entropy is inextricably linked to energy. Recalling the line in the old song *Love and Marriage*, "You can't have one without the other," I chose the title *Energy and Entropy: A Dynamic Duo.*[1]

Rudolf Clausius, who conceived entropy, acknowledged the importance of *both* energy and entropy, writing that *the energy of the universe is constant, and the entropy of the universe continually increases.* As *any* macroscopic physical process occurs, e.g., an automobile engine running, a star in the cosmos exploding, or any animal breathing, there is a continual redistribution (spreading) of internal energy, and the mysterious thing called the *entropy of the universe* continually increases. According to Clausius, it *cannot* decrease.[2]

Internal energy and entropy are properties of any object large enough to be seen with the naked eye. Numerical values of entropy for elements and compounds have been tabulated in thermodynamics databases. Because of its non-decreasing property for an isolated system, entropy has puzzled many students, teachers, and researchers. People have tried to find a simple way to describe and understand entropy. One that has caught on is that entropy is a measure of *disorder*, which unfortunately misses the energy–entropy linkage, and is deficient in other ways (see Sec. 7.3). In what follows, I discuss other entropy metaphors and stress that any acceptable metaphor should reflect the interconnection of energy and entropy.

---

[1] This linkage is challenged in some cases. Superfluid ^{4}He at temperatures below 2.172 K has entropy ~ 0 but nonzero zero-point internal energy.

[2] From Clausius onward, the principle of entropy increase has been applied to the universe, which is *not* in thermodynamic equilibrium. Strictly speaking, this requires a definition of non-equilibrium entropy, which Clausius did not have.

My target audience is (i) teachers of physics, chemistry and engineering; (ii) college science and engineering students; (iii) scientists and engineers who seek a fresh view of thermal physics concepts, and (iv) interested readers who have sufficient maths and physics backgrounds. To help make the manuscript more accessible to a diverse audience, I use images, graphs, and analogies to help get important ideas across. To minimise the need to work through the mathematics by those who have little interest in that, I have included *Key Points* throughout each chapter, emphasising important concepts and results. Where possible, I emphasise *concepts*, recalling the dictum, "Physics should be presented as simply as possible, but no simpler."[3]

The first two chapters, which introduce the concept of energy and how it leads to the need for entropy, should be accessible to the full target audience. Chapters 3 and 4 entail more substantial mathematics, especially Sec. 4.8 on the Jarzynski equality, while Ch. 5, on radiation and the photon gas, Ch. 6, on numerical entropy, and Ch. 7, on language in thermodynamics, are less mathematical, albeit with some sophisticated concepts. Some of Ch. 8 on working, heating and cooling will be familiar to many readers. Parts of Ch. 9, on the sanctity of the second law of thermodynamics, are largely qualitative and accessible to a wide audience. For readers seeking a roadmap, Ch. 1 through Sec. 3.3 plus Chs. 7 and 9, contain the main ideas and some applications.

My own road to this topic was a rocky one, and I understand how frustrating thermodynamics can be. Entropy was not even mentioned in my first college introductory physics textbook, *University Physics* (1955) by Francis Sears and Mark Zemansky. Entropy *was* covered (though alas, not *uncovered*) in a junior-level course on thermodynamics, and my first exposure to it produced an uncomfortable level of confusion.

In graduate school, I learned the mathematical framework of statistical mechanics, and was able to derive expressions for the energy and entropy of well-defined models of gases and solids. I still had a nagging feeling that more could be known about entropy. Later I learned about information theory, which enabled the replication of much of statistical mechanics, but with a twist: entropy appears as a measure of *missing information* or, in other words, *uncertainty*. This provided the first clue as to what entropy might *mean*. In the years since, I've pondered, taught and written about energy and entropy. In this monograph, I share some of the things I've learned that have helped me better understand energy and entropy. I truly hope this collection of thoughts will help readers gain a better understanding of the richness and beauty of thermodynamic entropy and its linkage with energy.

There exist a variety of books on entropy and/or energy, but none of them reveals in a compelling way the essential relationship between entropy and

---

[3]This is patterned after a quotation attributed by some to Albert Einstein. The point is that thermodynamics is rooted in mathematics. I found it impossible to present the topic solely qualitatively without oversimplification.

energy. This monograph is intended to fill that void, albeit *not* to replace other works. If you have questions or comments as you read this, please write me at hsleff@cpp.edu. I would love to hear from you.

Harvey S. Leff
Portland, Oregon
March 2020

Acknowledgments

I am grateful to the Department of Physics and Astronomy at Cal Poly Pomona for providing me with the opportunity to teach thermodynamics and statistical mechanics numerous times. It was primarily during that teaching that I puzzled over and digested many of the subtleties and nuances of those disciplines. I am also indebted to Reed College for the opportunity to be a Visiting Scholar in the Physics Department, 2010-2020, during which much of this book was written. The libraries at Cal Poly Pomona and Reed College were invaluable to me. I am grateful to many colleagues, too numerous to identify, with whom I discussed the key ideas in this book. I also wish to acknowledge the American Association of Physics Teachers (AAPT), whose *American Journal of Physics* and *Physics Teacher* taught me much and provided an opportunity to publish articles for teachers and students. I learned much and shared my findings at AAPT's local and national meetings.

I thank the following people, who bravely read various stages of the manuscript and provided me with invaluable suggestions for improvement: Richard Kaufman, Don Lemons, Carl Mungan, Andrew Rex, and Daniel Schroeder. I am indebted to my wife Ellen, who proofread the entire manuscript. Of course, I am solely responsible for any remaining errors or lack of clarity.

The manuscript was prepared with the superb Mac LaTeX editor Texpad, which helped ease the writing process.

CHAPTER 1

Energy is Universal

CONTENTS

1.1 MYSTERIOUS INVISIBLE ENERGY

Everything that *happens* in the universe entails energy. There are various forms of stored energy, namely, the energy of motion (kinetic energy), energy due to electromagnetic forces (EM energy), energy due to gravitation (gravitational energy), and energy associated with nuclear forces. Nonzero kinetic energy occurs for a moving object that travels from one place to another. It also is involved when an object spins and/or vibrates. Electromagnetic energy binds electrons to nuclei in atoms, and nuclear energy holds nuclei together.

Energy release is associated with chemical bonds when the product compounds in a chemical reaction have lower total energy. Examples include combustion of coal, natural gas, petroleum, wood, and candle wax. As I shall explain, the stored energy is *negative*, and energy release is the result of even more negative energies stored in the combustion products, primarily CO_2 (carbon dioxide) and H_2O (water).

Invisible energy is ubiquitous in the universe. It enables plants to grow and animals to live. It resides in solids, liquids, gases, plasmas, and in fact just about everything. Outside Earth, energy drives the birth, life, and death of each star, which stores and converts enormous amounts of energy into light, heat, neutrinos and other particles. Energy transformations run the universe.

Thermodynamics is the science devoted to the study of energy and its transformations within and between *macroscopic* bodies. These transformations typically entail spatial redistributions of energy. The term *macroscopic* connotes virtually any matter that has a very large number of molecules. Typically this number is of order 10^{20} or greater. Examples include a human hair, speck of dust, paper clip, horseshoe magnet, human body, the air in a room, an automobile engine, multi-story building, planet, star, galaxy, and even the whole universe.

Humans' digestive systems transform energy derived from their food and the oxygen breathed in, enabling their hearts to pump blood, their muscles to work, and their lungs to continually take in oxygenated air and expel carbon dioxide. Energy is delivered to our homes via electricity and (typically) fossil fuels, along with the oxygen in the air. That energy is used to perform functions that keep us warm in winter and cool in summer, keep foods cold, cook meals, run our computers, power our kitchen appliances and TV, and generally make our lives more comfortable.

Thermodynamics is extraordinarily important, not just to humans, but to all living things. All of these depend on a continual flow of energy *to* them from nutrients and oxygen needed to digest them, and solar radiation to sustain their lives, and *from* them as they dump energy to their surroundings.

1.1.1 Internal energy

Gaining an understanding of energy stored within matter is a fundamental goal of this book. Students of beginning physics learn about the *kinetic energy* of non-existent point-like particles (that occupy zero volume) and also about kinetic energy for objects such as disks and spheres, which can move from place to place and also rotate.

Typically there is very little, if any, attention given to the energy stored *within* materials. For example, the energy stored by a pot of water increases when it is heated on a stove top. We cannot see that energy, but we are aware that the object gets hotter. We can even *feel* that it is hot by bringing our hand near the pot. Although we do not seem to be touching anything, we sense warmth. What we feel is the effect of invisible energy, *infrared radiation* being emitted from the pot and traveling to our skin. Further, an electric stove's heating coil gets so hot that it glows, as illustrated in Fig. 1.1. In that case,

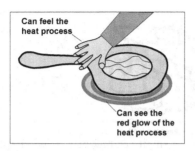

Figure 1.1: A pot of boiling water on an electric stove top. A nearby hand feels the heat, though seemingly touching nothing, and our eyes can see the glowing, red-hot burner coil.

our eyes *can* detect that there is energy being emitted by the red-hot coil. If the water were allowed to boil away, the pot itself might begin to glow!

Key Point 1.1 *Macroscopic bodies store internal energy, which is denoted here by E. Suppose a system has zero net energy flow through its boundary, and unchanging pressure, temperature and any other measurable variables. This quiescent state is called thermodynamic equilibrium. In such an equilibrium state, the internal energy typically can be written as $E(T, V, N)$, a function of the system's temperature T, volume V and number of molecules N.*

When a liquid like water is heated, its energy increases and some of the water molecules gain sufficient kinetic energy to break out of the liquid, to the air above it. At each temperature, the equilibrium water vapour pressure has a well-defined value. When the water temperature reaches $100\,°\mathrm{C}$, its vapour pressure equals the (assumed) normal atmospheric pressure, and boiling ensues. While the liquid water has a fixed volume for any given temperature and pressure, the water vapour takes on the shape of its container. Although we can see boiling, we never see the invisible energy.

If on the other hand, water at room temperature is cooled sufficiently, say, to $0\,°\mathrm{C}$, enough energy is removed from the system that the forces between the water molecules can bond strongly with nearby molecules, forming solid ice that maintains a fixed, rigid shape. And although we can see the difference between ice and water, we do not see the energy directly.

Yet that energy is responsible for important behaviour patterns of water. In order to understand this further, one needs to appreciate some of the rudiments of thermodynamics. Thermodynamicists are like detectives, trying to pin down properties that entail invisible energy. Simplified depictions of a solid and gas are in Fig. 1.2.

Key Point 1.2 *Solids, liquids and gases consist of many individual atoms. Those atoms and the molecular structures they form can jiggle, rotate, vibrate and move from place to place, and they store both kinetic and potential energy.*

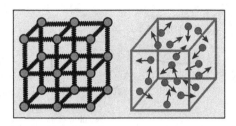

Figure 1.2: Depictions of small regions of a fixed-shape solid (left) and gas (right), which assumes its container's shape.

The sum of these two stored energies is the internal energy E that is postulated in thermodynamics.

The atoms of the solid behave roughly as if they are connected by springs. The forces on each atom come largely from attractions with nearest-neighbour atoms. If for example an atom moves slightly leftward, it gets closer to the atom to its left and farther from the atom to its right. Because of this, it is pushed rightward by the left atom and pulled in the same direction by the right atom. That is, it is repelled by the left atom and attracted by the right one. This results in each atom having a *home* location, a lattice site, about which it vibrates. Figure 1.3 shows the typical force between two atoms as a function of their separation distance.

The intermolecular force graph in Fig. 1.3(a) represents the force on a test molecule when it interacts with a fixed molecule at the origin of the coordinate system, where the separation distance is zero. For the very smallest positive separations (i.e., to the left of the dashed line), the force on the particle of the curve is positive, namely, to the right. This is a *repulsive* force, pushing the molecule of interest away from the molecule at the origin. For larger distances, corresponding to points on the green part of the curve, the force is negative, namely, to the left, or *attractive*. At the point that separates the negative and positive force regions, the force between molecules is zero. The repulsive and attractive regions are shown in Fig. 1.3(b). The atoms of a gas are free to move throughout the container. If the gas is *dilute*, the atoms interact with other atoms only occasionally. For sufficiently low temperatures, the attractive part of the force is responsible for *condensation*, where the gas becomes a liquid. This is responsible for the common atmospheric phenomenon of rain.

Key Point 1.3 *In a system with many molecules, the net force on any molecule comes from the sum of all the forces of multiple molecules near it. The result is that at sufficiently low temperatures and small enough volumes, each molecule tends to remain near the point where the forces on it add to zero. It then jiggles in a small region, labeled "jiggle region" in Fig. 1.3(a). This picture leads to a rough understanding of how liquids and solids can exist. In contrast, at sufficiently high temperatures and low enough densities (i.e., large enough*

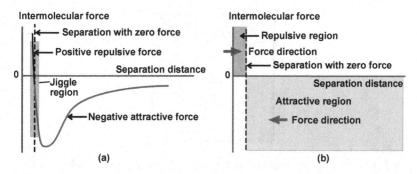

Figure 1.3: (a) Intermolecular force on a test molecule whose position is to the right of the vertical (force) axis, by a molecule at the origin of the coordinate system. (b) Attractive, repulsive and zero-force regions for the test particle.

volumes), molecules can move much farther from one another, and form a gas. The attractive forces between them become weaker as the separation distance increases.

Suppose you place a piece of iron in the flame of a fire. Of course, it gets hot as its internal energy increases. If you remove it from the fire and place it in a pot of cool water, the iron cools and the water heats up. The internal energy of the metal has decreased and that of the water increased. After a while, the metal, water, and surrounding air all have the same temperature. Their internal energies have adjusted to achieve thermodynamic equilibrium.

Is it possible to explain this familiar chain of events in an understandable way? Historically, neither internal energy nor energy itself were understood. Rather, it was thought that when hot and cold objects were put near one another, there was a spreading of an invisible fluid called *caloric*. I discuss caloric and energy in more depth in Secs. 1.2–1.7.

1.1.2 Brownian motion

The invisible energy stored within a liquid like water comes to life, at least in part, through what is called *Brownian motion*. Water molecules undergo a constant jiggling motion which is more intense as the temperature is increased. Some of the most energetic water molecules leave the liquid and become water vapour molecules in the air. This happens at any temperature, and is more pronounced at higher temperatures. In this way, water generates a vapour pressure at its water-air interface that increases as the water is heated.

The fact that water molecules jiggle, was discovered by Robert Brown in 1827. He suspended pollen grains in water and observed them with a microscope. The grains could be seen jiggling around, which suggested that they were being bumped by water molecules that themselves, were moving in unpredictable, i.e., *random*, ways. This is depicted in Fig. 1.4. The random

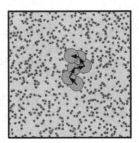

Figure 1.4: Depiction of motion of a grain of pollen (large circles) caused by continual jostling from collisions with numerous smaller water molecules. The arrows show a grain's path as time evolves.

jiggling of atoms is called *Brownian motion*. Yet, Brown was not the first to observe such motion. Jiggling of particles of coal dust on the surface of alcohol had been observed in 1785 by Dutch scientist, Jan Ingenhousz.

Even more notable, in 60 BC, Lucretius had a similar idea that related to dust particles in air, and he made the earliest known prediction of atoms. He observed that when sunbeams enter a building, one can see tiny particles jiggling, which he attributed to movements of "atoms" that we cannot see. Although the motion of dust particles can be influenced by air currents, the visible *dancing* of small but visible dust particles is from those particles being jostled by air molecules. This is Brownian motion of the dust particles.

In 1905, Albert Einstein applied his physical intuition and mathematical skills to Brownian motion. This was the *same* year he published his monumental contributions on the theory of special relativity and the photoelectric effect. Surprisingly to some people unfamiliar with Einstein's work, his contribution to the photoelectric effect is what brought him the Nobel Prize in 1921. The reason evidently is that it was in that work where he introduced the concept of *light quanta* or *photons*. Thus in 1905, Einstein not only postulated the existence of photons, but his work on Brownian motion solidified the idea that elements consist of *atoms*. Not bad for a year's work! Einstein's work led him to a relatively simple equation for the number of Brownian particles per unit volume as a function of time.[1]

1.2 CALORIC: A SEDUCTIVE IDEA

As early as the 1620s, Sir Frances Bacon and Galileo Galilei independently hypothesised that heat was a consequence of the microscopic motion of the invisible particles that made up the hot body. However there was no known way to relate this motion to any Newtonian dynamic quantity. Further, even

---

[1]J. Bernstein, "Einstein and the existence of atoms," Am. J. Phys. **74**, 863–872 (2006).

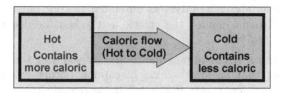

Figure 1.5: Caloric was assumed to be an indestructible fluid that flowed from hotter to colder objects. It could neither be created nor destroyed.

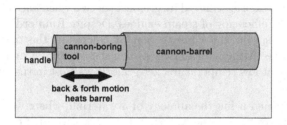

Figure 1.6: Boring a cannon seemingly produced an arbitrary amount of caloric.

after thermometers were invented, it was unclear whether temperature and quantity of heat were one and the same, or were distinct physical concepts.

The Scottish chemist Joseph Black was the first to make a distinction between them, viewing temperature as the intensity of heat within a body. Bodies in thermodynamic equilibrium would have the same temperature, independent of size while the heats in those bodies could differ. Equally important, Black viewed heat as a measurable quantity, recognising that the time required to boil water depended on its mass; the greater the mass, the longer the heating time, and the more heat supplied. Black seems to have understood the essence of internal energy, but did not separate it from a heat process.

Understanding the need for quantitative measurements, Black developed calorimeters, and perfected the method of mixtures in which two masses of liquid at two different temperatures were mixed and a final intermediate temperature was reached. This led to the concept of specific heat capacity, defined as *the quantity of heat (proportional to heating time) per unit mass required to raise the temperature of a body by one degree.* Black interpreted his findings as proof that heat was a conserved substance; e.g., when mixing two liquids the amount of heat lost by one equaled that gained by the other.

The concept of conservation was becoming attractive at the end of the 18th century, strongly supported by Antoine Lavoisier, who named the assumed conserved substance that flowed from hotter to colder objects *calorique*, i.e., *caloric* in english. As a substance, caloric, depicted flowing in Fig. 1.5, was massless, invisible, and evidently was unrelated to Newtonian mechanics. In 1820, Nicolas Clement defined the calorie as the heat needed to raise water from 13.5 °C to 14.5 °C.

Count Rumford (Benjamin Thompson) revived the theory of heat as motion, showing that the work done in cannon boring produced limitless amounts of heat, as depicted in Fig. 1.6. This destroyed the idea that heat was a conserved substance. He also proved that caloric was by necessity massless. Oddly, Rumford had little influence on the development of thermal physics because he did not envisage a new conservation law to replace that of caloric.

In some ways caloric achieved its greatest success when Sadi Carnot applied it to analyse the efficiency of steam engines. Despite Rumford's work, Carnot based his analysis on the conservation of caloric in the operation of his heat engine cycle. Although caloric was transferred from a high temperature reservoir to one at low temperature, zero caloric was lost in the cycle, yet work was done.

Carnot reasoned using the analogy of a waterfall, where work is done, yet no water is destroyed. For the waterfall, the work depended on the difference in height, while Carnot's engine depended on a difference in temperature. Notably, Carnot's caloric had units of work per degree of temperature (similar to the units of entropy), quite different from Black's caloric, whose thermal unit was akin to today's calorie. Although Carnot's work suggested a relation between heat and mechanical work, no such connection was made until the independent efforts of Mayer, Joule, and Colding.

Why would caloric fluid flow as envisaged? Perhaps because a hot object is more dense with caloric than a cool one, and caloric tends to *spread* until the caloric densities, and resulting temperatures, of all objects involved are equal? This seemed to make sense; i.e., caloric tended to spread until it was distributed uniformly. One belief about caloric was that it consisted of particles that *repel* one another, which is consistent with a tendency to spread spatially.

We know that *spreading* phenomena occur in other contexts. For example, if there is an oil spill in an ocean, that oil tends to spread over time to distant regions, and the density of oil tends to become more nearly uniform. And if a large truck containing a noxious gas overturns and gas is released, that gas tends to spread in all directions. When you pour milk into coffee, the milk disperses, i.e., spreads, over time, mixing throughout the coffee.

As I already explained, despite the beauty and simplicity of this mental picture, the caloric idea was found to be incorrect. The reason, which goes back to the experiments of Count Rumford, seems obvious nowadays. One can heat an object simply by rubbing it. For example, we rub our hands together to warm them in winter. The more we rub, the more warming, so if caloric exists, it cannot be a conserved quantity. It would be *generated* whenever we slid a chair on a floor, hammered a nail, stopped a car (as two parts of the brake rubbed together), and even combed our hair. Caloric would be created regularly.

Figure 1.7: The death of caloric led to a more general principle, the conservation of energy, which can be transferred via any of three processes: heat, work, and transfer of material.

Key Point 1.4 *The death of caloric, Fig. 1.7, came with the discovery of energy and conservation of energy. Unlike caloric, energy can be transferred not only by heating or cooling, but also by work. Energy is also transferred whenever material that stores energy is moved from one place to another, e.g., when petrol is transferred from a vendor's storage tank to a car's gas tank. The cannon-boring experiments of Benjamin Thompson can be explained in energy terms. Work done by friction forces increased the cannon barrel's and cannon-boring tool's internal energies and temperatures, which led to subsequent heating of the cannon's environment, increasing its internal energy.*

Although caloric has been dead for over 170 years, remnants of its seductive nature remain in our language. The term *heat* is commonly treated as if it were a substance like caloric. For example, when a block slides along a surface and comes to rest, what happens to its kinetic energy? The common, but unsatisfactory, answer is, "It went to heat." That answer does not jibe with thermodynamics, where *heat* is a process and internal energy is the energy stored within matter. An Internet search of "heat as motion" found: *Heat is the energy an object has because of the movement of its atoms and molecules which are continuously jiggling and moving around, hitting each other and other objects. When we add energy to an object, its atoms and molecules move faster increasing its energy of motion or heat.* This too is inconsistent with the language of thermodynamics.

One easily finds more incorrect statements, for example, "The motion of atoms and molecules creates a form of energy called heat or thermal energy" should be "The motion of atoms and molecules creates a form of energy called *internal energy*." Also, "Light from the sun is converted to heat as the sun's rays warm Earth's surface" should be "Energy from the sun is absorbed by the atmosphere and other matter on Earth's surface, where it resides as internal energy." These examples are part of *the language problem of thermodynamics*, which is discussed further in Ch. 7.

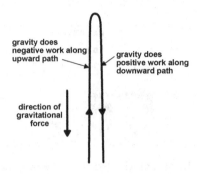

Figure 1.8: Gravitational work done when a ball is thrown upward.

1.3 ENERGY TRANSFERS: WORK, HEAT, MASS

Work

Work comes from the application of a force on an object, applied through a nonzero displacement. If the force is unvarying and applied in one direction along a straight line, the magnitude of the work is defined simply as

$$\text{work} = \text{force} \times \text{distance.}^2 \tag{1.1}$$

It is important to understand that if a force is applied, but *not* through a distance, then zero work is done. If you rub one hand against the other, you do work. But if you hold a heavy barbell over your head without moving it, you are doing zero work on it! Of course, your heart would be pumping faster than normal, doing work pushing your blood through your circulatory system. But that is not work on the barbell *per se*.

If you throw a ball upward vertically, as in Fig. 1.8, it slows because the downward force of gravity *opposes* the upward motion. Gravity is doing negative work on the ball, slowing it to zero velocity at the highest point, whereupon the ball starts moving downward, with gravity doing positive work that speeds it up.

Earth's gravity acts on you constantly, but unless you change your vertical position, it does zero work on you. If you walk down a flight of stairs, gravity does positive work on you. It's positive because the gravitational force points downward and your vertical displacement is also downward. And if you walk up a flight, gravity does *negative* work on you because the directions of the force and displacement are opposite one another. The negative work makes walking up a stairwell more strenuous than walking down.

It is a bit more difficult to understand the combination of work processes involved when you arise from a seated position (see Fig. 1.9). Getting up

---

[2]If the force varies over the path, then (1.1) is replaced by $work = \int_\pi F(\vec{r}) \cdot d\vec{r}$, where $\vec{F}(\vec{r})$ is the vector force (magnitude and direction) as a function of the displacement vector $\vec{r}$, and the specific path is connoted by π.

Figure 1.9: During rising, various body regions, e.g., the circles, are lifted to higher positions because of the upward force of the floor, which provides the person's upward momentum needed during rising.

from a chair causes most of the molecules in your body (upper legs, hips, abdomen, chest, arms, shoulders, neck, head...) to move upward. Accordingly, your *centre of gravity* rises. Energy for this rising is provided by the body's digestion of food, which enables muscle action. Your centre of gravity and gravitational potential energy rise with the help of *internal* work done by your muscles, but that is not enough. The force of the floor as you rise from a seated position acts through zero distance. Although it does zero work on you, according to Newton's second law, that force is needed to increase your momentum. Thus, the force of the floor $= dp/dt = ma$. Here momentum = mass $\times$ velocity of the centre of mass. The floor's nonzero force, which exceeds your weight during rising, is required to increase your centre of mass velocity and momentum.

Key Point 1.5 *There is a stark difference between momentum and energy. A nonzero external force is needed to change an object's momentum even if that force does zero work, while an energy source is needed to increase its energy.*

Similarly, if you stand on roller skates and push yourself and the skates away from a wall, as shown in Fig. 1.10, the force of the wall on you does zero work because it acts only at the *fixed* wall. This supports the observation that *external* work is not necessary to increase an object's speed. The energy that makes that possible comes from the digestion of food. That the energy comes from an *internal* process should be no surprise. Obviously, inanimate objects like books do not rise spontaneously. But humans and other animals can do so. Even flowers open in the sunlight and close at sundown. These are driven by internal processes. In contrast, you do positive work on a book whenever you lift it from a table and put in on a shelf that's higher than the table, and you do negative work when you move it to a lower shelf. This last part is true because while lowering the book, you exert an upward force (less than the book's weight) as the book moves downward.

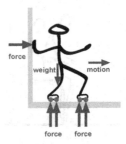

Figure 1.10: The external force of the wall on the roller skater increases his/her momentum. Zero external work is done. The skater's kinetic energy increase comes from stored bodily energy from prior digestion of food, which is released to energise his/her muscles.

Key Point 1.6 *Work, the first of the three major energy transfer methods discussed here, requires a force acting on an object through a nonzero displacement, and can be positive, negative, or zero.*

Heat

Energy accounting for the process of rising from a seated to a standing position shows that our bodies lose energy, part of which increases the gravitational potential energy of body parts moving farther from ground level. A host of bodily chemical reactions keep us at a body temperature of approximately 37 °C or 98.6 °F, and when this body temperature is higher than the air around us, we continually transfer energy to the air, heating it.

This energy transfer occurs because of contact between our skin and the surrounding air, and also via evaporation of body-temperature water through skin pores. This water vapour then becomes a part of the air and warms it when the air is cooler than the body. We also heat the surrounding air by radiation. Every object radiates energy at a rate that depends on its absolute temperature and surface area.[3] Finally, because the air near us becomes warmer than that farther away, and expands, this causes convection currents that move the less dense warmer air upward and away from us, and cooler air toward us. Although we are not always aware of these phenomena, in extremely cold outdoor weather or under an inappropriately heavy blanket in summer, we feel either too cold or too warm.

Other examples of radiation are sunlight heating the Earth and an incandescent lamp lighting, and also heating, a room. We understand this radiant heating to be part of the electromagnetic (EM) spectrum, which includes phenomena from radio and TV waves to x-rays and gamma rays. If the wavelength of EM radiation is in the visible part of the spectrum,

---

[3]See Sec. 2.5 for an overview of temperature.

Table 1.1: Some Common EM Radiation Types & Their Wavelengths

Radiation Type	Wavelength Range (nanometres)
Gamma Rays	less than 0.01
X-Rays	0.01 − 1.0
Ultra-Violet	1 − 400
Visible	400 − 700
Infrared	750 − 100, 000
Microwave	100, 000 − 100, 000, 000
Radio	greater than 100, 000, 000

$400 - 700$ nanometres (nm), we can see it, as illustrated in Fig. 1.1. For larger wavelengths, in the infrared (IR) region, we *feel* it as *heat*. Radiant heat is similar to light, except that the wavelengths involved are much longer, namely $750 - 100,000$ nm. As our skin absorbs IR radiation, we sense the temperature rise. Wavelengths just beyond the IR region are microwaves, which are used to heat food in modern microwave ovens. These familiar parts of the EM spectrum, plus other higher and lower EM radiation are summarised in Table 1.1. Examples of heat processes and other energy transformations are in Sec. 1.7. For comparison, the diameter of a hydrogen atom is roughly 0.05 nm, the diameter of a uranium atom is about 0.35 nm, and the relatively large biological molecule ribosome has a diameter of about 20 nm.

In the 19th century, a number of researchers, including Julius Robert von Mayer, James Prescott Joule, and Ludvig Colding, carefully observed processes where heat was produced by *work* processes. These three men were born within four years of one another. Independently, each determined a numerical value that connects mechanical work with heat. Mechanical work had been expressed in units of force × distance (nm), and heat was measured in calories (cal).

Key Point 1.7 *The result of the monumental research on work and heat was the seminal result that both heat and work are distinct processes with energy transfers Q and W, which can change a system's internal energy, ΔE, with no net energy change in the universe.*

The most accurate and extensive experiments done to obtain a numerical connection between heat and work were done by Joule, whose apparatus had a mass m moving through a distance h, with the gravitational field doing work mgh, turning a set of paddle wheels in the process, as depicted in Fig. 1.11. Joule measured the temperature increase of the water and equated the number of implied calories to the work mgh to get what was called the mechanical equivalent of heat, $4.160 \, \mathrm{N\,m\,cal^{-1}}$. Previously, it had not been appreciated that heat and work were both transfers of a more general *thing* (a noun) called energy. The currently accepted value of the mechanical equivalent of

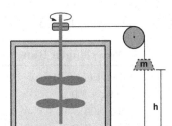

Figure 1.11: Depiction of Joule's paddle wheel experiment.

heat is $4.186\,\mathrm{J\,cal^{-1}}$. In modern parlance, this is simply a conversion factor that changes from one energy unit to another.

In honour of Joule's brilliant work, the newton-metre was named a joule, the unit of energy adopted in the worldwide System of International Units. The issue of whether Joule or Mayer found the mechanical equivalent of heat first led to substantial acrimony. Furthermore, the Danish Colding did work along similar lines, which seems to be largely unknown. Hermann Helmholtz, another pioneer in the formulation of the principle of energy conservation, wrote that it was J. R. Mayer who first discovered the law in 1842, and that one year later, Colding announced the same law. At roughly the same time, Joule's experiments in England led to the same conclusion. Helmholtz observed, "We often find, in the case of questions to the solution of which the development of science points, that several heads, quite independently of each other, generate exactly the same series of reflections.[4]"

Key Point 1.8 *Learning the connection between work and heat was much more than simply finding a number connecting the units joules and calories. The efforts of Joule, Mayer, Colding, and others was instrumental in showing that there is a universal physical entity, energy, that can neither be created nor destroyed. This was a great path-breaking advance.*

These developments propelled energy, and in particular, the notion of conservation of energy, into prominence as one of the most fundamental principles of physics. In fact, a common definition of *physics* is *the branch of science concerned with the nature and properties of matter and energy.*

Material transfer

The third method of energy transfer is to move material that bears energy from region A to region B. We think of food as *containing* energy for our bodies, and petrol as similarly *containing* energy to power our vehicles. Although

---

[4]H. Guerlac, Selected Readings in the History of Science, vol. II, part II, Ch. XVll (Cornell, 1953).

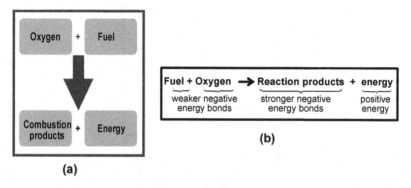

Figure 1.12: (a) Foods and fossil fuels do not *contain* extractable energy. Rather they must combine chemically with oxygen to form combustion products. (b) Those products have stronger bonds with more negative energies. Positive energy must be is released to conserve energy.

this is not entirely accurate, we *are* transferring energy when we eat, fill our gas tanks, or bring a loaf of bread from the grocery to our homes. The phrase *containing energy* is not accurate because, strictly speaking, foods and petrol do not release energy that they store. Rather, in each case, they must combine with oxygen from the atmosphere and with that oxygen, they engage in chemical reactions that release energy, as in Fig. 1.12. While some authors state that oxygen plays a role in the energy release during combustion, most do not address the following observation:

Key Point 1.9 *Bond energies are negative and it takes positive external work to separate the atoms that constitute a molecule to a distance approaching infinity. In the infinite separation limit, with the atoms assumed to be at rest, the interparticle potential energy is defined to be zero. Stronger bonds are more negative than weaker ones, and the main point is this: If the bonds that hold reactant molecules together are weaker than those that hold the combustion product molecules together, then energy is released in the process.*

An example is the combustion, i.e., rapid oxidation, of hydrogen. This is depicted in Fig. 1.13 Energies shown are bond energies associated with the initial diatomic molecules. The chemical reaction in the top line reads, two diatomic hydrogen molecules (H_2) combine with one diatomic oxygen molecule, (O_2), to give two water molecules (H_2O). The total energy of the reactants is $-436 - 436 - 498 = -1370$ kJ mol^{-1}, while the total bond energy of the combustion products is lower, $-926 - 926 = -1852$ kJ mol^{-1}. Weaker bonds have been replaced by stronger, more negative energy bonds, and compensation is needed to avoid violation of the conservation of energy. This is the 482 kJ mol^{-1} of energy released as *heating energy*.

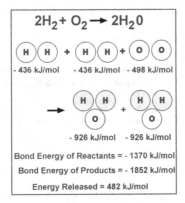

Figure 1.13: Depiction of energy changes during the combustion of hydrogen (diatomic molecule H_2) by 'oxidation', i.e., combination with oxygen (diatomic molecule O_2).

Key Point 1.10 *During combustion, energy called the heat of combustion is released. That energy was not stored by the fuel, but results from the combination of the fuel with oxygen. In the process, weaker chemical bonds have been replaced by stronger bonds.*

Another relevant illustration is provided by a burning candle, as shown in Fig. 1.14. Suppose the system is taken to be the candle plus surrounding air in an imaginary cylinder with twice the diameter and twice the height of the candle itself. Consider a time interval of a few seconds. If we ignore the movement of air upward by *convection* currents induced by the temperature difference between the flame and surrounding air, the energy transfer through the imaginary cylindrical system boundary is $Q < 0$. This energy, *radiated* from the high temperature flame, is a loss of energy. If we include any work W done *on* the system, and also the energy M gained by air movement, then

$$\Delta E = Q + W + M \text{ (first law of thermodynamics).}^5 \qquad (1.2)$$

Because hot air carries energy out through the system's boundary, $M < 0$. As air within the cylindrical boundary moves upward toward the system boundary, gravity does negative work on it, and $W < 0$. The negativity of W, Q *and* M assure that the system loses energy, i.e., $\Delta E < 0$, as expected.

This unconventional introduction of the first law of thermodynamics is inspired by the fact that a burning candle is an excellent example of an energy system that entails combustion of a fuel, along with heat, work, and mass transfers. The interesting example of a burning candle is discussed further in Examples E1 and E2 in Sec. 1.7.

---

[5]This presupposes that the total internal energy E is well defined initially and finally.

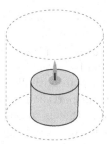

Figure 1.14: A thermodynamic system (dashed lines), consisting of a burning candle and the air surrounding it.

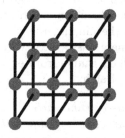

Figure 1.15: Depiction of a rigid body's molecules, held fixed in place by rigid connecting rods of (assumed) negligible mass. This is an imagined system that does not exist in nature.

1.4 IMAGINED SYSTEMS WITH E = CONSTANT

1.4.1 Rigid bodies

In Newtonian classical mechanics, one normally deals with two types of model systems, point particles and rigid bodies. Both are imagined idealisations, the first consisting of point particles that are assumed to have mass, but to occupy zero volume. The second consists of point particles with mass, arranged in a rigid lattice, depicted in Fig. 1.15. The name rigid body, which is reminiscent of *rigour mortis*, means that the point particles, which are intended to represent atoms, cannot even jiggle. In reality, real solids consist of atoms that can jiggle, and they are not rigid. For example, a copper rod's length increases when it is heated, as the average distance between atoms increases. This could not happen with a rigid body.

Because a point particle occupies zero volume, it cannot rotate or vibrate (say, like a jiggling drop of water). All it can do is move from place to place, and classical mechanics predicts how forces affect that motion. Point particles are mental models that are not real, but can sometimes exhibit behaviour that *approximates* real systems. For example a dilute gas model gives rise to an equation for pressure as a function of temperature and volume that approximates real dilute gases quite well.

Rigid bodies have finite volumes, can move from place to place, and also can rotate. A mental model could be an envisaged football-shaped object that is assumed to be rigid (unlike a real football, which is deformable), but can spin and/or wobble. When a point particle moves, it possesses *kinetic energy*. The same is true for a rigid body that is tossed across the room. If it rotates as it moves across the room, each of its point particles, has an extra motion and thus an extra *rotational kinetic energy*.

A property shared by both a point particle and a rigid body is that neither can store energy internally. For a point particle, this is true because it occupies zero volume and there is nothing *internal* about it. For a rigid body, because its point masses cannot jiggle, there is no energy stored within it. Two energy types typically encountered the kinetic energy associated with translational motion, rotational kinetic energy associated with spinning (for rigid bodies, but not point particles, which cannot rotate), and gravitational potential energy because of the pull of Earth.

Key Point 1.11 *Like idealised point particles, rigid bodies do not exist in nature. They are idealisations used to simplify mathematical studies of motion. In some cases, they provide good approximations to the behaviour of real physical systems. Neither a point particle nor a rigid body possesses internal energy. They cannot engage in energy spreading, and are outside the domain of thermodynamics.*

1.4.2 Frictionless surfaces

An idealisation that is often used in physics is that of a frictionless surface. Thus, beginning physics students often study rigid blocks that slide on assumed *frictionless* surfaces. For such imagined objects, there is no internal energy and thus no spreading to or from their interiors is possible. As with point particles and rigid bodies, these idealised entities lie outside the domain of thermodynamics and thermodynamic entropy.

All *real* surfaces have some degree of roughness and when a block slides over, say, a horizontal table, there is a horizontal force that slows the block. Both the block and table become warmer as the internal energy of each increases, and energy spreads internally within them. A model for this entails a mental picture of both surfaces containing small bumps or spikes that deform when they encounter one another. As these bumps push one another, they do work. In this process, atoms become jostled. As the bumps deform one another, the internal energies of both objects increase.

In this case there is not necessarily a heat process between block and table. Rather, the work process generates warm regions near the surface of each object which trigger *internal* heat processes within the block and table,

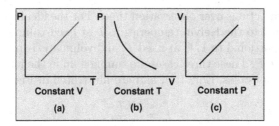

Figure 1.16: Dilute gas graphs at (a) constant volume, (b) constant temperature, (c) constant pressure.

and external heat processes with the air. The net result is increased internal energy of the block, table, and surroundings. This increases the entropy

Students in introductory physics learn that if a block of wood is pushed across a horizontal table and the coefficient of friction is denoted by μ, the external force required to keep the block sliding at constant velocity is $\mu \times$ the block's weight. However, it turns out that the work done by the external force is *not* equal to the work done by friction of the block on the horizontal surface. This is because, the little bumps do not push each other over the full distance that the external force acts. Rather, the friction work is the external force $\times$ an effective displacement that is less than the block's displacement. The work done by the external force is the work done by friction plus the gain in internal energy of the block-surface system. This is a case where thermodynamics really has a significant effect, and classical mechanics alone is inadequate. Friction is discussed further in Sec. 7.1.5.

In a similar vein, it is common to neglect air resistance for objects moving through air, e.g., baseballs and rockets. However, in real life air resistance cannot be neglected. For example, a parachute would not function without air resistance and as a parachutist falls vertically at nearly constant velocity, work is done on the air by the parachute and *vice versa*. Put differently, increased internal energy spreads through the parachute's material and the air's molecules. An extreme example of air resistance is during reentry of a space vehicle to Earth's atmosphere. The vehicle's nose cone can experience temperatures of $7,000\,^{\circ}\text{C}$ or more. Clearly, air resistance is not negligible!

1.5   DILUTE GAS MODEL: IDEAL GAS

An ideal gas is a simplified model of a monatomic gas, with the major simplifying assumption that the particles are point particles that do not exert forces on one another. The total interparticle potential energy is zero, and all the energy is kinetic. Particles can bounce off the container walls elastically (i.e., with no kinetic energy loss). A nonzero pressure on the walls results from the average force from collisions by many molecules with a distribution of speeds and directions over observation times. For the ideal gas, the pressure P is proportional to the Kelvin temperature T at fixed volume V.[6] Similarly, pressure is proportional to $1/V$ at fixed T, and volume V is proportional to T at fixed pressure P. These three cases are summed up in the ideal gas equation of state for N molecules, Eq. (1.3), and are illustrated in Fig. 1.16.

$$P = \frac{NkT}{V}. \tag{1.3}$$

Here k is the Boltzmann constant, which relates temperature to energy, with kT having energy units. This is discussed at length in Sec. 2.5.

Key Point 1.12 *For a dilute gas,*

1. *For fixed volume (isochoric process), when temperature increases, molecules have higher average energies, and a larger momentum transfer rate at the container walls, so $P \propto T$.*

2. *Increasing V at fixed T (isothermal process) means longer transit times between collisions with walls, decreasing the momentum transfer rate, and $P \propto 1/V$.*

3. *For a fixed value of pressure (isobaric process), the momentum transfer rate at the walls is fixed. If T is increased, in order to keep the average force fixed, a V increase is necessary. This increases the average collision time, which offsets the temperature increase, and $V \propto T$ at constant P.*

4. *The internal energy, $E = \frac{3}{2}NkT$ can be interpreted as each "degree of freedom," x, y and z, contributing energy $\frac{1}{2}kT$ on average to the internal energy of each particle.*

5. *For a classical ideal gas, temperature can be defined to be proportional to the average kinetic energy per molecule. However the average energy per molecule for a quantum ideal gas is not generally proportional to temperature.[7] Temperature is not proportional to the average kinetic energy per molecule in general.*

6. *For a classical nonideal gas, the average energy per molecule is E/N. In that case, E is the average kinetic energy E_k plus an average potential energy. The latter energy involves pairs of interacting molecules, and the meaning of potential energy per molecule is unclear. Temperature T is not generally proportional to E/N for a nonideal gas.*

7. *Two types of volume change are of particular interest for gases. One is an isothermal change, where the gas temperature does not vary. The other is an adiabatic change, with zero energy exchange other than work.*

---

[6]The Kelvin temperature scale is defined in Sec. 2.5.

[7]For a quantum mechanical gas, $E/N \propto T$ in the classical limit, but not otherwise.

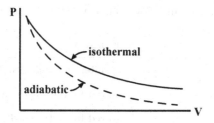

Figure 1.17: Graph of pressure P vs container volume V for slow *isothermal* and *adiabatic* expansions, starting in the same state.

Isothermal and adiabatic expansions are graphed in Fig. 1.17. The gas pressure drops for both an expansion at constant temperature and an adiabatic process (with $Q = 0$) because any expansion lowers the collision rate of molecules with the walls. During the isothermal expansion, as the gas does work expanding against an external load, energy is transferred to it from the hot reservoir to keep temperature constant. This energy input explains why the gas pressure during isothermal expansion is greater than for the adiabatic expansion, where there is no receipt of heating energy.

8. There is a tendency for students to use the ideal gas equations for any and all systems. My admonition, "The world is not an ideal gas" worked for some of them. In Sec. 5.3, I introduce another ideal gas, the photon gas, which has very different pressure, energy and entropy behaviour than a classical ideal gas of molecules with nonzero mass. In Sec. 4.2.3, I cover the van der Waals model of a non-ideal gas, which I hope will counter any tendency to view the world as an ideal gas.

1.6 ENERGY DEFINITIONS, UNITS

Although energy is ubiquitous, it is nevertheless true that it does not have an easy definition. Rather, it must be approached by identifying different energy types. A list of some common stored energy types is shown in Table 1.2.

The kinetic energy of an object has a simple formula. We can guess what it might depend on, namely, how fast the object moves and how much matter is doing the motion. Indeed, the formula is simply,

$$kinetic\ energy = \frac{1}{2} \times mass \times (speed)^2 = \frac{1}{2}mv^2. \tag{1.4}$$

A bird with mass 1.0 kg, flying with speed $10\,\mathrm{m\,s^{-1}}$ has kinetic energy

$$\frac{1}{2} \times 1.0\,\mathrm{kg} \times (10\,\mathrm{m}\,s^{-1}) \times (10\,\mathrm{m}\,s^{-1}) = 50\,\mathrm{N\,m} = 50\,\mathrm{J}.$$

The joule (J), named after James Joule, is the standard energy unit used by scientists all over the world. It is a relatively small amount of energy,

Table 1.2: Some stored energy types and examples

Type	Examples
kinetic energy (energy of motion)	sprinter running, grand prix racing car
gravitational potential energy	stored by the Earth–moon system
elastic potential energy	stored by a stretched or compressed spring
chemical energy	(negative) energy stored within the bonds that hold molecules together
internal energy	stored by matter because of molecular motion and intermolecular forces

Table 1.3: Comparison of the joule and other energy units

Energy Unit	Joules	Conversion
food calorie, Cal or kcal	4186	4.186 kJ/kcal
Btu (British thermal unit)	1,055	1.055 kJ/Btu
kilowatt hour	3,600,000	3.6 MJ/kWh

and the familiar food calorie, written Cal or kcal, equals $4,186$ J. A 60 watt incandescent lamp transforms $216,000$ J of energy each hour. A human heats his/her environment at about the same rate as a 60 watt bulb, but at a much lower temperature, over a much larger surface area. Table 1.3 shows how the joule compares with other common units.[8] United States Energy Information Administration (EIA) estimates that in 2013 the entire world *used*, i.e., transformed, about 5.6×10^{20} J of energy from all fossil, nuclear, and renewable energy sources. The EIA refers to "energy consumption," which really means "fuel consumption." In 1875 seventeen nations signed the international Treaty of the Metre, the first common measurement system for international trade and the global exchange of ideas. The system of units was based on artefact standards, i.e., physical objects that were stored in the International Bureau of Weights and Measures in France. That system of units became the *International System of Units*, abbreviated SI, in 1960. Since that time, the system has evolved. A major change was the adoption of the frequency (called $\Delta\nu_{Cs}$) of microwave radiation released by a transition

---

[8] See, e.g., URL: http://en.wikipedia.org/wiki/Conversion_Of_units

Table 1.4: Defining constants of the SI system

Defining constant	Symbol	Numerical value	Unit
hyperfine frequency of Cs	$\Delta\nu_{\mathrm{Cs}}$	9 192 631 770	Hz
speed of light in vacuum	c	299 792 458	m s^{-1}
Planck constant	h	6 626 070 15 $\times 10^{-34}$	J Hz^{-1}
elementary charge	e	1.602 176 634 $\times 10^{-19}$	C
Boltzmann constant	k	1.380 649 $\times 10^{-23}$	J K^{-1}
Avogadro constant	N_A	6.022 140 76 $\times 10^{23}$	mol^{-1}
luminous efficacy	K_{cd}	683	lm W^{-1}

Source: URL https://www.nist.gov/si-redefinition

between the two hyperfine levels of the ground state of the caesium 133 atom.[9] This *atomic clock* frequency is now the basis of the second, replacing the antiquated definition $1\,second = 1/86,400$ mean solar day.

In 1983, it was agreed at yet another international General Conference on Weights and Measures that the believed-to-be unvarying speed of light c would be used to define the metre. The adopted speed value $c = 299,792,458$ metres per second agreed well with prevailing measurements of c. Thus, the metre is now defined as the distance light travels in $1/(299,792,458)$ seconds. In a real sense, a *ruler made of light* has replaced the previously-used standard metre bar in France.

This left only the kilogram as a fundamental unit based on a physical object in France, a candle-sized cylinder of platinum-iridium. Although it was kept in a vault under carefully controlled conditions, and was used only every 40 years to calibrate copies, there is evidence that it was changing. Perhaps this is because of ultra-slow, but finite-rate *sublimation* from solid to vapour (The reverse transformation from vapour to solid is called *deposition*). In any case, a new standard was proposed in 2018, and is scheduled to take effect May 20, 2019, the day I am writing this paragraph!

The new definition of the kilogram utilises an electromechanical measuring instrument called a Kibble balance, proposed by Bryan Kibble in 1975. That device measures the weight of a test object to high accuracy, using an electric current to produce a compensating magnetic force to balance the gravitational force. The definition entails use of the now fixed numerical value of the Planck constant, along with the definitions of the *metre* and *second*, defined respectively in terms of the speed of light and the hyperfine transition

---

[9]The term *hyperfine* refers to a very small splitting of an electronic energy level because of interactions of nuclei with internally generated electric and magnetic fields.

frequency of the caesium atom alluded to earlier. The idea is simple in principle, but the Kibble balance is a highly advanced metrological instrument (few exist in the world) and the detailed physics is complex and not needed here.

In addition to the three constants, speed of light, the Planck constant, and caesium's hyperfine frequency, four more constants are used to define *all* physics units. The elementary charge magnitude (for an electron and proton) is used to define the ampere, the Boltzmann constant is used to define temperature, the Avogadro constant is needed to define the mole, and the luminous efficacy is used to define the candela. The defining constants adopted in 2020 are listed in Table 1.4. The Planck constant is necessary for the *energy* of a photon, $h\nu$. The speed of light is an upper bound (albeit never reached) on the speed of a massive particle, and it would presumably require infinite *energy* to reach it. The elementary charge is a strength parameter for electric potential *energy*. The Boltzmann constant links energy to temperature, and luminous efficacy is a measure of the visible light flux in a small frequency interval, relative to the rate per unit area at which that light carries energy. (A related quantity is the spectral luminous efficiency in Fig. 5.10.)

Key Point 1.13 *Notably, with the exception of the Avogadro constant, each of the other fundamental constants in Table 1.4 has an explicit connection with one or more aspects of energy, reinforcing the already known fact that energy is a central concept of physics, indeed, THE central concept.*

For an object on or near Earth, gravitational potential energy is mutually shared by the Earth and the object, and has a simple approximate form if the object is sufficiently close to Earth's surface. Here we must agree to define the potential energy to be zero when the object is at ground level.[10]

$$\text{Gravitational potential energy (GPE)} = mgh. \tag{1.5}$$

Note that the GPE depends on how much matter, measured by mass m (kg), the object has, the strength ($9.8\,\text{N metre}^{-1}$) of the gravitational field at Earth's surface, and h the distance above the surface of the Earth. For a 100 kg (roughly 220 lb) person 0.5 metres above the ground, the gravitational potential energy is 490 J. This is not the potential energy of the person or of the Earth, but rather of the person-Earth system.

Key Point 1.14 *Gravitational potential energy is stored by the gravitational field. It is a mutual potential energy that causes an object and Earth to be attracted toward one another. Because Earth is so massive compared to objects on Earth, a falling object's motion is obvious while Earth's motion through a relatively tiny displacement is difficult to detect.*

---

[10]The full gravitational potential energy $-GmM_{earth}/(R_{earth}+h) \to 0$ when mass m is taken infinitely far from Earth ($h \to \infty$), and is negative for all finite separations.

Table 1.5: Food calorie breakdown for peanut butter

Calories	180 Cal
Total Fat	15 g
Total Carbohydrate	6 g
Protein	7 g

Fuels like paraffin release energy when they combine with oxygen from the air. The release for paraffin is about 40 MJ of energy per kilogram. This is equivalent to 9.6 $kcal\,g^{-1}$. Paraffin is largely fat, which releases about 9 $Cal\,g^{-1}$ when burned. Combustion of a carbohydrate or protein releases about 4 $Cal\,g^{-1}$, less than half that for fat. A label on a jar of peanut butter in my kitchen shows the information in Table 1.5. Using the guidelines above, this implies 9 $Cal\,g^{-1}$ × 15 g+4 Cal/g×(6 g +7 g) = 187 Cal, which compares reasonably well with the stated caloric total of 180 Cal. The 3.9% difference comes from roundoff error; i.e., the 9 $Cal\,g^{-1}$ and 4 $Cal\,g^{-1}$ figures for fat and protein or carbohydrate are approximations, not exact values.

1.7 ENERGY TRANSFORMATION EXAMPLES

To conclude this chapter, here are thirteen illustrative everyday examples, labeled E1 to E13:

E1. We feel warm when our hand is near, but not touching, a burning candle (Fig.1.18). The reason is *not* evident to many. Energy is transferred away from the flame in three ways: radiation, convection, and conduction. We feel warmth primarily from electromagnetic radiation, which travels at the speed of light (slightly slower in air) away from the flame in all directions. The hotter the flame, the more energy is transferred each second. Radiation is also why we feel warmer when we are in direct sunlight, and why our hand warms when it gets near an incandescent light bulb or hot oven. We also feel cold near a cold wall, when our skin transfers energy at a rate that exceeds that at which it receives radiative energy from the wall.

Returning to the candle, convection sets up air currents that carry expanding warm air upwards, which is less dense than unheated air, and is pushed upward by the higher-density cooler air. A simple exercise with a candle and flashlight illustrates the phenomenon of convection very clearly. Light a candle on a table. Darken the room as much as possible, and then shine a flashlight at the burning flame, onto a wall or piece of white paper. You will see evidence of air currents moving upwards because of the variations of the air's density, which are visible shadows depicted in Fig. 1.18b. Conduction and convection warm the air that is in contact with the flame just as warm water warms your hands when washing them, or an ice cube cools your hand when

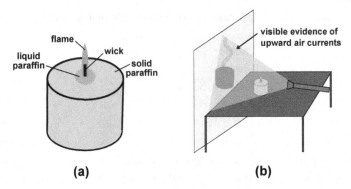

Figure 1.18: (a) A burning candle. (b) Air currents above a burning candle made visible.

you pick it up. When the latter combination of *heat processes* from a candle warm an object, the net added energy (traditionally called Q) increases the energy the object stores internally (called E).[11] This is summarised by the equation

$$Q = \Delta E. \tag{1.6}$$

This reads: Added heating energy = change in internal energy. Energy is conserved. Again, this presupposes that E is well defined initially and finally.

E2. A burning candle is an excellent model of an energy system.

Referring to Fig. 1.18a, a burning candle has a fuel, namely the candle wax (paraffin), typically derived from a fossil fuel. It is solid at room temperature, but it melts above approximately normal body temperature ($37\,°C = 98.6\,°F$). The visible puddle at the base of the wick of a lit candle is liquid paraffin. Although neither the solid nor liquid is flammable, vapour paraffin will burn. The wick acts as a pipeline that carries liquid fuel to the flame. That liquid becomes hot enough to vaporise, mix with air, and burn.

The speed at which the paraffin fuel is transported adjusts automatically to keep the candle burning well. This occurs by a beautiful *negative* feedback mechanism. If the fuel rate gets too high, there is not enough replenished oxygen to maintain the rate of combustion and the burning rate slows. If the burning rate gets too low, the wick's top becomes relatively dry and capillary action speeds up, bringing fuel to the area more rapidly. A candle is a self-regulated energy system!

Suppose a candle burns for 900 s and its mass loss is 1 g = 10^{-3} kg. For paraffin, the amount of energy released during combustion is 40 MJ kg^{-1}. Thus, the energy release in 900 s is 40 MJ kg^{-1} × 10^{-3} kg = 40 kJ. Power is defined as the energy released per unit time–i.e., the *rate* at which energy is used, and the unit of power is therefore J s^{-1}, which is the familiar watt (W). Thus, for

---

[11]Note: Many authors use U rather than E for internal energy.

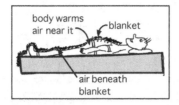

Figure 1.19: Person sleeping under a blanket (artificially transparent for illustration purposes).

the candle the average output power over the measurement time is

$$Power = \frac{40\,\text{kJ}}{900\,\text{s}} = 42\,\text{W}. \tag{1.7}$$

This result might seem surprising because a candle does not seem as bright as a 40 W incandescent light bulb. Indeed, it is not. The reason is that the flame temperature is much lower than that of the filament of an incandescent light bulb–about 1800 K = 1527 °C *vs* 2600 K = 2327 °C, where K signifies the kelvin absolute temperature scale.[12]

Both the candle and light bulb are hot enough to radiate infrared radiation (*heat*) along with visible light. The candle's temperature is much lower so it glows less brightly with a *greater* fraction of the emitted E-M energy in the invisible infrared, and a *smaller* fraction in the *visible* region. The candle gets smaller as it burns because its paraffin fuel is being transformed into both carbon dioxide, CO_2, and water, H_2O. A common paraffin is $C_{25}H_{52}$, and the combustion reaction is $C_{25}H_{52} + 38\,O_2 \rightarrow 25\,CO_2 + 26H_2O$.

E3 At night we keep warm by snuggling under a blanket.

We heat ourselves under a blanket. Our bodies process chemical energy derived from our food interacting with the oxygen we breathe in, providing energy to keep our hearts beating and maintaining nearly constant body temperature of about 37°C. Under a blanket, as depicted in Fig. 1.19, your body radiates and conducts energy to the initially cooler air between you and the blanket, which slows energy exchange with the cooler room air above it. The air beneath the blanket becomes warm, which slows the transfer of energy from your body, making you feel warmer.

E4. When we walk barefoot on a carpet, it feels comfortable, but when we step onto a vinyl floor at the same temperature as the carpet, it feels cold.

Our bodies detect the *rate* at which energy is removed from our feet. When the rate is relatively high, our feet feel cold. A carpet is a good insulator and poor thermal conductor, and even though it is quite cold, the carpet removes

---

[12]The degree size of the kelvin is the same as that for the familiar Celsius (formerly centigrade) scale, 1.8 times the size of the Fahrenheit degree. However, the kelvin scale is shifted such that the low temperature limit in nature is zero, 0 K = −273.15° C = −459.67° F.

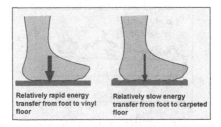

Figure 1.20: Conduction of energy from a warmer foot to a colder vinyl floor and a carpet at the same temperature.

energy from bare feet slowly. In contrast, a vinyl floor is a much better thermal conductor that removes energy more quickly from our feet, so it feels colder. This is depicted in Fig. 1.20. Clearly our feet are *not* good thermometers. That is, they do not detect temperature very well. They are better at detecting the rate at which–i.e., how fast–a heat process occurs.

E5. When we vigorously operate a manual bicycle air pump, repeatedly pushing its handle up and down, the pump warms.

We exert a force through a distance, moving the pump handle, which moves a piston through the bike pump's cylinder. The piston compresses air in the cylinder, forcing it through a small opening. When the air in the pump cylinder is compressed quickly, it gains energy and has insufficient time to transfer this energy through the cylinder's walls. Thus, it heats up, also heating the metal pump casing. As the pump's piston moves through the cylinder repeatedly, friction heats the piston and cylinder walls, which adds to the heating effect.

E6. Automobile tire pressure increases when driving at high speeds.

As an automobile tire turns, the bottom outer part of the tire flattens against the road (see Fig. 1.21). As segments of the tire flatten out and then become curved again, this heats the rubber, similar to bending and straightening a portion of a metal coat hanger many times in succession, making it warm. (The force you apply to do the bending and unbending does *work* on the metal, adding energy to it.) Additionally, there is likely some slippage of the tire against the road. Both effects heat the tire by mechanical work being transformed to internal energy. The effect is greater at higher speeds. The tire pressure increases with temperature, at nearly fixed volume, as faster air molecules hit the inside of the tire wall with greater force.

E7. Scientists use a double-walled container called a dewar (which is the basis for glass-lined Thermos jugs that keep coffee warm), to keep liquids such as nitrogen and helium from warming and vaporising.

Energy transfer by conduction and convection is slowed by a double-walled shell that is evacuated of most of the air. The air pressure between the two walls is very low, and less air means a lower the rate of energy transfer by conduction and convection. Radiation losses are cut down by using silvered

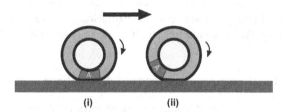

Figure 1.21: Tire region A is shown at two times, (i) with A at the tire's bottom, and (ii) slightly later, with A displaced a bit.

Figure 1.22: Depiction of a *Dewar* bottle, invented by James Dewar to hold low temperature liquids. Energy losses are slowed as described in the text.

surfaces on three of the wall surfaces, inside the outer wall and both inside and outside inner walls, as illustrated in Fig. 1.22. Silver reflects a substantial fraction of the radiation that hits it, and absorbs relatively little energy. Most of the radiation between the two walls simply bounces back and forth between the silvered surfaces.

E8. A steel nail head gets hot when we hammer it into a wall.

The act of hammering distorts the nail's shape and imparts energy to its atoms. Repeated hammering amplifies this effect. This work process increases the energy stored internally by the nail's atoms. Following reasoning similar to that leading to Eq. (1.6), we can approximate the energy transformation in this example over a short time period during hammering by the work

$$W = \Delta E \text{ net work on nail} = \text{change in internal energy.} \qquad (1.8)$$

In a state with zero net energy flow and unchanging measurable thermodynamic variables such as temperature, pressure and volume, the nail's internal energy is E. As seen earlier, such a quiescent state is called *thermodynamic equilibrium*. The net work done on the nail is W. This holds approximately for only a short time period because the warmed nail also transfers heating energy to the cooler air and the wall holding the nail. Recalling Eq. (1.2), a more complete equation is

$$W + Q = \Delta E \quad \text{(first law of thermodynamics).} \qquad (1.9)$$

In words, energy transfer via work + energy transfer via heat to a system = change in internal energy of the system. Equations (1.2) and (1.9) are forms of the *first law of thermodynamics.*, and (1.8) is a special case that applies only when $Q \approx 0$.

Key Point 1.15 *Here are some important observations.*

1. *The first law is a statement of* conservation of energy. *Energy can neither be created nor destroyed; it can only change its form.*

2. *The first law recognises that the internal energy stored within matter can be represented by a well-defined mathematics function called E.*

3. *For homogeneous macroscopic substances, E has the important property of being an extensive function, i.e., twice the amount of material under the same thermodynamic conditions will have twice the original internal energy:* $E(2V, 2N) = 2E(V, N)$.[13]

4. *Temperature does NOT scale with a thermodynamic system, but rather is invariant under scale changes of the number of molecules and volume, and is called an intensive function or variable.*

5. *A mathematical generalisation for homogeneous systems is that for any positive integer* λ, $E(T, \lambda V, \lambda N) = \lambda E(T, V, N)$ *and* $T(\lambda E, \lambda V, \lambda N) = T(E, V, N)$.[14]

6. *For a specified set of thermodynamic conditions, e.g., temperature, pressure, and mass of material, E has a specific numerical value. We normally do not know that value and are content to measure changes that occur. Notably, W and Q exist only during a process, and it is internal energy E that is stored prior to and after all processes.*

E9. We need food and oxygen in order to live and be active.

This is an elaboration of a point made earlier. The number of food calories in a food can be measured by burning it in a closed container and measuring the energy that is released in the process. A common example is the sugar form, glucose. Each glucose molecule contains six atoms of carbon, 12 atoms of oxygen, and 6 atoms of oxygen, and is denoted by $C_6H_{12}O_6$. When we digest our food, we *burn* it very slooooooooowly, not at the temperature of fire, but rather, at normal body temperature. More accurately, we oxidise it, combining it with oxygen according to the chemical reaction,[15]

$$C_6H_{12}O_6 + 6O_2 \rightarrow 6CO_2 + 6H_2O. \tag{1.10}$$

---

[13]This property does not hold, say, for a gas in a gravitational field, which has a density gradient, is *not* homogeneous and has non-extensive energy.

[14]Restricting λ to integers assures that λN is an integer. Mathematically, λ can be any positive real number.

[15]Our digestion is more complex. A sequence of reactions has the net effect of Eq. (1.10).

Under constant-pressure conditions, it is useful and common to define the state function *enthalpy*, whose *name* has been attributed to Heike Kamerlingh Onnes, and is suggested by the Greek word for *heat*, $\theta\alpha\lambda\pi o\varsigma$ *(thalpos)*.[16]

$$H \equiv E + PV. \qquad (1.11)$$

Key Point 1.16 *When a process occurs in a constant-pressure atmosphere, the enthalpy change accounts for the work (positive, negative, or zero) done by the atmosphere on the target system.*

1. *If the constant-pressure atmosphere does work* $-P_a\Delta V$ *(positive when* $\Delta V < 0$*) and an external source of work does work* W_{ext} *on the system, then the total work on the system is* $W = -P_a\Delta V + W_{\text{ext}}$*, and the first law, Eq. (1.9) can be recast as* $\Delta H = Q + W_{\text{ext}}$*.*

2. *The upshot is that when there is no external work other than atmospheric work, then* $W_{ext} = 0$ *and the latter equation becomes the useful relation*

$$Q = \Delta H. \qquad (1.12)$$

Notably, in this case, Q is a path dependent energy transfer that equals the change in a path independent state function.

3. *It is* ΔH *that equals Q, and not H itself. It is incorrect to view H itself as "heat" because H is a state function, and Q is the energy transferred in a heating process. Perhaps Kamerlingh Onnes was inspired to use the letter H by his first name, Heike, rather than by heat.*

4. *Equation (1.12) makes possible data compilations of* ΔH_f *for the formation of specific compounds, calling it the enthalpy of formation.*

5. *We expect* ΔH_f *to be correlated with the strength of the bonds that form diatomic hydrogen. However, other existing energies and energy changes, e.g., zero-point vibrational energy for* H_2 *and the kinetic energy change during the process, weaken that correlation.*[17]

Returning to glucose combustion in Eq. (1.10), the energy released is, $|Q| = |\Delta H| = 2.8 \, \text{M J mol}^{-1}$.[18] Of course, when glucose or any other foodstuff is digested in the body, it is not heating that is involved, but rather a slower process involving *free energy*, explained in Ch. 3. The digestion process *does* keep our bodies warm, but it also fuels our muscles, brain, and the like, keeping our hearts beating involuntarily and our minds working. It is *oxidation* such as that in Eq. (1.10) that makes this possible.

---

[16]I. K. Howard, "H Is for Enthalpy, Thanks to Heike Kamerlingh Onnes and Alfred W. Porter," J. Chem. Ed **79**, 697–698 (2002).

[17]A detailed discussion of these and other effects is given in Treptow, R. S., "Bond Energies and Enthalpies," J. Chem. Ed. 72, 497-499 (1995).

[18]The enthalpy change, $\Delta H = \Delta H_f(\text{products}) - \Delta H_f(\text{reactants}) < 0$.

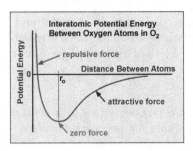

Figure 1.23: Potential energy of diatomic oxygen molecule for variable oxygen atom separation. Typically the separation distance is near r_0, where the interatomic force is zero.

In the combustion process, the glucose, with relatively weak bonds, is replaced by compounds with relatively strong bonds (and *lower* potential energies. This enables a positive energy release. It is through oxidation that energy is released in what chemists call an *exothermic* chemical reaction. Because one mole of glucose has a mass of 180 g (12 g × 6 from carbon, 1 g × 12 from hydrogen, and 16 g × 6 for oxygen), the energy released per gram is 2.80 MJ/180 g = 15.6 kJ g^{-1}, or 3.72 Cal g^{-1}. This is in good agreement with the guideline mentioned earlier that carbohydrates (of which glucose is one) releases approximately 4 Cal g^{-1}.

I use the common convention that when the constituent particles are pulled apart by forces doing positive work, becoming infinitely far from one another, the potential energy of interaction of the system approaches zero. This implies that the bond energies of these compounds is negative. A sketch of a typical potential energy of a diatomic oxygen molecule is shown in Fig. 1.23. This figure is yet another way to understand the force *vs* separation distance graph in Fig. 1.3.

The potential energy is defined such that the slope of the curve equals the force between oxygen atoms, but with opposite algebraic signs. See Key Point 1.17. The regions for which the interatomic force is repulsive–pushing the atoms apart–and attractive–pulling the atoms together–are labeled in the figure.

Key Point 1.17 *The force always points in the direction of decreasing potential energy, just as Earth's gravitational field is downward, in the direction of decreasing gravitational potential energy. For $r < r_0$, the force is rightward (repulsive, away from the origin, $r = 0$; and for $r > r_0$, the force is leftward (attractive, toward the origin).*

When the separation is a bit larger than r_0, the atoms are pulled closer together, and when the separation is a bit smaller, than r_0, they are pushed farther apart. Thus, some of the energy stored in an O_2 molecule is vibrational, from oscillations about its mechanical equilibrium (zero force) position r_0.

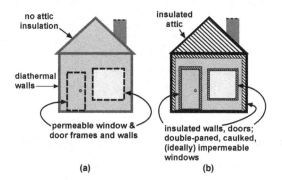

Figure 1.24: Depiction of two similar houses, (a) a *leaky* house that allows air infiltration, and considerable energy exchanges with the environment; (b) a well-insulated house with well-sealed window and door frames, drastically reducing energy and particle exchanges. Both cases assume some air flow through the chimney, via the fireplace damper; colder outside air flows in and warmer inside air flows out.

E11. Boundaries that allow heat processes are called *diathermal* or *diathermic*. Those that pass matter are permeable. These can be understood using a residential house as a model.

Referring to Fig. 1.24, the diathermal outer walls are connoted by solid lines, and some permeable regions around the windows that enable air leaks are illustrated by dashed lines. A house *cannot* be totally impermeable to air exchanges with the outdoors, or sufficient oxygen for the inhabitants would be depleted.

E12. Explain whether the following are adiabatic (perfectly insulated such that $Q = 0$), diathermal, permeable, or some combination of these. (a) Your skin. (b) The walls of a thermos jug. (c) The walls of a pressure cooker.

(a) Your skin is a semipermeable membrane, which passes water vapour and other things out, but lets very little in. It continually transfers energy because of temperature differences, so it is diathermal. (b) Ideally a thermos jug would be adiabatic and impermeable to all matter. In real life, thermos jugs do allow a very slow energy transfer through their walls, i.e., they are diathermal over long time intervals, but can be approximated as adiabatic over shorter time intervals. They are permeable to the extent that they leak around their stoppers. (c) A pressure cooker has diathermal walls, and is impermeable to matter, unless the inside pressure gets too high, whereupon a pressure-release safety valve releases water vapour.

E13. A windmill enables us to capture wind energy to generate electricity.

Wind is a system of air molecules that travels in a preferred direction. Wind exerts force and does physical work when turning the blades of a windmill, converting its kinetic energy to rotational kinetic energy of the

Figure 1.25: Windmill with three rotating blades.

blades. Wind-electricity converters are coupled to an electric generator that converts rotational energy into electrical energy. In some cases, windmills are coupled mechanically to a pump that can distribute water for irrigation purposes.

The power generated by wind is roughly proportional to the wind speed v to the third power; i.e., *Power* $\propto v^3$, so a 30 mph wind generates about $(3)^3 = 27$ times as much power as a 10 mph wind. Because wind speeds normally increase with altitude, windmills usually are mounted on towers.

Windmills are not 100% efficient because (a) only their blades intercept the wind directly and thus, they cannot capture all the energy in a circular cross section (shaded circular area in Fig. 1.25); and (b) when the wind collides with the windmill blades, there is dissipation of work to internal energy of the air and windmill. This happens as the air that collides with the blades encounters friction, the blades become deformed, and the air develops irregular (turbulent) flow. It is through the dissipation of mechanical work to internal energy of the air and windmill that thermodynamics becomes relevant. Typical modern windmill efficiencies are about 40%.

In 2018, the countries with the largest installed (not necessarily used) wind-generating capacities were: China, 211 GW; US, 97 GW; Germany, 59.3 GW; India, 35.1 GW; Spain, 23.5 GW; and the UK, 21.0 GW.

CHAPTER 2

Energy is Not Enough

CONTENTS

2.1   THE WORK-ENERGY THEOREM

2.1.1   Conservation of energy

In examples E1 to E13, energy is conserved. The everyday examples entail considerations of temperature and internal energy, concepts that are outside the realm of Newtonian mechanics. Work, heat, internal energy, and temperature are all concepts that appear in, and are central to, thermodynamics. The simple graphical summary in Figure 2.1 shows that

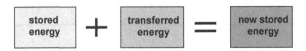

Figure 2.1: Symbolic diagram of a system's stored energy being modified by a transferred energy.

Figure 2.2: A tennis racket hitting a tennis ball, deforming the netting and the ball.

energy can be transferred to or from a system, changing the system's stored energy. The transferred energy can be either *to* or *from* the system of interest.

2.1.2 Inadequacy of work-energy theorem

In classical mechanics, there is a work-kinetic energy theorem, which is

$$\text{Net work on an object} = \text{Change of kinetic energy of the object.} \quad (2.1)$$

If a baseball is pitched, the work done by the pitcher becomes kinetic energy of the ball, as in Eq.(2.1). When a high speed baseball hits a batter's bat, the bat exerts a force on the ball, reversing its direction of motion. The bat is in contact with the ball for a fraction of a second, and pushes the ball through a distance, giving it a new kinetic energy. In this case Eq.(2.1) holds only to a rough approximation.[1] This is because in the collision between bat and ball, some energy goes into generating the sound that one hears. The "crack of the bat" comes from energising air surrounding the bat, sending sound waves in all directions.

Additionally during the collision both ball and bat deform temporarily because they are *not* rigid bodies. In the process of deformation their internal energies increase a bit. A similar effect for a tennis racket hitting a ball is more dramatic visually, as illustrated in Fig. 2.2. The racket strings and ball both distort noticeably, as they each absorb some of the energy of the work done by strings on the ball and *vice versa*. In this case, Eq.(2.1), which comes from classical mechanics, must be replaced by an energy balance equation that accounts for internal energy changes.

---

[1] The work-kinetic energy theorem is exact for *point* particles, but the bat and ball are *macroscopic* objects.

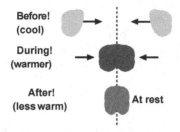

Figure 2.3: Two pieces of putty have a head-on collision, during which all the kinetic energy is lost.

Before generalising Eq.(2.1), I will introduce an even more striking example, a head-on, *perfectly symmetric* collision of two pieces of putty, as in Fig. 2.3. Initially, each has the same kinetic energy, which we write for each piece of putty, $K_0 = \frac{1}{2}mv^2$, where m and v mass and speed. After the collision, the resulting blob of putty is at rest, so *all* the kinetic energy, namely $2K_0$, is lost in the collision. Applying Eq. (2.1), we get Net Work = Final kinetic energy − Initial kinetic energy = $0 - 2K_0 = -2K_0$ or

$$\text{Net work} = -2K_0 \quad \text{(a nonsense result)}. \tag{2.2}$$

This simple-looking result makes no sense because the left side of Eq.(2.2) is zero while the right side is a negative number. The net work is zero because all contact between the two putty pieces occurs at the fixed vertical plane of symmetry, the dashed line in the figure. The force of each putty piece on the other acts through zero displacement, so zero work is done by each piece.

$$\text{Net work} = \text{Work by left piece} + \text{Work by right piece} = 0 + 0 = 0. \tag{2.3}$$

Key Point 2.1 *There are no* external *forces acting on the putty pieces, so the net work* definitely *is zero. Obviously, it is not possible for Eq.(2.2) to hold because its left side is zero and its right side is a negative number. The difficulty is that it appears that energy is not conserved in the collision because all the initial kinetic energy vanishes, and classical mechanics allows for no other energies. Indeed, Eq. (2.1) is inapplicable here, because it does not account for energy being transferred to the atoms that comprise the putty, i.e., the putty's* internal *energy.*

In reality, when two pieces of putty collide, they distort, energy is transferred to the individual atoms, and the putty warms. The combined putty piece then cools back to room temperature in a quiescent state of thermodynamic equilibrium.[2] The collision alters the kinetic *and* internal

---

[2]Quiescence is relative to measurable macroscopic changes, while unceasing invisible molecular activity is in play.

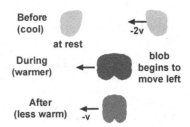

Figure 2.4: Head-on collision of the 2 putty pieces of Fig.2.3, but observed from a frame of reference with rightward velocity v relative to the laboratory.

energies of the putty pieces, and the correct conservation of energy equation is *net work* $= W = \Delta K + \Delta E$. Because $W = 0$,

$$\Delta E = -\Delta K = 2K_0. \tag{2.4}$$

I have ignored the heat process that would accompany the envisaged head-on collision. Were it included, Eq.(2.4) would be replaced by $Q = \Delta K + \Delta E$, where Q is the heating energy transferred TO the putty (which is negative here) from the environment, or

$$\Delta E = -\Delta K - |Q|. \tag{2.5}$$

Key Point 2.2 *For the head-on putty collision described above, the internal energy change of the putty pieces equals the lost kinetic energy, less the heating energy transferred to the environment.*

It is often of interest to investigate whether physics entities depend on the coordinate system used to analyse the system. In the head-on collision example, the laboratory frame is also the *centre of mass frame*. If instead, an observer is in a coordinate system moving rightward with speed v relative to the laboratory system, the left piece initially has zero speed and kinetic energy, as shown in Fig. 2.4. The right piece of putty has velocity $-2v$ and thus kinetic energy $4K_0$. After the collision, the final blob of putty with double the mass of the individual pieces, moves leftward with velocity $-v$, so its total kinetic energy is $2K_0$. The kinetic energy change is $2K_0 - 4K_0 = -2K_0$.

Although the individual kinetic energies differ in the new coordinate system, the change in kinetic energy is the same as in the *centre of mass* frame. The force of each putty piece on the other acts through a *nonzero* displacement, and the work done by each on the other is nonzero. However, Newton's third law assures equal and opposite forces, and the two forces act through the *same* displacement. Thus, the total work $W_{\text{tot}} = 0$ and thus, the change in kinetic energy is $\Delta K_0 = -2K_0$. The change in internal energy $\Delta E = 2K_0$, Eq.(2.4), holds also in this second frame of reference.

Key Point 2.3 *More generally, the internal energy change is the same in all inertial (non-accelerated) frames of reference.*[3] *Internal energy change is independent of the relative motion between the centre of mass of a system and the observer. It reflects a truly* internal *property of a system.*

This putty collision is an example of an *inelastic* collision, in which kinetic energy of two macroscopic objects is transformed into internal energy. The increase in internal energy is generally accompanied by a temperature increase, making the object(s) warmer than their environment, thereby triggering a heat process with the environment, namely, the nearby air, furniture, floor, walls, and ceiling. This is depicted by the shading change from black to dark grey between the second and last segments of Figs. 2.3 and 2.4.

2.2 HEAT DEFINED IN TERMS OF WORK

Classical mechanics is a field of study that is about 400 years old. In comparison, thermodynamics is only about 170 years old. As discussed in Sec. 1.2, the history of thermodynamics changed dramatically when the caloric theory was overthrown and it was realised that heat was an energy transfer. The first law of thermodynamics relates the two energy transfers, heat Q and work W with the internal energy change ΔE. That law, including the fact that internal energy E is a state function that depends only on the system's equilibrium state, is sufficient to define heat in terms of measurable work.

Key Point 2.4 *As we saw in Key Point 1.1, a thermodynamic equilibrium state of a system with zero net flow of energy through its boundary has unchanging measurable thermodynamic parameters such as temperature and pressure. The existence of equilibrium states enables a critical link between classical mechanics and thermodynamics.*

Envisage two different scenarios. First, imagine that a gas is allowed to slowly expand against a piston *isothermally* (constant temperature) as in Fig. 2.5(a) , labeled path *eg* in Fig. 2.5(c). Then a two-part path also connects *e* and *f* in Fig. 2.5(c). The first part is a pure work, adiabatic process *ef* with $Q = 0$. The second is path *fg*, a constant-volume work process, accomplished with a rotating paddle wheel inside the gas chamber. This raises the gas temperature and pressure at constant volume V_0.

Call the work done *on* the gas along path *eg* $W_{\text{isothermal}}$ and the work done *on* the gas along *ef* W_{ef}. The work done *on* the gas by the paddle wheel is denoted by W_{fg}. The total adiabatic work for path *efg* is

$$W_{\text{adiabatic}} = W_{ef} + W_{fg} \text{ (and } Q_{efg} = 0). \tag{2.6}$$

Along the isothermal path *eg*, the change in internal energy for the gas is

$$(\Delta E)_{eg} = Q_{eg} + W_{\text{isothermal}}. \tag{2.7}$$

---

[3]H. S. Leff & A. J. Mallinckrodt, "Stopping objects with zero external work: Mechanics meets thermodynamics," Am. J. Phys. **61**, 121–127 (1993).

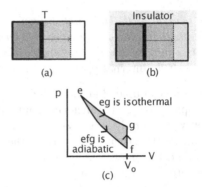

(c)

Figure 2.5: (a) Isothermal expansion of a gas from volume V_e to volume V_0, doing work W. (b) Adiabatic expansion of a gas with the same volume change, followed by (c) adiabatic energy increase at constant volume to state g, doing work W_{ad}. Paths eg and efg have the same initial and final equilibrium states.

For the adiabatic path efg, the net internal energy change is

$$(\Delta E)_{efg} = 0 + W_{\text{adiabatic}}. \tag{2.8}$$

Because for both paths, $\Delta E = E_g - E_e$, we may equate the right hand sides of Eqs. (2.7) and (2.8) to get the result

$$Q_{eg} = W_{\text{adiabatic}} - W_{\text{isothermal}}. \tag{2.9}$$

Key Point 2.5 *Heat energy Q can be defined solely in terms of work by comparing adiabatic and non-adiabatic processes that connect specified initial and final states, i and f, as in Eq. 2.9. More generally, we can write for any non-adiabatic process connecting i and f, $\Delta E = Q + W$, while an adiabatic process connecting the same two states has $\Delta E = W'$. Equating the two expressions for ΔE leads to $Q = W' - W$. This defines Q in terms of the in-principle measurable work quantities W and W'.*[4]

2.3 ENERGY IS NOT SUFFICIENT

In the development of thermodynamics, it became clear that the powerful principle of conservation of energy was not enough to explain physical behaviour of macroscopic systems. For example, energy conservation does not explain why a cup of hot coffee cools down to approximately room temperature, but a cup of room temperature coffee does not heat up

---

[4]It is not obvious that *any* two equilibrium thermodynamic states can be connected by an adiabatic path. The existence of an adiabatic path, either from i to f or *vice versa* is assured within the mathematical framework of Lieb and Yngvason, discussed in Sec. 10.2.1.

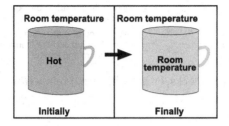

Figure 2.6: Hot cup of coffee cooling (from left to right).

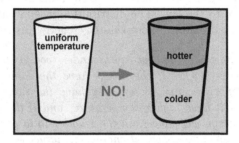

Figure 2.7: An impossible process where a uniform temperature distribution spontaneously becomes one with hotter and colder regions.

spontaneously. Figure 2.6 shows pictorially spontaneous, irreversible cooling. This is an example of an *irreversible* process, i.e., the reversed process is not observed. Similarly, the process shown in Fig. 2.7, where a system at uniform temperature spontaneously forms a colder and hotter region, is not observed. After many years of *never* observing this process, suggests that it is impossible. Energy conservation does not rule out impossible processes such as that in Fig. 2.7, which *can* conserve energy. The impossibility of such a process can be related to nature's propensity for energy to be distributed in an equitable way. This is the focus of the next section.

2.4 DISSIPATION, ENERGY SPREADING, EQUITY

2.4.1 Energy exchanges & equity

For the cooling of the hot cup of coffee in Fig. 2.6, the final temperature of the room and coffee is slightly higher than the original room temperature when the coffee was hot. The increase in the room's temperature occurs as the coffee releases energy to the room, which decreases the coffee's temperature significantly while the room's air and furnishings experience relatively small temperature increases. Energy has spread from the relatively small cup to the relatively large room, redistributing energy.

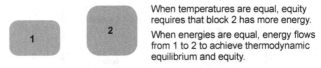

When temperatures are equal, equity requires that block 2 has more energy.

When energies are equal, energy flows from 1 to 2 to achieve thermodynamic equilibrium and equity.

Figure 2.8: Two copper blocks, either with equal temperatures or equal internal energies.

In a sense, relative to the surroundings, the cup had *too much* energy for its size and contents, while the surrounding room air, furniture, and walls had too little energy relative to the cup. The distribution was initially an *inequitable* one, but then nature brought a more *equitable* energy distribution.

Key Point 2.6 *The equity concept can be understood for the example of two different sized copper blocks. If the blocks have the same temperature, the larger one stores more energy. When they have the same internal energy, the smaller block will have higher temperature, and if the blocks are put in thermal contact, energy will flow from the smaller to larger block, 1 → 2, in Fig. 2.8. This phenomenon is spatial energy spreading. More generally, when energy is distributed inequitably among subsystems that are able to exchange energy by a heat process, the inequity is rectified by energy spreading. Equity exists only when the subsystems have the same temperature. The equity idea is based on nature being fair, and can be a helpful way to understand familiar thermodynamic behaviour.*

In some contexts the equity concept seems clear and is *intuitive*. Based on equity in nature, we expect the energy contained in one cubic meter (1 m^3) of air in a room with a uniform temperature to be one tenth of that in 10 m^3 of the air at the same temperature and pressure.

The concept of equity is inspired by the realisation that the energy spreading phenomenon cannot be explained solely using conservation of energy. Energy tends to spread only in certain ways, namely, ways that lead to increased equity. In addition to the first law of thermodynamics, which is based solely on defining internal energy and on energy conservation, another law, which can be related to the concept of equity, is needed. Although equity is no more than a concept at this point, it is given a quantitative basis in Sec. 3.3.3. What kind of law will assure that some processes are not observed in nature, while others are? Rudolf Clausius proposed a way in 1854, with the following statement, albeit with no mention of equity:

Key Point 2.7 *Clausius statement: Heat can never pass from a colder to a warmer body without some other change, connected therewith, occurring at the same time. (Note: This is a standard usage of the term heat, which I use in respect for Clausius. I prefer "Energy can never pass...")*

As rationale, Clausius wrote, "Everything we know concerning the interchange of heat between two bodies of different temperatures confirms

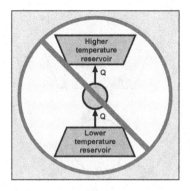

Figure 2.9: Graphic display showing that a violation of the Clausius statement of the second law of thermodynamics is not observed in nature. The small central circle represents a cyclic device whose physical state is the same before and after the (impossible) process illustrated.

this; for heat everywhere manifests a tendency to equalise existing differences of temperature, and therefore to pass in a contrary direction, i.e. from warmer to colder bodies." An equity-based interpretation is that a higher temperature object has excess energy relative to a lower-temperature one, and if the two objects can exchange energy, the higher temperature one will send energy to the lower temperature one, reducing inequity. If an energy transfer from the colder to hotter object occurred, it would lead to an even more inequitable distribution of energy. This suggests that the second law of thermodynamics is a statement that the physical world behaves in a way that promotes equity.

A graphic representation of the Clausius statement is shown in Fig. 2.9. Earlier, in 1851, Lord Kelvin (William Thomson) proposed a very different-looking statement of the second law of thermodynamics: *It is impossible, by means of inanimate material agency, to derive mechanical effect from any portion of matter by cooling it below the temperature of the coldest of the surrounding objects.* The term "mechanical effect" refers to what we now call *work*. A more modern, and perhaps more clear version of Lord Kelvin's statement was proposed by Max Planck in 1897: *It is impossible to construct an engine which will work in a complete cycle, and will produce no effect except the raising of a weight and the cooling of a heat-reservoir.* The Kelvin-Planck statement of impossibility is illustrated in Fig. 2.10.

Can this be interpreted in term of the equity concept? With only one reservoir at uniform temperature, there is no heat process induced by an inequity. Energy would have to be released *spontaneously* from a reservoir, and somehow be converted to pure work. Without inequity, there seems to be no reason for such an energy release, and no reason for the heat-to-work process to occur. The equity concept suggests that in order to generate work from heat, there must be at least two reservoirs, with energy flowing from the

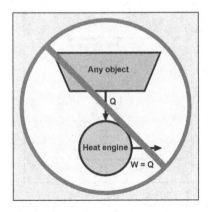

Figure 2.10: Graphic display showing that a violation of the Kelvin-Planck statement of the second law of thermodynamics is not observed. The shaded circle represents a heat engine whose physical state is assumed to be the same after the illustrated process as it was prior to that process. Recall that the same assumption was made in the Clausius statement, Fig. 2.9.

hotter to the colder, reducing the inequity, with some amount of the energy from the higher temperature getting converted to work. This describes a *heat engine.*

The term "engine" refers to a hypothetical system that consists of a *working substance*, e.g., a gas, that undergoes expansion, compression and heat processes, always beginning and ending in the same equilibrium state. The meaning of the Kelvin-Planck statement is that it is not possible to extract energy from a single object, thus cooling it, with the sole result of producing an equal amount of work. The conversion of *heating energy* to *external work*, cannot be done without a compensating effect, which is typically the unintended heating of something else, e.g., the surroundings. An automobile is a real-life heat engine that derives energy from high temperature combustion of a petrol-air mixture. Because it uses new fuel each cycle it is not a *closed* cycle that has a working substance that continually is brought back to its initial state. The cycles commonly considered in thermodynamics are highly-simplified idealisations of reality.

2.4.2 Carnot cycle & reversibility

A common idealised model of a heat engine is a gas, the "working substance," enclosed in a cylinder with a piston that is linked mechanically to a pulley that can lift a weight. The system can receive energy from a hot constant-temperature reservoir. For example, this might be a large body of hot water, say as large as the Atlantic Ocean, whose temperature does not

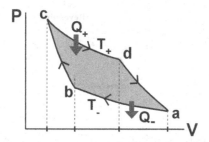

Figure 2.11: Reversible Carnot cycle consisting of two isothermal paths, alternated with adiabatic paths.

change perceptibly during the envisaged heat processes. Similarly, I assume the gas can release energy to a cold-water reservoir without changing that reservoir's temperature measurably.

A reversible Carnot cycle is illustrated on a pressure-volume diagram in Fig. 2.11. It has *isothermal* energy exchanges along paths ab and cd, alternated with a pure-work (adiabatic) expansion along da and an adiabatic compression along bc. The initial and final thermodynamic states of the gas are identical, so for each cycle, the first law of thermodynamics implies, $\Delta E = 0 = Q_+ + Q_- + W$ or $W = -(Q_+ + Q_-) = -W_{by}$. Note that $W < 0$ is the negative work *on* the working substance, while W_{by} is the positive work *by* the working substance. All processes are done infinitely slowly, so it is possible to run the cycle *backwards*, in which case it is not a heat engine, but rather is a refrigerator (or heat pump, depending on usage).[5] The reversible Carnot engine is named after Sadi Carnot, whose theoretical research on heat engines and in particular, their maximum possible efficiency, was of fundamental importance in the history of thermodynamics.

Key Point 2.8 *For the infinitely slow reversible Carnot cycle, the nonzero energy changes that occur during each cycle are confined to $Q_+ > 0$ from the hot reservoir and $Q_- < 0$ to the cold reservoir, and the work $W_{by} = Q_+ + Q_- > 0$ by the working substance on an external load.[6] Because the process is infinitely slow, the power output (time rate at which work is done) is zero. These properties hold for all reversible cycles.*

An example of reversible and irreversible heating processes is depicted in Fig. 2.12(a), where a system is heated sequentially using n reservoirs. In the limit $n \to \infty$ and $T_{j+1} - T_j \to 0$, this process is reversible. Figure 2.12(b) shows a system being heated irreversibly from T_1 to T_n, i.e., through a single finite temperature difference.

---

[5]This is explored in Ch. 8.
[6]The work on an external load could raise weights hanging from a pulley.

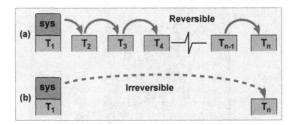

Figure 2.12: (a) Reversible heating, when $|T_{j+1} - T_j|$ approaches zero for all j, and $n \to \infty$. (b) Irreversible heating through a finite temperature difference.

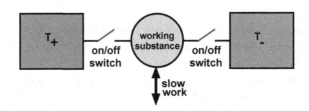

Figure 2.13: A thin insulating filament and switches enable ultra-slow energy transfers by opening the switches briefly, pulsing the flows. (Reproduced from H. S. Leff, "Reversible and irreversible heat engine and refrigerator cycles," Am. J. Phys. **86**, 344–353 (2018), with the permission of the American Association of Physics Teachers.)

Key Point 2.9 *The general definition of a reversible process is one for which infinitesimal changes in variables like temperature, pressure or volume can induce a reversal of the paths followed by both the system and environment.*

Rarely do we envisage heating processes that are both irreversible and quasistatic, but such processes are possible in principle as limiting cases of ultra-slow heating processes.[7] An infinitely slow process can be envisaged using a thin, insulating filament that maintains an arbitrarily slow energy flow rate through a finite temperature difference. The energy transfer rate can be lowered further by pulsing the transfers using *on/off* switches, as in Fig. 2.13, which shows a heat engine with reservoirs at T_+ and T_-. The engine operates quasistatically, but irreversibly with temperature differences between the working substance and the reservoirs. For example, a reversible Carnot cycle with temperatures T_h and T_c where $T_- < T_c < T_h < T_+$ could have irreversible energy exchanges with the reservoirs. (This is done in Sec. 8.2.5.)

Key Point 2.10 *The plot in Fig. 2.11 shows the Carnot cycle paths, but does not make clear just where or how the cycle performs external work. A simplified*

---

[7]An ultra-slow friction work process that converts work to internal energy can be irreversible and quasistatic. Reversal of the process converts more work to internal energy.

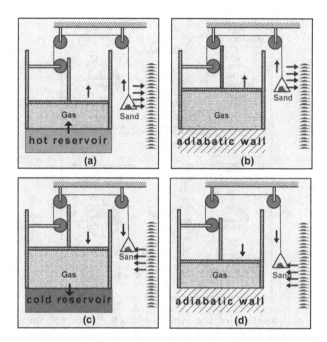

Figure 2.14: Idealisation of a heat engine: a gas in a cylinder with a movable piston linked by pulleys to sand grains on a pan. Processes (a, b, c, d) comprise a cyclic process for the gas, during which it does net positive work. This constitutes a model heat engine.

model in Fig. 2.14 is coupled to a pan and pulley to lift sand grains from lower to higher shelves. The linkage between the piston and pan is designed so the piston and pan move either upward or downward simultaneously.

The four steps of the sand-coupled cycle are:[8] (a) Sand is moved grain by grain horizontally from the pan to the vertical shelves while the gas is in contact with the hot reservoir. The gas, which has the temperature of the hot reservoir, does positive work and the piston and pan rise until the net force on the piston (from gas pressure and gravity) equals the weight of the sand-filled pan.[9] (b) The gas is isolated from its surroundings using a non-conducting (adiabatic) wall, preventing energy exchange by a heat process. Simultaneously, grains are moved from pan to shelves, raising the pan and piston. The gas does positive work during this segment as it loses energy and its temperature drops to the temperature of the cold reservoir. (c) The gas is put in contact with the cold reservoir and sand grains are moved from shelves to pan, lowering

---

[8]Details for this model and other cycles are in H. S. Leff, "Heat engines and the performance of external work," Am. J. Phys. **46**, 218–224 (1978).

[9]I assume the pulleys operate without slippage or dissipation, and that the frictionless piston's mass is appropriate for the envisaged processes.

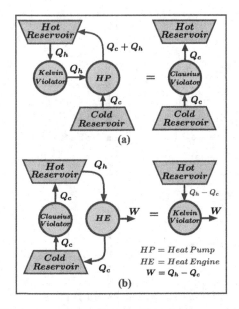

Figure 2.15: (a) Violation of the Kelvin-Planck statement implies violation of the Clausius statement. (b) Violation of the Clausius statement implies violation of the Kelvin-Planck statement.

the pan and piston. The gas does negative work at the temperature of the cold reservoir. (d) The gas is isolated using an adiabatic wall, and grains are shifted from shelves to pan, lowering the pan to its original height. The gas does negative work adiabatically and its temperature rises to the temperature of the hot reservoir. The gas now has its original temperature and volume. It has returned to its original thermodynamic state.

Steps (a), (b), (c) and (d), done infinitely slowly, result in positive net work done on the sand. Most discussions of Carnot cycles are abstract, with no indication of how external work is accomplished. With infinite slowness, the process can be reversed by reversing the movements of the sand grains, with more grains moved downward than upward. The sand's gravitational potential energy decrease provides the work to accomplish the reversed process.

Figure 2.15(a) shows that if the Kelvin-Planck statement (no heat-to-work with one and only one reservoir) is violated, *and* if the first law of thermodynamics holds, then the Clausius statement is violated too. The left side shows an impossible Kelvin-Planck heat engine delivering work to a refrigeration cycle, which removes Q_c from the cold reservoir and returns $Q_c + Q_h$ to the hot reservoir. The net effect is that energy Q_c is moved from the cold to the hot reservoir, in violation of the Clausius statement of the second law of thermodynamics.

Similarly, Fig. 2.15(b) shows that If the Clausius statement (no heat process from cool to warm without any other effect) is violated, *and* if the first law of thermodynamics holds, a heat engine can be combined with the Clausius-violating device such that the combination is equivalent to energy coming from a single reservoir and being converted entirely to work $W = Q_h - Q_c$. This would violate the Kelvin-Planck statement of the second law of thermodynamics.

Key Point 2.11 *Despite the fact that the Clausius and Kelvin-Planck statements of the second law of thermodynamics appear very different, each one implies the other, as shown in Fig. 2.15(a) and (b).*

2.5 AN OVERVIEW OF TEMPERATURE

2.5.1 International temperature scale

Temperature is so familiar to us that it might seem unnecessary to devote a section to it. However, it turns out that temperature is a rather subtle concept generally, and this section is devoted to clarifying how one can define and measure it. Thermodynamic systems can be hot, warm, cool, or cold, which we can detect qualitatively with our subjective senses. To quantify temperature, we humans have invented various types of thermometers and standardised temperature scales. The zeroth law of thermodynamics (ZLT) focuses on the phenomenon of thermal equilibrium:

Key Point 2.12 *ZLT: If systems, A and C, are each in thermal equilibrium with a third system, B, they are in thermal equilibrium with each other.*

This means that when placed in thermal contact with one another, systems A and C suffer no changes in their macroscopically observable properties. If system B is a glass tube thermometer with a narrow capillary containing a small amount of mercury, when it is in contact with A or C, the length of the mercury column will be the same. If B is put in contact with a hotter or colder object, the length of its mercury column will change, and is an indicator of temperature. Such a thermometer can be used to arrange a collection of systems according to their degree of hotness, with a longer mercury column taken to mean a higher temperature. This works provided the systems being observed are not so hot or cold as to expand the mercury column beyond the glass length or to freeze the mercury.

Note that the zeroth law of thermodynamics does *not* address the process of equilibration, namely, the *approach* to equilibrium, but only the case where two or more objects are *already* in thermodynamic equilibrium with one another. The study of equilibration processes lies within the domain of the second law of thermodynamics.

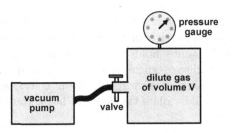

Figure 2.16: Schematic diagram of a constant-volume gas thermometer, which uses pressure to measure temperature.

For systems that are cold enough to freeze mercury, i.e., $< -38.8\,°C$, we can do a similar ordering by choosing B to be another suitable device, e.g., a dilute gas whose volume is kept constant. For such a constant-volume gas thermometer the pressure of the gas can be used to measure temperature. A symbolic view of such a thermometer is in Fig. 2.16. The reason this type of thermometer is considered is that all gases behave similarly if they are sufficiently dilute, even at temperatures near $0\,K$.

A constant-volume gas thermometer consists of a gas that is dilute enough that (a) intermolecular forces and energies are negligible because molecules are relatively far enough apart on average, and (b) quantum mechanical effects that occur when individual particle wave functions overlap are inconsequential. For such a gas, as the gas is made more dilute by pumping out molecules, the product $Pv \to constant$ at any temperature, where $v \equiv V/N$. The constant is larger when the gas is hotter, and thus, temperature can be measured, and even defined,[10] to be proportional to the product Pv using a constant-volume gas thermometer. As the gas becomes more dilute, it approaches a classical ideal gas that satisfies the equation (1.3), $Pv = kT$.

Generally, with a suitable definition of temperature, a gas has a pressure-volume-temperature functional relationship of the form[11]

$$\frac{Pv}{kT} = 1 + B(T)v^{-1} + C(T)v^{-2} + \dots \qquad (2.10)$$

Measuring the gas pressure for a highly dilute gas, where the terms beyond the first one in Eq. (2.10) are negligible, enables a definition of temperature in terms of a measurement of pressure. To understand this, put the gas in contact with a chosen *reference system* in a specific *reference state* labeled ref with a temperature called T_{ref}, and denote the measured gas pressure by P_{ref}.

---

[10]See the discussion of the ITS-90 temperature scale later in this section.

[11]This series was first investigated by Dutch physicist, Heike Kamerlingh Onnes, who was trying to fit experimental data for various substances. He called the series a *virial expansion*. The term *virial* had been introduced earlier by Rudolf Clausius, borrowing from the Latin *vis*, which relates to force or energy. The virial expansion was derived later using statistical mechanics, as explained in Sec. 4.2.

Now put the constant-volume gas thermometer in contact with a system at an unknown temperature T, wait for equilibration, and measure the new gas pressure p. Retaining the first two terms on the right of (2.10) to see the dependence on the degree of dilution, we see that the ratio P/P_3 is

$$\frac{(P/T)}{(P_3/T_3)} = \frac{1 + B(T)v^{-1}}{1 + B(T_3)v^{-1}} \approx 1 \text{ for sufficiently large } v. \tag{2.11}$$

It is common to use the triple point of water as a reference state, and to define the temperature of that state as $T_3 = 273.16$. Then, rearranging Eq. (2.11), the temperature of the system whose temperature was unknown is

$$T \equiv 273.16 \left(\frac{P}{P_3}\right). \tag{2.12}$$

Key Point 2.13 *The constant-volume gas thermometer enables a measurement of temperature in terms of the gas pressure using the triple point temperature of water, defined to be $T_3 = 273.16$ K, as a reference. This temperature is referred to henceforth as the Kelvin temperature.*

In practice, when a constant-volume gas thermometer is put in contact with a system in thermal equilibrium, the measured ratio P/P_3 approaches a constant value as more and more molecules are pumped from the gas thermometer. This enables the definition (2.12). Despite its attractive nature for teaching, the constant-volume gas thermometer is too cumbersome for laboratory use.

The International Temperature Scale (ITS-90) is the main defining system used by scientists. It uses an elaborate set of 17 fixed points for which temperature on the kelvin scale is assigned specific values. These fixed points are either triple points, such as the triple point of water, or melting points. The ITS-90 system spans a temperature range from 0.65 K to $10,000$ K, using four defining instruments: (1) Helium vapour pressure thermometers for the range $0.65 - 5.00$ K. (2) A constant-volume gas thermometer for $3.0 - 24.5561$ K. (3) Platinum resistance thermometers for the relatively large interval $13.8033 - 1234.93$ K. (4) At temperatures above 1234 K, radiation pyrometers are used to collect thermal radiation energy from an object in the visible and infrared. That energy is focused on a detector that converts it into an electrical signal that signifies the object's temperature. The latter conversion can be accomplished by a *thermopile* (an energy collector with thermocouples) or much more quickly, with a photomultiplier tube.

Notably, the ITS-90 system provides no standard for temperatures under 0.65 K. This is remarkable because researchers have routinely worked at lower temperatures in recent years, e.g., in the work of David Lee, Douglas Osheroff, and Robert Richardson on exotic phases of liquid helium-three (^{3}He) that exist only below approximately 2.6 mK.[12] These physicists shared the

---

[12]This is discussed in Sec. 8.6.4.

1996 Nobel Prize in Physics for their work, despite the fact that it entailed a temperature domain for which there is no international standard. The lack of such a standard is even more notable given the fact that adiabatic demagnetisation refrigerators are used routinely to bring nuclear spin systems into the microkelvin region and lower. A proposed extension of the ITS-90 would use four fixed points related to the melting curve of ^{3}He together with eight more fixed points where specific materials become superconductors. If adopted, this will provide an international standard that goes below 1 mK.[13,14] Low temperature phenomena are examined in Sec. 8.6.

Key Point 2.14 *The existing international temperature scale and its extensions all define temperature in terms of the measurement of another quantity, such as a gas pressure, electric resistance, or electric current.*

2.5.2   What is temperature?

While energy and entropy are the hallmark functions of thermodynamics, temperature has a different status. It arises in various ways.

Dilute gases. The kinetic theory of dilute gases combines Newtonian mechanics and some elements of probability theory to provide an expression for the average kinetic energy per molecule $\frac{1}{2}m\overline{v^2}$, and the kelvin temperature of the gas is defined as

$$T \equiv \frac{m\overline{v^2}}{3k} \text{ with } k = 1.381 \times 10^{-23} \text{J K}^{-1} \text{ (Boltzmann constant).} \qquad (2.13)$$

This is a common way of introducing temperature to students, but it should be noted that although temperature in (2.13) is proportional to the average molecular kinetic energy, that is not true for all systems, as clarified in Sec. 6.4.4. In any case, Eq. (2.13) is consistent with Eq. (2.12).

Key Point 2.15 *Generally, when quantum physics plays a role, temperature is not proportional to the average kinetic energy per molecule.*[15]

Carnot cycle. One of the most important properties of the reversible Carnot cycle, which I will examine in more detail later is that the energies exchanged with the hot and cold temperature reservoirs satisfy

$$\frac{|Q_+|}{T_+} = \frac{|Q_-|}{T_-} \quad \text{or} \quad \frac{Q_+}{T_+} + \frac{Q_-}{T_-} = 0. \qquad (2.14)$$

---

[13]In the absence of a standard, researchers still measure ultra-low temperatures and document their measurement techniques, which other researchers can verify and build upon.

[14]Recently, German researchers reported on single-atom quantum probes for accurate measurements of temperature in the 10^{-7} K range for small systems: Q. Bouton *et al.*, "Single-atom quantum probes for ultracold gases boosted by nonequilibrium spin dynamics," Physical Review X **10**(2020).

[15]H. S. Leff, "What if entropy were dimensionless?," Am. J. Phys. **67**, 1114–1122 (1999).

The temperatures T_+ and T_- are measured on the Kelvin scale. The (nonnegative) magnitudes $|Q_+|$ and $|Q_-|$ appear in Eq. (2.14), as do the algebraic values $Q_+ > 0$ and $Q_- < 0$. The existence of temperature is implicitly assumed in the very definition of a Carnot cycle, which requires two isothermal paths. Equation (2.14) can be used to *define* the kelvin temperature scale. We may choose either T_+ or T_- to be that of a reference system, say the triple-point temperature of water, $T_{\text{tr}} = 273.16$ K. The other reservoir temperature could be determined by measuring $|Q_+|$ and $|Q_-|$.

Key Point 2.16 *A Carnot thermometer could in principle be used to measure an unknown temperature. However, reversible Carnot cycles and thermometers exist only in our minds, and are useful only in thought experiments.*

Although many are comfortable with their understanding of temperature, fewer seem comfortable with entropy. This is true in part because we can feel when objects have higher or lower temperatures, but we cannot sense entropy directly. To the extent that one's comfort with temperature stems from the incorrect belief that temperature is always proportional to the average kinetic energy per particle, there is a false sense of security.

Key Point 2.17 *Internal energy increases with, but is not always proportional to, temperature. An intuitive view is that relative temperature is an indicator of a system's tendency to transfer energy via a heat process.*

Thermodynamic definition. The *thermodynamic definition* of temperature involves a partial derivative of the internal energy E with respect to entropy S, holding the volume and number of molecules fixed,

$$T = \left(\frac{\partial E}{\partial S} \right)_{V,N}. \tag{2.15}$$

Key Point 2.18 *The thermodynamic definition of temperature shows it to be a link between the variation of energy with entropy. Although this linkage takes on more meaning in Sec.3.3.1, after introducing the concave relationship between entropy and energy, it is not as intuitive as hotness and coldness.*

Key Point 2.19 *Practical ways to measure temperature entail the detection of pressure, electrical resistance, and current, which provide no clue of what temperature actually signifies physically. Theoretical definitions like Eq.(2.15) are abstract. Neither the types of thermometers nor the abstract theoretical definitions help us understand what temperature means physically. Perhaps the most physically meaningful way to understand temperature, is as an indicator of the direction of energy flow in a heating process, from hotter to colder.*

2.6 CONNECTING ENERGY & ENTROPY

Figure 2.17: Rudolf Clausius, who conceived the concept of entropy. (Courtesy of The Linda Hall Library of Science, Engineering and Technology.)

2.6.1 Clausius's main contributions

Rudolf Clausius (1822–1888), shown in Fig. 2.17, was a major contributor to the development of thermodynamics as we know it. His work followed that of Sadi Carnot, whose *Reflections on the Motive Power of Fire* (1824) showed that there is a maximum efficiency for any heat engine operating between two reservoirs with temperatures T_+ and $T_- < T_+$. Subsequently, an expression for that maximum, now called the *Carnot efficiency*, was found to be

$$\text{Maximum efficiency} = \frac{\text{work generated}}{\text{heating energy input}} = 1 - \frac{T_-}{T_+}. \qquad (2.16)$$

Notably, this depends only on the two reservoir temperatures. Carnot's work was based on caloric theory, before the concept of energy was appreciated.

Clausius approached the subject differently and his work, along with the contributions of James Joule, William Thomson, and others in the mid-19th century, led to our present understanding of the role of energy in thermodynamics. Among the most important ideas to emanate from the work of Clausius, are:

(1) Thermodynamic systems store internal energy E, the *average* total kinetic and potential energy stored within the system's boundary. It is an average, because thermodynamic systems continually exchange small amounts of energy with their surroundings, i.e., the total system energy fluctuates. We now know from statistical mechanics, whose development followed Clausius, that fluctuations are much smaller than E itself, and the time-averaged value is sufficient for most purposes. The existence of internal energy obviated the need for caloric to explain heat phenomena. The overthrow of caloric theory brought into question all results that were derived assuming its existence. It was fortuitous that Carnot's main results, which he obtained assuming the existence of caloric, are consistent with results using the energy paradigm.

(2) The internal energy E is a function of each thermodynamic equilibrium state and does not depend on how that state was achieved. In order to pin down a specific value of internal energy, we must specify its equilibrium state, e.g., by giving its temperature and pressure, and define the energy of a reference state, as was done with temperature. This is not done typically. Rather, we restrict our attention to energy changes, given by the first law of thermodynamics, Eq. (1.9), $\Delta E = Q+W$, where neither Q nor W is a function of the state of the system, but depend on the specific thermodynamic *process*. Remarkably, the sum of heat and work in the first law equation is the change ΔE in the state function E, which is independent of the path followed. In words, the internal energy has no history dependence; it is determined fully by the thermodynamic state, regardless of how a system got there.

(3) There are two laws (and not just a single law, as had been believed) of thermodynamics.[16] The second law of thermodynamics can be framed in terms of a new thermodynamic function of state called entropy. Furthermore, the entropy of any adiabatically isolated, closed system[17] within the universe cannot decrease

Key Point 2.20 *Heat and work are dependent on the specific path followed in a thermodynamic process from state i to state f. However, the change in internal energy is the same for all paths from i to f, and is the sum of the process-dependent energy transfers, Q and W.*

Key Point 2.21 *That there are two separate laws of thermodynamics, not one law, was another revolutionary idea. The first law of thermodynamics is clearly an energy-based law that combines the concepts of energy storage and transfer with energy conservation, plus the existence of a state function called internal energy. The second law of thermodynamics extends this, addressing which energy exchanges are and are not possible. An infinite variety of processes that satisfy energy conservation are possible, yet do not occur, e.g., the spontaneous flow of energy from a cooler to a warmer object.*

Key Point 2.22 *In contrast with energy, which is conserved, the total entropy value for the universe (isolated because there is nothing else) cannot decrease. Entropy cannot be destroyed globally, and continually increases as irreversible physical processes occur in the universe.*

Key Point 2.23 *An important point that is often left unsaid is that any object in thermodynamic equilibrium with its surroundings, i.e., unchanging in a measurable way, has a numerical value of entropy. The very existence of measurable numerical values of entropy for various materials takes entropy from the mysterious to the concrete. Tabulated data exists for the entropy of*

---

[16]Later, in the beginning of the 20th century, it was discovered that there is also a *third* law of thermodynamics, which is explained in Sec. 6.3.

[17]A closed system is one that cannot gain or lose material particles like atoms, electrons, neutrinos, and the like through its boundary.

the elements and many compounds. These are useful in doing a variety of calculations in chemical physics, biophysics and physical chemistry.[18]

2.6.2 Clausius entropy & entropy increase

Clausius gave a straightforward definition of entropy change due to a reversible energy exchange between a system of interest and one or more other systems. An example is simply heating a system with a tiny amount of energy dQ joules when it has temperature T, changing the system's entropy by

$$dS = \frac{dQ}{T}.$$ (2.17)

Notice that dQ is the notation used for an *infinitesimal* energy transfer and dS is used for the corresponding infinitesimal entropy change. Similarly, I shall use the symbol Q for *finite* energy transfers during a heat process and S for the value of entropy in an equilibrium state.[19] If a large number N of such tiny energy transfers and measurements is made, the total entropy change is defined to be the sum of the individual tiny processes,

$$\Delta S = \frac{dQ_1}{T_1} + \frac{dQ_2}{T_2} + \frac{dQ_3}{T_3} \cdots + \frac{dQ_N}{T_N}.$$ (2.18)

The number N can be enormous and the individual steps dQ_i correspondingly tiny so that

$$Q = dQ_1 + dQ_2 + dQ_3 + \cdots + dQ_N \quad \text{where } N \to \infty.$$ (2.19)

The point is that although each individual step brings tiny (treated as infinitesimal) energy and entropy changes to the system of interest, the sum of many such steps can amount to a finite change.

Key Point 2.24 *Remarkably, using his definition of entropy, Clausius was able to deduce the following sweeping statement of the second law of thermodynamics: "The energy of the universe is constant; the entropy of the universe tends toward a maximum."*

Equations (2.17) and (2.18) imply that the units of entropy are J K^{-1}, which is explored in Sec. 6.4. Despite the fact that the entropy of the universe cannot decrease, the entropy of a non-isolated object, such as a glass of water, *can*

---

[18] An example of such data is in Fig. 6.3.

[19] Traditionally in calculus, d is used to connote a tiny quantity called a *differential*. Here I use the symbol d for a tiny quantity of an energy *transfer*, of which heat and work are the prime examples. It is sometimes referred to as an inexact differential because it applies to energy transfers that are path-dependent. In contrast, a small change in entropy S is denoted by dS, a quantity that is independent of the path, being determined solely by the initial and final equilibrium states.

Universal Entropy Clock
(turns faster when more or larger physical processes occur)

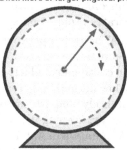

Figure 2.18: The universal entropy clock, an imagined device that rotates clockwise only. It rotates faster when something *big*, e.g., a supernova explosion, occurs. It *cannot* rotate counter-clockwise, which would correspond to a forbidden entropy decrease for the universe.

decrease. But as it does, the principle of entropy increase assures that the entropy of other non-isolated objects *must* increase by at least as much. For example if an ice cube is placed in a glass of room temperature water, the water becomes colder and its entropy decreases. However the melting of the ice and heating of the water increases the entropy more,[20] and the entropy increase exceeds the decrease. The total entropy of the universe will increase.

Key Point 2.25 *The total entropy of the universe continually increases as stars are born and die, tectonic plates within Earth's crust collide, car engines run, plants grow, water flows, musicians play, singers sing, and, in fact, whenever anything happens in the universe. All activity creates entropy. The entropy of a non-isolated part of the universe can decrease, but its total entropy cannot decrease.*

The continual entropy increase of the universe whenever any physical process occurs is depicted with an imagined entropy clock in Fig. 2.18. This has a single needle that moves clockwise when anything happens, i.e., any physical process occurs. I have seen tee shirts and buttons that say, "Entropy Isn't What It Used To Be." Indeed, the entropy clock keeps increasing and will never return to the value it had when you began reading this sentence.

2.6.3 Systems not in equilibrium

Not everything is in equilibrium with its environment. Certainly, living things are not. An important example is *you*. Your body temperature is typically

---

[20]I've ignored the effects on the glass and surrounding air. Also if we wait a while there will be a further entropy increase as the water comes to equilibrium with the room air.

different from room temperature. Indeed, when the room and you have the same temperature, you feel uncomfortably warm because you cannot get rid of the energy your body generates rapidly enough.

The maintenance of normal body temperature (about $98.6°F = 37°C = 310K$) is achieved by a combination of two things. One is the bodily chemical reactions that occur continually, which warm you and prevent your temperature from lowering below normal body temperature. These *exothermic* chemical reactions heat your body and its surroundings. The other is the transfer of energy from your body, keeping it from getting too warm. Part of this is by transpiration, the evaporation of water through your skin pores. The evaporation part helps because it takes a relatively large energy to vaporise water. This is why we feel cool when we emerge from a swimming pool or bath soaking wet. When the outdoor temperature exceeds body temperature, we perspire. Cooling occurs because we lose energy by vaporising the water released through our skin. Each of these processes is accompanied by an increase in entropy of the universe.

Key Point 2.26 *Humans maintain a temperature different from their surroundings. They continually transfer energy and water through skin pores to the environment, and regularly ingest food and dump waste liquids and solids to the environment. Humans are* not *in thermodynamic equilibrium. This is true for all living animals and plants. The only animal or plant that is in thermodynamic equilibrium with its surroundings is a dead one, and even then only after all internal chemical reactions and exchanges of energy and matter with the surroundings have ceased. Pretty gory, but true.*

2.6.4 Disgregation

Prior to his introduction of entropy in 1865, Rudolf Clausius published a series of memoirs that show his evolving ideas that led him to entropy. He referred to molecular motion of atoms as "heat" H.[21] In modern language, what Clausius called H is the average kinetic energy of a system's molecules. He believed that heat loosened the connections between molecules and increased their average separation distances. This involved both internal work by intermolecular forces, together with any external work on the system. To quantify this tendency to separate molecules he postulated the function Z, the "disgregation," and assumed the total work separating the molecules was proportional to temperature. By a multistep argument employing the first law of thermodynamics, he arrived at the result $dQ/T = dH/T + dZ$. The left side is the change dS in the Clausius entropy, dH is the change in "heat content" and dZ is the corresponding change in disgregation.

---

[21]This should not be confused with the state function *enthalpy*.

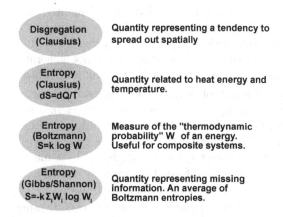

Figure 2.19: Chronological view of the evolution of entropy, beginning with Clausius's early introduction of the now defunct *disgregation* and ending with Shannon's information theory.

Martin Klein pointed out that Clausius believed disgregation had more physical significance than entropy itself.[22] Classical statistical mechanics shows that dH arises from changes in molecular momenta and dZ comes from configurational, i.e., positional, changes as molecules separate.

Disgregation as a measure of particles moving away from one another is suggestive of spatial spreading of molecules. Those molecules carry energy, so as they spread, so does energy. Because Z is determined by particle configurations through the volume V, it alone cannot describe entropy, which changes for constant-volume heating when $dZ = 0$, and molecules do not become more widely separated. An entropy chronology that puts disgregation in perspective, is shown in Fig. 2.19.

2.6.5 Entropy as missing information

The concept of entropy arises in a completely different context, namely, information theory, developed in the 1940s by Claude Shannon. His focus was communication theory, not physics. He was interested in the process of sending and receiving messages along communication channels such as long cables. Inevitably, errors are introduced and there is uncertainty about the accuracy of received messages. Shannon introduced a mathematical function that gives a measure of uncertainty, and upon the advice of John von Neumann, called it entropy. A more apt term is *missing information*, which I'll call *MI*.

Mathematically, Shannon's *MI* turns out to be identical to the entropy functions that arose earlier in the development of statistical physics by

---

[22]M. J. Klein, "Gibbs on Clausius," Hist. Stud. Phys. Sci. **1**, 127–149 (1969).

Boltzmann and independently by Gibbs. Subsequently, in 1957, Edwin T. Jaynes showed in detail how statistical physics can in fact be derived using Shannon's framework. Information theory provides a different view and interpretation of statistical mechanics. It does not generate any results for equilibrium statistical mechanics that Boltzmann and Gibbs did not know.

One important result of Jaynes' work is establishing a *bona fide* view of thermodynamic entropy as a measure of *missing information*. Shannon's information entropy provides a pathway for obtaining the probabilities of individual macrostates under specified conditions such as constant temperature or constant energy. Those probabilities are found by a mathematical process that maximises Shannon's information entropy function, MI. Boltzmann obtained the same results by maximising probabilities.

The missing information metaphor can be useful, especially in understanding that descriptions of macroscopic matter necessarily discard enormous amounts of information about system details, working ultimately only with a small number of macroscopic variables such as pressure, volume, and temperature. Our systems are assumed to have 10^{20} or more molecules. With overwhelming probability, a system with so many molecules will evolve to the *macrostate* with the largest number of possible microstates. For us observers, the microstates are invisible. We can *only* observe the macrostate, and the specific microstate can change continually over time without us being aware of it.

Jaynes showed that the missing information for a system's state j is proportional to $-\ln(W_j) = \ln(1/W_j)$, where W_j is the probability of the system being in state j. The higher the probability W_j, the smaller the missing information. That is, the more certain you are that the system is in microstate j, the less uncertain you are. If the probability for any state is 1, then $MI = 0$ for that state because $\ln(1) = 0$. Such a state would occur with certainty.

Key Point 2.27 *For an isolated system with energy E, if there are Ω microstates, then information theory tells us that the states are equiprobable, $W_j = 1/\Omega$. Thus $MI \propto \ln(1/W_j) \propto \ln \Omega$. The uncertainty about the specific microstate increases with the number of microstates.*

Key Point 2.28 *For a non-isolated system in contact with a constant-temperature reservoir, Jaynes showed that the MI is proportional to the average of $\ln(1/W_j)$, namely, $\sum_j W_j \ln(1/W_j) = -\sum_j W_j \ln(W_j)$. He linked this to the thermodynamic entropy S, and evaluated the probabilities W_j. He found that if state j has energy E_j, then $W_j \propto e^{-E_j/\tau}$, where τ is a shorthand notation for $k \times$ temperature T. We will see in Sec. 3.4 that this agrees with Boltzmann's findings. The factor $e^{-E_j/\tau}$ is the much-used "Boltzmann factor."*

2.6.6 Confusion about entropy

The role of Helmholtz and Boltzmann. In 1882, Hermann Helmholtz described entropy with the German word *Unordnung*, which translates to *disorder*.[23] In 1898, in his *Lectures on Gas Theory*, Ludwig Boltzmann wrote, that the world is enormously complicated, and becomes more so as time marches on. To explain this, he concluded that initially, the world must have been in a highly improbable, *ordered* state and proceeded to a highly *disordered* one.

Key Point 2.29 *Given Boltzmann's status as a luminary, textbook writers and researchers adopted his oversimplified one-word disorder metaphor to describe entropy. Despite substantial criticisms of the term disorder, some dictionaries still define entropy as a measure of disorder, and many textbook authors have adopted that term. I say more about this dilemma in Sec. 7.3.1.*

Entropy and heat engines. Clausius demonstrated the principle of entropy increase using the mathematical model of a heat engine cycle as a convenient tool that entails both work and heat.[24] Many modern day introductory physics books develop entropy using heat engine cycles, so perhaps unsurprisingly, students incorrectly think that entropy has some deep connection to heat engines. Entropy is a function of the thermodynamic state of a system and has no *special* linkage with heat engines.

Confusion in textbooks. Confusion has surrounded entropy since its inception. In early editions of his book, *Theory of Heat*, James Clerk Maxwell followed a suggestion of Peter Tait, and took entropy to *decrease*(!) during spontaneous processes in an isolated system. He corrected this in subsequent editions, saying it introduced "great confusion into the language of thermodynamics."

In his book, *Entropy; or Thermodynamics from an Engineer's Standpoint*, James Swinburne wrote: "As a young man I tried to read thermodynamics, but I always came up against entropy as a brick wall that stopped my further progress. I found the ordinary mathematical explanation, of course, but no sort of physical idea underlying it. No author seemed even to try to give any physical idea." In the realm of thermodynamics, this statement could have been written yesterday. However, it was actually written prior to the publication date of Swinburne's book in 1904, well over a century ago! Earlier (and likely unknown to Swinburne), James Clerk Maxwell, Ludwig Boltzmann, and Josiah Willard Gibbs developed a statistical framework that provided a more *physical* view of entropy, accounting for the kinetic and potential energies of a system's molecules.

---

[23]R. Baierlein & C. A. Gearhart, "The disorder metaphor," Am. J. Phys. **71**, 103 (2003).
[24]Clausius's argument is given in Sec. 8.1.2.

CHAPTER 3

Entropy: Energy's Needed Partner

CONTENTS

3.1 COMPOSITE SYSTEMS

No system that we deal with on Earth is completely isolated from its environment. Familiar systems like the amount of air in a house can change because of small pores in walls and cracks around windows and doors. The air can gain energy from solar inputs through walls and windows, and can lose energy in winter. The water in a lake can decrease by leakage through small

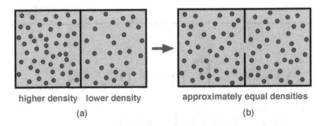

higher density   lower density   approximately equal densities

(a)   (b)

Figure 3.1: (a) Partitioned container with higher particle density on left. (b) With a small opening in the partition, the distribution equalises.

pores in the Earth or by evaporation, and can increase because of runoff from the ground and rainfall or snow.

Even the best insulated container cannot keep its interior hot or cold indefinitely. A well-insulated wall simply increases the amount of time it takes for energy to penetrate it. Insulation *slows* the energy transfer rate by heating processes, but over sufficient time, the temperatures outside and inside even a well-insulated container will become the same.

We think of sealed cartons, houses, and other closed containers as having unchanging volumes, but small volume changes can be induced by atmospheric pressure changes. Similarly, as the temperature varies, materials expand and contract. A dramatic example is a blown up, tied off balloon that is put in a freezer. Cooling reduces the air pressure *inside* the balloon, and the atmospheric pressure compresses the balloon and the inside air.

To assess why such phenomena occur, consider a container filled with say, N molecules of a gas. The volume is divided into two equal-volume chambers. The left chamber contains, say N_L molecules, and the right chamber contains N_R molecules, with $N_L + N_R = N$. Suppose $N_L > N_R$, i.e., there are more molecules in the left chamber, as in Fig. (3.1a). In Fig. 3.1b, a small opening is made in the dividing partition, and molecules redistribute, leading to an approximately equal number in each chamber. This happens because before the hole is opened in the partition, more left chamber molecules will collide with the partition each second than will right chamber molecules, because the density on the left is higher. The initial number moving from left to right each second will exceed that from right to left, equalising the densities. Even with equal densities, the numbers of molecules in the chambers can fluctuate.

Key Point 3.1 *For the two-chamber system in Fig. 3.1a, it is more probable initially for a left → right passage through the partition opening than for a right → left passage. The physical behaviour is statistical, requiring probabilities. When the numbers of molecules in the two chambers become equal, the probabilities for passage in either direction become equal, with (typically negligible) temporal fluctuations about equilibrium. Large fluctuations, though possible, are very low probability events.*

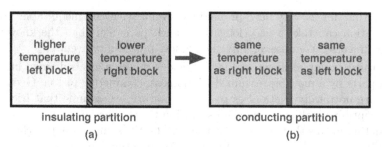

Figure 3.2: (a) Equal-volume blocks, separated by an adiabatic wall, with higher temperature in the left block. (b) The partition is replaced by a conducting one, enabling temperature equilibration.

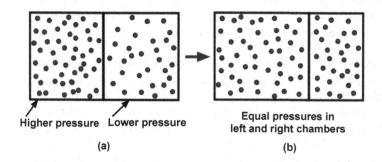

Figure 3.3: (a) Container with fixed conducting partition separating equal chamber volumes. (b) The partition becomes a freely moving frictionless piston that repositions.

Next, consider the two-chamber system in Fig. 3.2(a). Each chamber holds a gas, liquid or solid. The important point is that the temperature of the left side is initially higher than that of the right side. Initially, the wall separating the systems is taken to be a very good insulator, so good that over a period of hours, no temperature change can be detected by our very best thermometers. Then the separating wall is replaced by one that is a very good conductor, and temperatures quickly equalise, as in Fig. 3.2(b). For simplicity, suppose the left and right materials are identical except for their initial temperatures. In this case, the internal energy is greater in the left chamber initially.

Key Point 3.2 *When an insulating wall is replaced by a conducting wall, energy can flow in either direction. Nature tends to work toward a more equitable distribution of energy, and energy flows from hot to cold until equity exists. The two systems will then have equal temperatures. An inequitable energy apportionment is possible while still conserving energy, but is extremely improbable, and this has never been observed.*

Waiting for a very low probability event with an inequitable energy distribution can take a very long time, perhaps longer than the known age of our universe! We observe only the highest probability events.

A third example, shown in Fig. 3.3, is one where a gas is divided into two equal parts by a movable, assumed-frictionless, conducting piston. Given that there are no totally frictionless pistons in nature, we assume that friction is imperceptibly small over typical observation times. The fact that the piston is conducting means that the temperature will be the same in the two chambers. Initially, the chamber volumes are equal. Assume there are fewer molecules in the right chamber so the pressure on its walls is lower. Once the constraint holding the piston is removed, the larger pressure in the left chamber pushes the piston rightward. It might overshoot the point where the pressures are equal and oscillate a little, finally settling down to an equilibrium position, albeit with some jiggling, i.e., pressure fluctuations. With overwhelmingly high probability, the pressures will be measurably equal.

Key Point 3.3 *In the latter examples, either an impermeable wall, insulating wall, or fixed conducting wall separates two subsystems of a composite system initially. Two things deserve emphasis: (i) Upon removal of a constraint, either opening a hole to allow diffusion; replacing an insulating wall with a conducting one; or releasing a fixed piston, enabling it to move frictionlessly, the system redistributes energy. (ii) The energy exchanges lead to a more equitable distribution of energy, and the process culminates when the most equitable distribution exists. This is the most probable thermodynamic state. For a work process, Newton's laws and thermodynamics both apply, leading to pressure equilibrium. Small, typically negligible, fluctuations occur and large fluctuations are extremely improbable.*

3.2   ENTROPY & PROBABILITY

3.2.1   Why probabilities?

The discussions above have taken us into the realm of *highly likely*, as opposed to *definite* events. How can we use this realisation to make the physics of macroscopic matter more meaningful and productive? In particular how can these ideas lead from energy to entropy? I continue to examine composite systems with two systems that can share energy. As we have seen in Figs. 3.1, 3.2 and 3.3, such sharing of energy can occur via diffusion of matter, a heating process, or a work process. In each case, energy and/or material leaves one region and enters another.

Commonly a system exchanges energy with its environment, which maintains an approximately constant temperature and atmospheric pressure. This *environment* consists of the surrounding air, room furnishings and equipment, walls, floor, and the like. To account for the surroundings, we

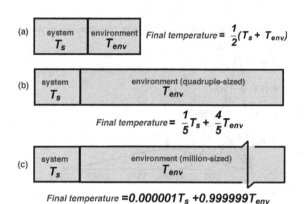

Figure 3.4: (a) A system in contact with a twin identical system as the environment. (b) The environment is four times larger than the system. (c) The environment is one million (10^6) times larger than the system.

can envisage a succession of ever larger model *environments*, with the system and environment having initial temperatures T_s and T_{env} respectively, as in Fig. 3.4. The environment in Fig. 3.9(a) is chosen to be identical with the system, leading to a final equilibrium temperature of $T_f = \frac{1}{2}(T_s + T_{env})$. This is obviously *not* a typical environment, whose temperature variations would be imperceptible. In (b) the environment's volume and mass are four times the system's, leading to a final temperature $T_f = \frac{1}{5}T_s + \frac{4}{5}T_{env}$, and in (c) where the environment's size is 10^6 times that of the system, $T_f = 0.000001T_s + 0.999999T_{env} \approx T_{env}$. Because our actual environment is so enormous, when it is put in thermal contact with the system, energy spreads and is shared until the final temperature of both system and environment are measurably equal, but the temperature of the environment has not changed measurably.

Although we know from experience that the latter picture describes reality, what determines the extent to which energy is shared and the system's volume adjusts? How does nature *know* when to stop sharing energy? *Any* mix of system and environment energies that adds to the original total energy would seemingly work, but that is not what is observed. The previous examples suggest that the physical law that governs this is probabilistic in nature.

One of the first to incorporate probabilities in physics was James Clerk Maxwell, most well known for his Maxwell equations of electromagnetism. In the late 19th century, he derived the probability distribution of velocities in a dilute gas. In equilibrium at some specific temperature, a relatively small fraction of the molecules have very low or very high speeds. The probability distribution is peaked at an intermediate speed, the most probable speed. Denoting the probability distribution by $f(v)$, $f(v)dv$ is the probability that a randomly chosen molecule has speed v in the interval $(v, v + dv)$. Because

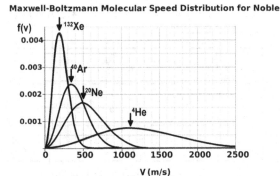

Maxwell-Boltzmann Molecular Speed Distribution for Noble

Figure 3.5: Speed distributions in gases at the same temperature peak at higher speeds for smaller mass atoms.

each molecule must have *some* speed, the sum of the probabilities must equal one, i.e., $\int_0^\infty f(v)dv = 1$. The Maxwell-Boltzmann speed distribution is

$$f(v) = \left(\frac{m}{2\pi kT}\right) 4\pi v^2 e^{-(\frac{1}{2}mv^2/\tau)} \quad (m = \text{molecular mass}; \tau \equiv kT). \quad (3.1)$$

The distribution $f(v)$ in Eq. (3.1) is shown in Fig. 3.5 for four noble gases, helium, neon, argon and xenon, at temperature 298.15 K. The lightest of these, helium-4 has the lowest atomic weight and the highest most probable speed, 12900 ms⁻¹. This makes sense because for a dilute gas, the average kinetic energy is proportional to the molecular mass, i.e., *average kinetic energy per molecule*, $\overline{E}/N = (\frac{1}{2}m\overline{v^2}) \propto \tau$. For given τ, the average $\overline{v^2} \propto m^{-1}$, consistent with the most probable speed decreasing with increased molecular mass. Although the Maxwell-Boltzmann speed distribution goes back to Maxwell's 1860 heuristic derivation and subsequent work by Ludwig Boltzmann, a detailed experimental analysis of its validity was not done until 1955.[1]

3.2.2 Boltzmann, probability & entropy

In 1877, Ludwig Boltzmann extended the use of probabilities to the field of statistical mechanics, which he pioneered. He realised the need to appeal to a probabilistic approach. His intuition and creativity led him to consider a composite system comprised of two bodies, with different temperatures.[2] His main idea was that, with very high probability, energy would become distributed in a specific way that corresponds to *thermal equilibrium*. Boltzmann introduced his own version of the entropy function, linking it

---

[1] R. C. Miller & P. Kusch, "Velocity Distributions in Potassium and Thallium Atomic Beams," Phys. Rev. **99**, 1314–1314 (1955).

[2] One of these subsystems could be the constant-temperature *environment*.

Figure 3.6: The author and Boltzmann's tombstone at the Central Cemetery in Vienna, Austria in 1998. (Reprinted with the permission of Ellen Leff.)

with the latter probability. He agreed with Clausius that the entropy of the composite system must increase or stay the same upon release of an internal constraint in the composite system. For Boltzmann, that increase reflected the composite system evolving to the state of maximum probability. Despite the fact that the work of Clausius was in the *exact* science of *thermodynamics* and Boltzmann's theory was *statistical*, it turns out that these two approaches predict energies and entropies consistent with one another.

On Boltzmann's tombstone in the Vienna Central Cemetery is written the now famous Boltzmann entropy (see Figure 3.6)

$$S = k \log W(E) \quad \text{the } E \text{ dependence is not on the tombstone.} \quad (3.2)$$

Of particular interest is a *target system* plus a constant-temperature (and pressure) reservoir. Then $W(E)$ represents the probability of a particular target system *energy E* and environment energy $(E_0 - E)$. It is inspired by the German word for *probability, Wahrscheinlichkeit.*

Boltzmann used the notation "log," which now connotes the *common* logarithm to the base 10, but evidently meant it to mean the *natural* logarithm "ln" to base e.[3] I was fortunate to be able to visit Boltzmann's gravesite (see Fig. 3.6). As I stood there, I had a feeling that I was at a sacred place. I still get goose bumps when I think of that visit.

Key Point 3.4 *Because $0 \leq W(E) < 1$, $S(E) = k \ln W(E)$ is negative, which is inconsistent with tabulated numerical values of entropy.*

One way to make the Boltzmann entropy non-negative is inspired by information theory, replacing $W(E)$ by $1/W(E)$ in Eq. (3.2). A second way is equivalent to adding a positive constant to Eq. (3.2).[4]

---

[3]Private communication with Don Lemons.
[4]Neither of these ways is the way history evolved.

In the first way, imagine an isolated target system with energy E, and suppose $\Omega(E)$ microstates have this energy. In information theory, when there are $\Omega(E)$ outcomes (microstates), with no constraints other than the sum of probabilities must be one, then the probability of the system being in each microstate is the same: $W(E) = 1/\Omega(E)$, and the missing information is $MI = \sum_j W_j(E) \ln(1/W_j(E)) = \ln \Omega(E)$. This suggests redefining Eq. (3.2) as $S = k \ln(1/W(E))$, in which case

$$S = k \ln \Omega(E) \geq 0 \quad \text{because } \Omega(E) \geq 1. \tag{3.3}$$

The second way to obtain a positive entropy expression is to consider a composite system with total energy E_0, comprised of a target system with energy E and a reservoir with energy, $(E_0 - E)$. Denote the number of accessible states for the target system by $\Omega(E)$ and the number of states accessible to the reservoir by $\Omega_r(E_0 - E)$. When the target system has energy E, the composite system has $\Omega_{\text{comp}}(E, E_0 - E)$ states, which can be written,

$$\Omega_{\text{comp}}(E, E_0 - E) = \Omega(E) \times \Omega_r(E_0 - E). \tag{3.4}$$

That is, for each target system state, there are $\Omega_r(E_0 - E)$ reservoir states and for each reservoir state, there are $\Omega(E)$ target system states. The total possible number of states is

$$\Omega_0 \equiv \sum_E \Omega(E) \times \Omega_r(E_0 - E). \tag{3.5}$$

Key Point 3.5 *The principle of equal a priori probabilities postulates that each of the Ω_0 microstates for the isolated composite system is equiprobable, with probability $1/\Omega_0$, i.e., there is no reason to favour one state over another.*

The probability $W_{\text{comp}}(E, E_0 - E)$ of $\Omega_{\text{comp}}(E, E_0 - E)$ states in Eq. (3.4) occurring is therefore,

$$W_{\text{comp}}(E, E_0 - E) = \frac{\Omega(E) \times \Omega_r(E_0 - E)}{\Omega_0}. \tag{3.6}$$

This implies the Boltzmann entropy for the composite system,

$$\begin{aligned} S_{\text{comp}}(E, E_0 - E) &= k \ln[W_{\text{comp}}(E, E_0 - E)] \\ &= k[\ln \Omega(E) + \ln \Omega_r(E_0 - E) - \ln \Omega_0] \\ &= S(E) + S_r(E_0 - E) + \text{constant}. \end{aligned} \tag{3.7}$$

In the second line, the first term is for the target system and the second term for reservoir. In the third line I've made the identification,

$$S(E) = k \ln \Omega(E) \text{ and } S_r(E_r) = k \ln \Omega_r(E_r). \tag{3.8}$$

Only the latter two terms in Eq. (3.7) depend on the energy of the system or reservoir. Retaining *only* the energy dependent terms (i.e., dropping the additive constant $\ln \Omega_0$), we have the standard thermodynamic additivity property for a composite system: $S_{\text{comp}}(E, E_0 - E) = S(E) + S_r(E_0 - E)$.

Key Point 3.6 *Equation* (3.8) *shows that* $S(E) \geq 0$, *and exemplifies the intimate linkage between energy and entropy. Boltzmann used a similar approach, where* $\Omega(E)$ *is the number of classical configurations of a gas with energy E. Here I take it to be the number of accessible quantum microstates.*

Two points of historical interest are: (i) Boltzmann never wrote the entropy equation as it is on his tombstone. Evidently that equation is due to Max Planck. (ii) Boltzmann never used the constant k bearing his name. Rather, k was introduced by Max Planck, who observed in his Nobel Prize lecture that to his knowledge, the constant referred to as the Boltzmann constant had never been used by Boltzmann. He stated that a possible explanation was that Boltzmann "never gave thought to the possibility of carrying out an exact measurement of the constant."[5]

Key Point 3.7 *The Boltzmann constant links energy, a concept from classical mechanics, with temperature, a central concept of thermodynamics. Its value,* $k = 1.38064852 \times 10^{-23}$ J K^{-1}*, provides a numerical connection between energy and temperature units. Notably, the Boltzmann constant is one of the defining constants of the SI system of units in Table 1.4 Without* k*, there would be no such connection. This explains the common appearance of* k *in the product* kT*, where* T *is temperature. I call this product tempergy, to be discussed in Sec. 3.4.*

Some properties of the logarithm are worth pointing out. The common or natural logarithm of a large number is a much smaller number. For example, $\log(1000) = 3 \ll 1000$. More dramatically, $\log(10^{100}) = 100$ while 10^{100} (one googol) = 1 followed by 100 zeroes. A main point is that the log and ln functions *tame* the arguments at which they are evaluated.

3.3 ENTROPY VS. ENERGY GRAPHS

3.3.1 Concavity

For a composite system, using Boltzmann's definition $S(E) = k \ln W(E)$, we find that when the energy sharing is *just right*, namely, when $W(E)$ and $\Omega(E)$ have their maximum possible values, the target system and environment are in thermodynamic equilibrium at the temperature of the environment. We focus attention on how the target system's entropy function varies with its internal energy E with the system volume fixed. Energy variations come solely from heating and not work processes: $dE = đQ$. For most systems of interest here, entropy has two important properties: (a) it increases with the system energy E; (b) is a *concave downward* function of E.[6]

---

[5]The Boltzmann constant could not be called *Planck's constant* because that term was used for the fundamental constant of quantum mechanics, h.

[6]Some exceptions that will be encountered later are constant-temperature reservoirs and gravity-driven systems, e.g., black holes.

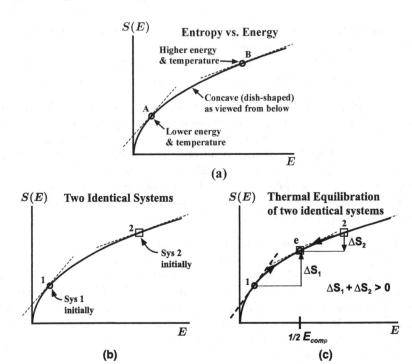

Figure 3.7: (a) Concave downward entropy *vs.* energy for a target system. States A and B have lower and higher temperatures, respectively. (b) Entropy *vs.* energy curve for two *identical* systems, with initial states 1 and 2 respectively. (c) When the latter two systems are put in thermal contact, the systems equilibrate to state e.

Properties (a) and (b) are evident in Fig. 3.7(a). As the energy E increases we expect temperature to increase. Notably, the slope $(dS/dE)_V = (\partial S/\partial E)_V$ of the curve $S(E)$ decreases with increasing energy, so it is tempting to suspect an inverse relationship between temperature and the slope of the entropy curve. We can pin this down further by considering two *identical* systems with different energies.

Because the envisaged systems are identical, they must have the *same* entropy *vs.* energy curve, as in Fig. 3.7(b), which shows system 1 with less energy than system 2 initially, an *inequitable* distribution. Our expectation is that system 1 also has a lower initial temperature than system 2. When the two systems are put into thermal contact, system 1 has *too little* and system 2 has *too much* energy for equilibrium. Energy will flow from system 2 to system 1, until each has half the total energy in state e, $E_e = \frac{1}{2}E_0$, as in Fig. 3.7(c).

At the final equilibrium state *e*, systems 1 (circle) and 2 (square) have the same state. In that state the slope of the entropy curve lies between the higher slope at state 1 and lower slope at state 2. System 1 has gained energy

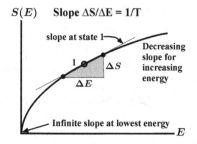

$S(E)$ **Slope $\Delta S/\Delta E = 1/T$**

slope at state 1

Decreasing slope for increasing energy

1   ΔS

ΔE

Infinite slope at lowest energy

E

Figure 3.8: Temperature at energy E is defined by $(\partial S/\partial E)_v = 1/T$.

and its temperature increased. System 2 has lost the same amount of energy and its temperature has decreased. In their final states, the two systems have the same slope and temperature. There is a strong suggestion that the slope of the entropy curve can be associated with inverse temperature. This idea is strengthened with the realisation that *if* the slope equals $1/T$, the Clausius expression for entropy emerges: $dS = (\partial S/\partial E)_V \, dQ = (1/T)dQ$, or

$$dS = \frac{dQ}{T} \quad \text{Clausius entropy expression.} \tag{3.9}$$

Thus, we adopt the definition

$$T \equiv (\partial S/\partial E)_{V,N} \quad \textit{thermodynamic definition of temperature.} \tag{3.10}$$

Because the graph in Fig. 3.8 has a well-defined temperature and entropy at each point on the curve, Eq. (3.9) connotes entropy change along a *quasistatic* path. Clausius required that Eq. (3.9) holds only for a reversible path, where the system's and environment's paths can both be reversed by small changes in one or more thermodynamic parameters. Here each system is the *environment* for the other system, and quasistatic processes are sufficient for Eq. (3.10). In Ch. 8, I illustrate *irreversible* cycles that are quasistatic, and for which (3.9) holds, point by point, along quasistatic, irreversible paths.

A notable and satisfying property implied by the concave downward nature of the entropy *vs.* energy curve is that it assures that in an equilibration process (illustrated in Fig. 3.7) the entropy increase of system 1 exceeds the entropy decrease of system 2. The increasing nature of the total entropy holds generally. Suppose one of the systems is a gas in a container of volume V at initial temperature T_i and the other system is a reservoir (the environment) at temperature T_{env}, which is assumed to be constant. Temperature constancy means the entropy *vs.* energy curve for the environment is a straight line, as illustrated in Fig. 3.9.[7] The system's initial temperature T_i is lower than that of the environment because the slope at i exceeds that of the

---

[7]This non-existent, but useful, imagined reservoir is called a Boltzmann reservoir, and is discussed in Sec. 3.4.1

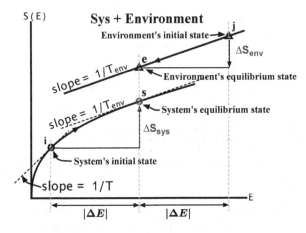

Figure 3.9: Thermal equilibration of the target system + environment. The initial state of system is labeled i, and the environment's initial state is j. Final states of system and environment are s and e respectively. The slope of the system's entropy curve at state s equals the slope of the linear entropy vs. energy line for the environment.

environment's straight line. The environment loses $|\Delta E|$ and the system gains an equal amount, bringing it to equilibrium temperature T_{env}. It is evident graphically that the system gains more entropy than the environment loses, i.e., $\Delta S_{sys} > |\Delta S_{env}|$. This assures that the total entropy of system plus environment increases.

3.3.2 Reflections on the entropy vs. energy curve

Notice that as E approaches its lowest possible value,[8] the slope of the entropy vs. energy curve approaches infinity, which signifies that temperature is approaching zero ($T \to 0$, $1/T \to \infty$). As the energy increases, the slope decreases toward zero. This signifies temperature continually increasing. Imagine starting near absolute zero temperature and slowly heating a system using pulses of small amounts of equal energy at regular intervals, measuring temperature at each step. The pulsed energy source is calibrated, and Fig. 3.8 shows how temperature changes with each added joule of energy.

Key Point 3.8 *The key points in this section are these:*

1. *The entropy vs. energy graph of a typical system when gravitational effects are negligible is an increasing, concave-downward function of its*

---

[8]The lowest energy is not necessarily zero. For quantum gases, it can be nonzero because of the Pauli exclusion principle, which is explained in Sec. 4.1.1; and also because of zero-point energy for crystalline solids, in accord with the uncertainty principle.

energy. Downward concavity assures that the entropy of a composite system is maximised at equilibrium. If the entropy vs. energy curve were concave upward for any or all energies, the above arguments would break down, and entropy would not be guaranteed to increase during temperature equilibration in a composite system.

2. *Defining inverse temperature as a slope $1/T = (\partial S/\partial E)_V$, is consistent with the Clausius entropy equation, $dS = dQ/T$. Reversibility of the constant-volume path used to define temperature is sufficient, and that path is necessarily quasistatic (or it could not be graphed). There is an inextricable linkage between entropy, internal energy, and temperature.*

3. *For a target system of interest that is in thermal contact with another system denoted by a prime ('), the entropy of the (assumed isolated) composite-system is the sum of the two system entropies. If the systems have different temperatures, they will exchange energy until they reach thermodynamic equilibrium, with maximum total entropy, $S + S'$.*

3.3.3 Equity revisited

The concept of equity was discussed qualitatively in Sec. 2.4.1. We can go a bit further using Boltzmann's entropy expression $S(E) = k \ln \Omega(E)$ in Eq. (3.3). Let two subsystems form an isolated composite system for which the total number of microstates of the composite system is $\Omega_{comp} = \Omega_1(E_1)\Omega_2(E_2)$, with $E_1 + E_2 = E_0$. The total composite system internal energy E_0 is assumed to be constant, so a small energy exchange $dE_1 = -dE_2$. The entropy of the composite system is $S_{\text{comp}} = k \ln \Omega_1(E_1) + k \ln \Omega(_2)$.

Suppose the two subsystems exchange a small amount of energy via a heat or a combined heat plus work process, with system 1 gaining energy $dE_1 > 0$ and system 2 losing an equal amount of energy, $dE_2 = -dE_1$. Because entropy increases with increasing energy, the number of microstates for the receiving system increases, i.e., $d\Omega_1 > 0$. Similarly, $d\Omega_2 < 0$ for the system that loses energy. Note that when $y = \ln x$, any change dx causes the change $dy = dx/x$. Applying this to S_{comp}, we have

$$dS_{\text{comp}} = \frac{d\Omega_1}{\Omega_1} + \frac{d\Omega_2}{\Omega_2} \equiv d\zeta_1 + d\zeta_2 \geq 0. \tag{3.11}$$

The infinitesimal fractions $d\zeta_1 \equiv d\Omega_1(E_1)/\Omega_1(E_1)$ and $d\zeta_2 \equiv d\Omega_2(E_2)/\Omega_2(E_2)$ are the fractional number of states gained or lost by the two subsystems, and the inequality follows from the second law of thermodynamics. If, as envisaged, $dE_1 > 0$ then $d\zeta_1 > 0$ and $d\zeta_2 < 0$. System 1 needs more energy and energy states to achieve equity. Only if $d\zeta_1 = -d\zeta_2$ is the energy distribution equitable.

Key Point 3.9 *An unchanging state of thermodynamic equilibrium exists between two subsystems only when the energy distribution is equitable, and small energy fluctuations have $d\zeta_1 = -d\zeta_2$. This means that a small energy transfer leads to **equal (and with opposite algebraic signs) fractional changes in the number of states accessible to the two subsystems.** When $d\zeta_1 - d\zeta_2$, one subsystem has excess energy and the other needs more energy to achieve equity.*

3.4 BOLTZMANN RESERVOIR & PROBABILITY

3.4.1 Boltzmann reservoir

The straight line in Fig. 3.9 for the constant-temperature reservoir is an idealisation. It is for a perfect "Boltzmann" reservoir, whose entropy expression is

$$S_{BR} = \left(\frac{1}{T}\right) E + constant, \tag{3.12}$$

where the subscript BR connotes *Boltzmann reservoir*.[9] Just as quasistatic and reversible processes do not exist in reality, neither do Boltzmann reservoirs. A closer look shows that Boltzmann reservoirs have some rather nonphysical properties, some of which are listed here.

1. Obviously, Eq. (3.12) satisfies the thermodynamic definition of temperature, (3.10). The property that the temperature is independent of the energy is certainly nonphysical.

2. If two Boltzmann reservoirs at the *same* temperature are put in thermal contact, either of them can have any fraction of the total energy. That is, there is no unique equilibrium state. If the reservoirs have energies E_c and E_h before thermal contact is established, then after contact, any pair of energies E'_c and E'_h with $E_c + E_h = E'_c + E'_h$ is possible.

3. Imagine putting two Boltzmann reservoirs with *different* temperatures in thermal contact, the higher temperature reservoir will give *all* its energy to the lower-temperature reservoir! In this example the lower-temperature reservoir absorbs heating energy but experiences zero temperature change. This implies an undefined heat capacity $C = \Delta Q/\Delta T$ because $\Delta T = 0$.

4. The entropy of a Boltzmann reservoir is linear in the energy E. It is *not* concave downward and is therefore non-physical, showing the bizarre behaviour in the previous item in this list. However, if a target

---

[9]This corresponds to $\Omega(E) = constant \times e^{E/\tau}$, where $\tau \equiv kT$, in Eq. (3.2).

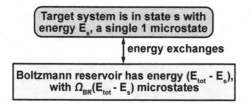

Figure 3.10: System in microstate s with energy E_s, in contact with a Boltzmann reservoir at temperature T.

system is put in contact with a constant-temperature environment, the final equilibrium state of the system occurs at an energy for which the system's entropy curve has the same slope (same temperature) as that for the Boltzmann reservoir.

Given these nonphysical behaviours, why introduce the Boltzmann reservoir? There are at least two reasons. First, the Boltzmann reservoir shows how one nonphysical behaviour can be accompanied by other ones; i.e., this is a good learning experience. The second reason is that constant-temperature reservoirs are very useful in thermodynamics, helping us focus on physical systems that do have properties similar to actual systems. A good example is in the next section on the important Boltzmann factor.

3.4.2 Boltzmann factor

Consider a system in contact with a Boltzmann reservoir at temperature T, as in Fig. 3.10. Let the total energy of the reservoir plus system be E_0 and suppose the target system is known to be in a particular quantum state s with energy E_s. The number of possible states for the composite system is that for the reservoir with energy $(E_0 - E_s)$, namely, $\Omega_{BR}(E_0 - E_s)$.[10] Using Eq. (3.12), we have

$$S_{BR}(E_0 - E_s) = k \ln \Omega_{BR}(E_0 - E_s) = \left(\frac{E_0 - E_s}{T}\right) + constant. \qquad (3.13)$$

For convenience, define,

$$\tau \equiv kT \quad \text{(tempergy)}. \qquad (3.14)$$

The name tempergy comes from τ being proportional to temperature, but with energy units.[11] Inversion of Eq. (3.13) results in,

$$\Omega_{BR}(E_0 - E_s) = \text{constant} \times e^{-E_s/\tau}. \qquad (3.15)$$

---

[10]This is because there is only one *microstate* s, which should not be confused with a *macrostate* for energy E, for which there are $\Omega(E) > 1$ microstates.

[11]Tempergy is examined in detail in Sec. 6.4.

The probability that a system that is in contact with a reservoir with tempergy τ is in microstate s with energy E_s is proportional to the number of ways that can occur, which is $\Omega_{BR}(E_0 - E_s)$, thus,

$$W_s(E_s) = Z^{-1} \times e^{-E_s/\tau} \text{ with } \sum_s W_s(E_s) = 1. \tag{3.16}$$

The proportionality factor Z^{-1}, called the *partition function*, generally is a function of the reservoir's tempergy and the target system's energy spectrum, which depends on the volume and total number of molecules, and the forces between those molecules. The exponential factor $e^{-E_s/\tau}$ is ubiquitous in thermal physics and is called the *Boltzmann factor*. If the energy $E = E_s$ is "degenerate," occurring for $\Omega(E)$ states, then the probability of the target system having energy $E = E_s$ is

$$W(E) = \Omega(E)W_s(E = E_s) = Z^{-1} \times \Omega(E)e^{-E/\tau} \text{ and } \sum_E W(E) = 1. \tag{3.17}$$

It follows that the partition function can be written

$$Z = \sum_s e^{-E_s/\tau} = \sum_E \Omega(E)e^{-E/\tau} \quad \text{(partition function).} \tag{3.18}$$

Equation (3.18) shows that the partition function can be evaluated either by summing over states or energies.

Example. A relatively simple, but educational, example is a system with two possible states, with energies E_1 and $E_2 > E_1$. From Eq. (3.16), it follows that the probabilities for the two states are

$$W(E_1) = Z^{-1}e^{-E_1/\tau} \text{ and } W(E_2) = Z^{-1}e^{-E_2/\tau}, \tag{3.19}$$

with $Z = e^{-E_1/\tau} + e^{-E_2/\tau}$. For temperatures $T \to 0$, $W(E_1) \to 1$ and $W(E_2) \to 0$, and the system spends nearly all its time in the lower-energy state 1. The constant-temperature reservoir cannot supply enough energy for the system to be excited to state 2. On the other hand, in the high temperature limit, $T \to \infty$, state 2 becomes as likely as state 1: $W(E_1) \to 1/2$ and $W(E_2) \to 1/2$. Note that in thermodynamic equilibrium, there is never a population inversion for which state 1 becomes more likely than state 2.[12]

Key Point 3.10 *Equations (3.16), (3.17) and (3.18) are the basic equations of statistical mechanics. Once the partition function $Z(\tau, V, N)$ is known for a system, there is a straightforward procedure for determining the target system's internal energy and entropy, which is outlined in the next section.*

---

[12]Such inversions can be *driven* and entail states with *negative* absolute temperatures (over short time intervals), which are discussed in Sec. 4.5.2.

3.4.3 Statistical mechanics

A system with temperature T, volume V and number of particles N can have any number of total states and each energy E' can have multiple, independent states, i.e., $\Omega(E') > 1$. From the partition function $Z(T, V, N)$, given by Eq. (3.18), one can calculate the system's average energy, entropy, pressure and the like. The average energy E is taken to be the system's internal energy. It entails multiplying the probability of the system having energy E' times E' itself, summed over all energies E', i.e.,

$$E \equiv \sum_{All\ E'} E' \times \left[\frac{\Omega(E') \times e^{-(E'/\tau)}}{Z(T, V, N)} \right]. \tag{3.20}$$

The quantity in square brackets is the probability $W(E')$ of the system having energy E', which is $\Omega(E')$, the *degeneracy* of energy E'. Once $Z(T, V, N)$ is known the entropy S can be expressed as

$$S(T, V, N) = k \ln Z + \frac{E}{T}. \tag{3.21}$$

Because the average energy E is identified as the thermodynamic internal energy E, $-kT \ln Z = E - TS$, a sum of the two terms that represent the energy-entropy duo. Finally, it is common to define the Helmholtz free energy

$$A(T, V, N) = E - TS, \tag{3.22}$$

and the Gibbs free energy

$$G(T, P, N) = E - TS + PV. \tag{3.23}$$

Combining Eqs. (3.22) and (3.21), we obtain,

$$A(T, V, N) = -kT \ln Z. \tag{3.24}$$

If a system is taken from an initial to final equilibrium state at the same temperature T, Eq. (3.24) implies that

$$\Delta A \equiv A(T, V_f, N) - A(T, V_i, N) = -kT \ln(Z_f/Z_i), \tag{3.25}$$

or

$$Z_f/Z_i = e^{-\Delta A/kT}. \tag{3.26}$$

We shall see in Sec. 3.4.3 that the Helmholtz and Gibbs free energy functions are helpful for understanding how energy and entropy work together to determine thermodynamic behaviour at constant-temperature (and also constant pressure for the Gibbs function) processes. In addition, Eq. (3.26) is central in the discussion of the Jarzynski equality in Sec. 4.8.

Key Point 3.11 *The Helmholtz and Gibbs free energy functions are particularly useful for phenomena that occur at the environment's temperature (Helmholtz function), and the temperature and pressure (Gibbs function). Given the ubiquity of such processes, the interplay between energy and entropy in both Eqs. (3.22) and (3.23) is of interest.*

3.5 HELMHOLTZ FREE ENERGY

3.5.1 Understanding free energy

It is common for a system to be in thermal contact with a constant temperature environment, for example the air in a laboratory. For such a constant-temperature system, the entropy of the system plus environment is maximised at equilibrium. But what about the system itself? I shall show that the principle of entropy increase for the system plus environment implies other inequalities for sub-systems with constant temperature or *both*, constant temperature and constant pressure.

Both the Helmholtz and Gibbs functions A and G defined in Eqs. (3.22) and (3.23) contain an energy term and an entropy term. Jeffrey Prentis observed that just as entropy can be viewed as a *spreading* function, we can (at least in some instances) view the Helmholtz free energy A as a "falling function," because of its tendency in a non-isolated system to fall in value. The two opposing tendencies that drive natural processes, the fall and spread of energy, are incorporated in the Helmholtz free energy, which decreases during a constant-temperature work-generating process that leads to equilibrium.

Prentis wrote that this "represents the exact compromise of nature to balance the conflicting tendencies to minimise the amount E of energy within the system and maximise the spread S of energy throughout the system." In this competition, the temperature T determines the winner. For relatively low T, the internal energy dominates because ΔE exceeds $T\Delta S$. For relatively high temperature, the reverse is true.

Consider the dilute gas in Fig. 3.11 in contact with a constant-temperature environment at temperature T. A freely-floating piston, i.e., one that can move vertically with zero friction forms the ceiling of the container. The piston's weight mg and its cross-sectional area $\mathcal{A}_p$ determine the equilibrium gas pressure, $P_0 = mg/\mathcal{A}_p$ so $P_0V_0 = (mg/\mathcal{A}_p)V_0 = mgh_0$ at equilibrium. I assume the ideal gas equations apply here, so $P_0V_0 = NkT$, and the internal energy of the gas is $\frac{3}{2}NkT = \frac{3}{2}P_0V_0$. I choose the target system to be the gas *plus* the gravitational field, and take the gravitational energy to be zero when the piston is touching the floor of the cylinder holding the gas, and denote the total energy (internal + gravitational) at equilibrium by $\mathcal{E}_0 = \frac{3}{2}P_0V_0 + P_0V_0 = \frac{5}{2}P_0V_0$.

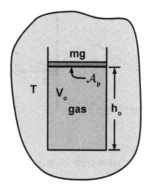

Figure 3.11: Dilute gas at temperature T, with freely-floating piston of weight mg and area $\mathcal{A}_p$. The equilibrium volume is V_0 and piston height is h_0.

Away from equilibrium, the piston height $h \neq h_0$ and the volume $V \neq V_0$. The relative volume ν_r is defined by

$$\nu_r \equiv V/V_0 = h/h_0 \quad \text{(relative volume)}. \tag{3.27}$$

The total energy is $\mathcal{E} = \frac{3}{2}NkT + mgh$. The second term can be recast as $mgh = mgh_0(h/h_0) = P_0 V_0 \nu_r$ and the first term as $\frac{3}{2}P_0 V_0 \nu_r$, so

$$\mathcal{E} = \frac{3}{5}\mathcal{E}_0(1 + \frac{2}{3}\nu_r). \tag{3.28}$$

In equilibrium, the relative volume $\nu_r = 1$, and the two terms add to $\mathcal{E}_0 = \frac{5}{2}P_0 V_0$. The entropy of the gas is given by the Sackur-Tetrode equation (4.4), $S = Nk[\ln(V) + 5/2 + K]$, where $K = constant$. Using the identity $V = \nu_r V_0$,

$$S = Nk\ln(\nu_r) + S_0, \tag{3.29}$$

with $S_0 = Nk[\ln(V_0) + 5/2 + K]$ being the equilibrium entropy. If the piston is forcibly moved such that the gas pressure $P \neq P_0$, so $\nu_r \neq 1$, equilibrium will be reestablished when the piston becomes free to move.

Key Point 3.12 *There is a competition between entropy S and energy $\mathcal{E}$. In classical mechanics, equilibrium means equality of the upward force on the piston by the gas and the downward force of gravity. The thermodynamic view is different. The gas pressure tends to maximise the gas volume, and thus, the entropy, while the total energy tends to be minimised because of gravity.*

This competition can be examined further using the Helmholtz free energy function, which is defined by

$$A \equiv E - TS. \tag{3.30}$$

For the current example, this can be written as

$$A = P_0 V_0[\nu_r - ln(\nu_r)] + \text{constant}. \tag{3.31}$$

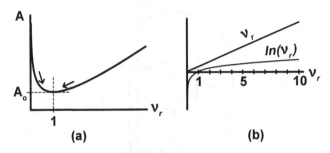

Figure 3.12: (a) Free energy with a free-floating piston *vs.* relative volume ν_r. (b) ν_r and $\ln(\nu_r)$ in A *vs.* ν_r.

A plot of A vs. ν_r is in Fig. 3.12(a), shows that A has its minimum value when the relative volume is one, namely, when there is mechanical as well as thermal equilibrium. In Fig. 3.12(b), the functions ν_r, which comes from the gravitational energy, and $\ln(\nu_r)$, which comes from the entropy, are plotted individually as functions of ν_r. Both functions have slope = 1 when $\nu_r = 1$. Because ν_r increases more rapidly than $\ln(\nu_r)$ for $\nu_r > 1$, the change in the term ν_r, which comes from the energy in Eq. (3.30), dominates for $\nu_r > 1$, namely, when the piston is *above* its equilibrium position. In contrast, the change in the term $\ln(\nu_r)$, which comes from the entropy in (3.30), dominates for $\nu_r < 1$, namely, when the piston is *below* its equilibrium position.

Mathematically, we have

$$(dA/d\nu_r) = P_0 V_0 [1 - 1/\nu_r], \qquad (3.32)$$

from which it follows that

$$(dA/d\nu_r) \begin{cases} < 0, & \text{for } \nu_r < 1, \text{ when entropy dominates.} \\ = 0, & \text{for } \nu_r = 1, \text{ where free energy has a minimum.} \\ > 0, & \text{for } \nu_r > 1, \text{ when energy dominates.} \end{cases} \qquad (3.33)$$

Key Point 3.13 *A number of important qualitative conclusions come from this rich example.*

1. *Figure 3.12(a) shows that when $\nu_r \neq 1$, the free energy is above the equilibrium free energy A_0. If free to move, the piston will move toward its equilibrium position.*

2. *When $\nu_r < 1$, the system is compressed relative to its equilibrium volume, and there is an entropic tendency for the piston to move toward its equilibrium position. Entropy dominates.*

3. *When $\nu_r > 1$, there is a tendency for the piston to fall because of gravity, thereby moving it toward the equilibrium position. Energy dominates.*

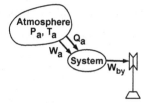

Figure 3.13: A system exchanges energy (W_a, Q_a) with a constant pressure (P_a) and temperature (T_a) atmosphere, and a pulley and weight.

4. *In either case the free energy is minimised. Equilibrium can be viewed as the state for which energy and entropy are apportioned in a manner that is consistent with the system's temperature—with neither "winning" the energy-entropy competition. The entropy effect is typically stronger at higher temperatures, which drives gas expansion, and energy dominates at lower temperatures when the piston tends to fall. This is an example of a powerful thermodynamics principle, the minimisation of the free energy function $A = E - TS$ for a system maintained at constant temperature. I explore this further in Sec. 3.5.2.*

3.5.2 Available energy and exergy

Available energy. Consider the following question: *Given a system that can exchange energy via work processes with both the atmosphere and a reversible work source, and also via heat processes with the atmosphere, how much work can be extracted from such a system?*

A target system of interest and its surrounding atmosphere, plus a reversible work source are depicted in Fig. 3.13. The system can exchange energy via heat and work processes with the atmosphere and a reversible work source, depicted as a weight and pulley. The atmosphere does work W_a and supplies heating energy Q_a to the target system. The work W_{by} is *by* the system on an external load (the pulley and weight).

Energy exchanges with the atmosphere are included because virtually all earthly processes are subject to them. For a process with the set of energy exchanges shown in Fig. 3.13, application of the first law of thermodynamics to the system gives $\Delta E = Q_a + W_a - W_{by}$ or

$$W_{by} = -\Delta E + Q_a + W_a. \tag{3.34}$$

The atmosphere is so huge that we treat it as a constant temperature T_a and pressure P_a reservoir. The work source is assumed to be reversible and to generate zero entropy. The atmosphere's entropy change is $\Delta S_a = -Q_a/T_a$, and it does work on the system, $W_a = -P_a\Delta V$. Writing the system's entropy change as ΔS, the total entropy change of the target system and reservoir is

$$\Delta S_{\text{tot}} = \Delta S - Q_a/T_a. \tag{3.35}$$

Solving Eq. (3.35) for Q_a and substituting it into (3.34), we obtain the interesting result

$$W_{\text{by}} = -\Delta(E + P_a V - T_a S) - T_a \Delta S_{\text{tot}}. \tag{3.36}$$

The operator Δ acts on E, V and S. It has no effect on the constant P_a or T_a. The second law of thermodynamics guarantees that $\Delta S_{\text{tot}} \geq 0$, which implies the inequality

$$W_{\text{by}} \leq -\Delta(E + P_a V - T_a S). \tag{3.37}$$

Equation (3.37) is useful for chemical reactions, where the constant pressure atmosphere can play an important role in the energy exchanges, e.g., when a reaction generates gases that have work done on them by the atmosphere. In that case, the system has temperature, pressure, volume and entropy $T = T_a, P = P_a, V$ and entropy S. The quantity in parentheses in Eq. (3.37) is called the *Gibbs free energy*, G:

$$G(T, p, N) = E - TS + PV. \tag{3.38}$$

Therefore, Eq. (3.37) can be written as

$$W_{\text{by}} \leq -\Delta G(T_a, P_a, N) = \text{available energy} \geq 0 \quad at \ T = T_a, P = P_a. \tag{3.39}$$

Key Point 3.14 *If a system undergoes a process at constant temperature T_a and pressure P_a, the amount of external work it can do during that process is limited by the negative of the change in its Gibbs free energy. In this sense, this change ΔG is the "available energy" for doing work in Eq. (3.39).*

In physics applications where the atmospheric work term $P_a \Delta V$ in Eq. (3.37) is typically negligible, the *Helmholtz free energy* A suffices.

$$A \equiv E - TS, \text{ the function introduced in (3.30)}, \tag{3.40}$$

and Eq. (3.39) is replaced by

$$W_{\text{by}} \leq -\Delta A = \text{available energy} \geq 0 \quad at \ T = T_a. \tag{3.41}$$

Key Point 3.15 *If a system undergoes a process at constant temperature T_a, the concomitant amount of external work it can do is limited by the negative of its Helmholtz free energy change, which is the available energy in Eq. (3.41). For a reversible process, Eq. (3.41) becomes an equality,*

$$W_{\text{by}} = -\Delta A \text{ or equivalently}, W = -W_{\text{by}} = \Delta A \text{ reversible process}. \tag{3.42}$$

Key Point 3.16 *If positive isothermal work W is done ON a system, the system's available energy increase is limited by $\Delta A \leq W$, with the equality holding for a reversible process.*

Example: Gas with floating piston

I return to the ideal gas with a freely-floating piston that led us to the Helmholtz free energy (3.30) and (3.31). At constant temperature, A is minimised at thermodynamic equilibrium, but suppose the system, excluding the gravitational field, is in equilibrium and the piston is forcibly moved upward or downward to a new relative volume $V_r' > 1$ or $V_r' < 1$. For the former, if released, the piston will exert a pressure higher than the equilibrium gas pressure at that volume and will compress the gas irreversibly to its new equilibrium position with $\nu_r = 1$. For the monatomic ideal gas, the work is limited by the available energy, i.e., $W_{\text{by}} \leq -\Delta A = A(T, \nu_r V_0, N) - A(T, V_0, N)$, so

$$W_{\text{by}} \leq -\Delta A = -NkT \ln(\nu_r) = \text{ available energy.} \qquad (3.43)$$

In Eq. (3.43), I used the fact that at constant temperature, $\Delta E = 0$ (for the ideal gas), and the entropy expression in Eq. (3.29) to evaluate ΔS.

Key Point 3.17 *For initial relative volume $\nu_r < 1$, upon release, the piston will rise, overshoot the equilibrium position, oscillate with decreasing amplitude, and stop in the equilibrium position with $\nu_r = 1$. The positive work done by the expanding gas is partially offset by negative work by the gas when the piston moves downward while oscillating. The net work by the gas $W_{\text{by}} < -\Delta A$, and because $\Delta E = 0$, $W_{\text{by}} < T\Delta S_{\text{gas}}$. The difference $T\Delta S_{\text{gas}} - W_{\text{by}} = T\Delta S_r'$, the energy sent to the reservoir because of the irreversibility. The entropy change of the universe is $\Delta S_{\text{univ}} = \Delta S_{\text{gas}} + \Delta S_r'$. In contrast, for a reversible expansion, $\Delta S_r' = 0$, and $W_{\text{by}} = T\Delta S_{\text{gas}}$.*

For $\nu_r > 1$, the piston will lower, oscillate, and settle in equilibrium. The work on the gas by the falling piston $W > \Delta A = -T\Delta S_{\text{gas}}$, and $W = T\Delta S_r' - T\Delta S_{\text{gas}}$. The work by the piston adds free energy $-T\Delta S_{\text{gas}} > 0$ to the gas by compression, and sends wasted energy $Q_r' = T\Delta S_r' > 0$ to the environment. In contrast, for a reversible compression, $T\Delta S_r' = 0$, and $W = -T\Delta S_{\text{gas}}$.

Exergy. Engineers define a state function B, called "exergy," to analyse thermodynamic processes in industrial applications. Exergy is related to a Gibbs energy change when the system is in thermal equilibrium with the atmosphere at pressure P_0 and T_0. When the system is in state i, B_i, the exergy, defined relative to a reference state r, is

$$B_i \equiv (E_i + P_0 V_i - T_0 S_i) - (E_r + P_0 V_r - T_0 S_r) \quad \text{(Exergy).} \qquad (3.44)$$

Figure 3.14 illustrates a process that entails the transfer of energy ΔE_i and the flow of exergy, which is a measure of Gibbs energy change. Exergy is important in that it represents the energy available for delivery as work.

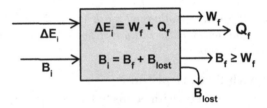

Figure 3.14: Symbolic view of energy transfers $(\Delta E_i, W_f, Q_f)$ and exergy flows $(B_i, B_f, B_{\text{lost}})$.

For example, for a hydrogen burning electricity power plant, B_i is the maximum possible work delivered by a known amount of hydrogen. The figure shows that the processes in the power plant decrease the exergy from B_i to B_f, the difference being $B_{\text{lost}} = B_i - B_f$. The electrical work W_f done delivered is limited by the remaining exergy: $W_f \leq B_f$. The first and second laws of thermodynamics can be viewed in the following ways.

- According to the first law, energy is conserved, with work W_{by} and heat Q being the two primary ways of transferring energy.

- In contrast with energy, exergy is the energy that is available to become work. It is always diminished in an irreversible process, where the entropy of the universe increases.

- The quantity $B_{\text{lost}} = B_i - B_f$ in Fig. 3.14 is sometimes called "lost work."[13] The term exergy does not add any new physics, but rather, provides a language with which to discuss processes.

In a process where the initial system has initial and final exergy B_i and B_f, the lost work is $B_{\text{lost}} = B_i - B_f = -\Delta G \geq W_{\text{by}}$.

A simple example is a monatomic ideal gas with initial volume V_i, in contact with a constant-temperature reservoir at T_0. We can in principle have the gas do work by expanding freely to final volume V_f, without any coupling to an external load; i.e., doing zero external work. Because the internal energy of a classical ideal gas depends only on temperature and is constant, the final temperature must be the same as the initial one.

The Gibbs function can be written as $G(T_0, P_0) = H - T_0 S$, and for an ideal gas, the enthalpy H is solely a function of the temperature, which is unchanged in the free expansion, and $B_{\text{lost}} = +T\Delta S = NkT \ln(V_f/V_i) = B_{\text{lost}}$. Had we coupled the system to an external load and done the expansion reversibly at constant temperature, the system could have done work $W_{\text{rev}} = B_{\text{lost}} = NkT \ln(V_f/V_i)$. For the free expansion, that work opportunity was lost entirely, and for any irreversible work process, $W_{\text{by}} < B_{\text{lost}}$.

---

[13]Some dictionary definitions of entropy relate it to the unavailability of a system's "thermal energy" for conversion into mechanical work. That statement connects entropy with the energy *transfer*, work, but not with the stored internal energy.

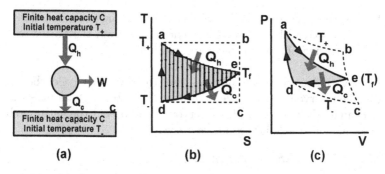

Figure 3.15: (a) Overview of a cycle using reservoirs with equal finite heat capacities (thermal inertias). (b) TS diagram of the cycle consisting of $\mathcal{N}$ sequential infinitesimal Carnot cycles. The limit $\mathcal{N} \to 0$ and width $\to 0$ is envisaged. (c) PV diagram of the cycle in (b), with Carnot cycles hidden. The dashed path $abcda$ is the reversible Carnot cycle between T_+ and T_-.

Key Point 3.18 *When a system is in thermal contact with a constant-temperature reservoir, thermodynamic processes occur spontaneously from higher to lower exergy. As exergy declines, the entropy of the universe and lost work increase.*

3.5.3 Available energy with finite reservoirs

Consider two identical finite reservoirs, which store energy and have initial temperatures $T_+ > T_-$. These finite reservoirs have finite, generally different heat capacities (thermal inertias). For simplicity, I assume the reservoirs have the same heat capacity C. As they exchange energy with other systems, the reservoir temperatures change. To find the available work that can be extracted from the finite heat source/sink pair, imagine an infinite sequence of infinitesimal, clockwise, reversible Carnot cycles that result in path *aeda* in Figure 3.15. Pane (b) shows a crude approximation using 18 Carnot cycles.

The time sequence of the envisaged physical process is that the upper and lower path segments of the left-most rectangle occur first, the next rectangle second, and so forth. The vertical adiabatic segments of successive Carnot cycles negate one another, leaving the left-most one, *da*, and the right-most one *e*, which is of zero length. In the limit of an infinite number of infinitely thin Carnot cycles,[14] cycle *aeda* is generated. The resulting paths, *ae* and *ed*, are variable temperature paths, and the shaded area *aeda* in Fig. 3.15 (b) and (c) equals the work W done by the cycle. Several points are of interest.

1. The reservoirs gain and lose energies $đQ_h = CdT_h$ and $đQ_c = CdT_c$ during each Carnot cycle, and have final temperature T_f.

---

[14]See H. S. Leff, "Available work from a finite source and sink: How effective is a Maxwell's demon?," Am. J. Phys. **55**, 701–705 (1987).

2. During the envisaged cyclic process, the working substance does work W on an external load, receiving energy Q_h from the hotter reservoir and delivering Q_c to the colder reservoir.

How much work does this cycle produce? For the reversible cycle, the entropy change of the reservoirs must be zero because reversibility guarantees that the entropy change of the system plus reservoirs is zero, and the entropy change of the working substance is zero for each cycle, i.e.,

$$\Delta S_{res} = \Delta S_h + \Delta S_c$$
$$= \int_{T_+}^{T_f} \frac{C}{T_h} dT_h + \int_{T_-}^{T_f} \frac{C}{T_c} dT_c \tag{3.45}$$
$$= \ln\left(\frac{T_f^2}{T_+ T_-}\right) = 0.$$

This reduces to

$$T_f = \sqrt{T_+ T_-}. \tag{3.46}$$

It follows that $W = Q_h - Q_c = C[(T_+ - T_f) - (T_f - T_-)]$. Using Eq. (3.46) and doing some algebra, we find that the work done by the working substance during cycle $aeda$, the maximum work possible, is

$$W_{max} = C\left(\sqrt{T_+} - \sqrt{T_-}\right)^2. \tag{3.47}$$

The input energy from the high temperature reservoir for the cycle is

$$Q_h = C(T_+ - T_f) = C\left(T_+ - \sqrt{T_+ T_-}\right). \tag{3.48}$$

Using the efficiency definition in Eq. (2.16), a little more algebra leads to the result that the efficiency η of cycle $aeda$ is

$$\eta \equiv \frac{W_{max}}{Q_h} = 1 - \sqrt{\frac{T_-}{T_+}}. \tag{3.49}$$

We shall see in Ch. 8 that Eq. (3.49) holds for a model of an irreversible heat engine called the Novikov-Curzon-Ahlborn cycle and other cycles.

Key Point 3.19 *The maximum work that can be obtained from the two identical finite reservoirs at initial temperatures T_+ and T_- depends on the heat capacity C of those reservoirs and their initial temperatures. The final temperature T_f is determined solely by T_+ and T_-.*

3.5.4   Entropic force

An entropic force *emerges* from the behaviour of many molecules, and is called an *emergent force*. This force is generated in a system with many ways to store energy or equivalently, many degrees of freedom. It arises from a statistical tendency to increase a system's entropy. An example is the force on the container walls of an ideal gas.

Ideal gas. For an infinitesimal reversible process, the first law of thermodynamics, $dQ = TdS$, and $dW = -PdV$ lead to $dE = TdS - PdV$. Writing $d(TS) = TdS + SdT$, we can write the differential of the Helmholtz free energy $dA = d(E - TS) = -SdT - PdV$, which implies

$$P = -\left(\frac{\partial A}{\partial V}\right)_T \text{ and } S = -\left(\frac{\partial A}{\partial T}\right)_V. \tag{3.50}$$

The entropy of a classical ideal gas is given in Eq. (4.4), implying

$$A(T, V, N) = E(T, N) - TS(T, V, N)$$
$$= \tfrac{3}{2}NkT - NkT\left[\ln\left(\frac{VT^{3/2}}{N}\right) + N \times \text{constant}\right]. \tag{3.51}$$

We can use Eq. (3.50) together with Eq. (3.51) to calculate the ideal gas pressure. The constant in (3.51) can depend on the particle mass m and the Planck constant h, but not on T, V or N. In the first line of Eq. (3.51), the internal energy $E(T, N)$ does not depend on the volume V for our ideal gas example. The second term, $-TS(T, V, N)$, is volume dependent, and the pressure can be written as

$$P = T(\partial S/\partial V)_T, \text{ an "entropic force" per unit area.} \tag{3.52}$$

Evaluation of Eq. (3.52) using Eq. (3.51) leads to

$$P = -(\partial A/\partial V)_T = +T(\partial S/\partial V)_T = NkT/V. \tag{3.53}$$

Key Point 3.20 *An entropic force has several notable characteristics. (i) It emerges from the behaviour of many molecules. (ii) It tends to change the volume (or length) in a way that increases entropy. (iii) It is proportional to the temperature.*

For the gas, this entropic force arises because molecules tend to spread their kinetic energies spatially as much as possible, thereby filling the container, maximising their entropy.

Polymers. The generation of *entropic forces* because of a system tending to maximise its entropy exists also in polymers like rubber. That system can be modelled as a chain of molecules connected by rigid, negligible-mass links that can rotate freely about connection points that act as frictionless ball

Figure 3.16: Model of a polymer fixed at both ends. Circles represent atoms. The vertical dimension is exaggerated here for what is intended to be a nearly one-dimensional chain.

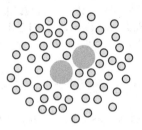

Figure 3.17: A small portion of a binary mixture of large and many more small spheres. Note the dearth of small spheres between the large ones.

joints, with a fixed atom at each end of the chain. This is depicted in Fig. 3.16. The end chains are assumed to be fixed, an example being a rubber band being held stretched with its ends motionless. The jiggling atoms have energies consistent with the system's temperature, and these energies generate a temperature-dependent entropic force at the fixed ends that tend to reduce its length. At constant temperature, the entropy increases with *decreasing* rubber band length. This is discussed further in Sec. 4.6.

Mixture of hard spheres. Another example is a mixture of large and small hard spheres.[15] When two or more large spheres are close enough together so as to exclude small spheres, more small spheres will collide with the outer edges of the large spheres as in Fig. 3.17, creating an effective attractive force between the large spheres. This is an entropic force.

Accelerating expansion of the universe. Albert Einstein published his theory of general relativity in 1916, in which the very presence of energy and matter alters spacetime, which determines how matter moves. Reciprocally, matter and energy alter the curvature of spacetime. It appeared that the universe began with a violent "Big Bang," that sent matter outward. Eventually, the spreading matter would presumably slow down and stop, and then collapse back to its Big Bang origin. To prevent such collapse, Einstein added a *cosmological constant* term Λ to his equations. However, Edwin Hubble's observations in 1929 discovered that galaxies were receding from Earth, with the most distant ones moving faster. The conclusion was that

---

[15]P. Taylor & J. Tabachnik, "Entropic forces–making the connection between mechanics and thermodynamics in an exactly soluble model," Eur. J. Phys **34**, 729–736 (2013).

space itself was expanding. This is not detectable locally because systems that are held together gravitationally (like Earth) do not recede from one other.

In 1998, two groups independently discovered the expansion of the universe was in fact accelerating. For this, Saul Perlmutter, Adam Riess, and Brian Schmidt shared the 2011 Nobel Prize in physics. Their finding begged for an explanation, and one that was appealing was Michael Turner's term *dark energy*. This is an assumed property of space, which has a constant energy density (energy per unit volume), and its total amount increases as the universe expands and its volume gets larger. The estimated value of the dark energy density is 1.67×10^{-27} kg m⁻³.

Despite what seems to be a tiny energy density, for the known size of the universe, the total amount accounts for 69.7% of the total energy of the universe. Dark matter is about 25.7% and ordinary matter represents 4.7% of the universe's energy. If the amounts of the latter two energies are fixed, their energy densities will decline as the universe expands, and the total amount of dark energy grows. The dark energy manifests itself with a seeming repulsive force that accelerates the expansion of the universe. Ironically, it puts a *cosmological constant* back in the equations of general relativity.[16]

Key Point 3.21 *The concept of dark energy filling new spatial volumes as the universe expands is suggestive of a form of energy spreading, and one might suspect that this is related to entropy increasing at an ever faster rate. However, dark energy is by no means thermodynamic internal energy, and furthermore, we do not know if thermodynamics actually holds on a cosmological scale.*

The *ad hoc* nature of dark energy and the lack of physical insight into its origin and meaning have led some researchers to seek an alternative model. I will describe another attempt to explain the accelerated expansion in terms of an entropic force. It is based on the *holographic principle*, which posits that the properties of a volume of space can be encoded onto the lower-dimensional surface bounding the volume. Jacob Bekenstein invoked this idea in the context of black holes in the 1970s, motivated by Stephen Hawking's theorem showing that the area $\mathcal{A}$ of a black hole's boundary, called its *horizon*, cannot decrease, and will increase in a dynamical process. Realising the latter property is reminiscent of the behaviour of thermodynamic entropy for an isolated system, Bekenstein postulated the entropy of a black hole to be proportional to $\mathcal{A}$. This is not proportional to the volume and is not extensive because of the strong influence of gravity.

$$S_{\text{bh}} = \eta k \mathcal{A}/\ell_{\text{pw}}^2 \text{ entropy of a black hole.} \tag{3.54}$$

---

[16]Other, competing theories explain the acceleration differently, and we do not know which, if any, will survive.

Figure 3.18: Symbolic depiction of a black hole's area, divided into $\mathcal{M}$ tiles, each with area ℓ_{pw}^2, where ℓ_{pw} is the Planck-Wheeler length in Eqs. (3.54), (3.55), and Key Point 3.23.

The black hole entropy S_{bh} has the same units as the Boltzmann constant k. The length ℓ_{pw} is defined by

$$\ell_{\mathrm{pw}} \equiv (hG/2\pi c^3)^{1/2} = 1.61 \times 10^{-35}\,\mathrm{m} \quad \text{(Planck-Wheeler length)}, \quad (3.55)$$

where G is the universal gravitational constant, h is the Planck constant, and c the speed of light in vacuum. Bekenstein wrote that η is a constant, presumed to be of order one, and Stephen Hawking found subsequently that $\eta = 1/4$.

The length ℓ_{pw} is the smallest length for which spacetime exists as a *smooth manifold* where all derivatives exist. For comparison, the radius of a proton is known to be a little smaller than $1\,\mathrm{femtometer} = 1 \times 10^{-15}\,\mathrm{m}$, 20 orders of magnitude larger than the Planck-Wheeler length. John Wheeler argued that for lengths $< \ell_{\mathrm{pw}}$ space-time is dominated by quantum mechanics and has a unique character. Indeed a black hole's entropy is very different from the thermodynamic entropy of normal matter, one difference being that the order of magnitude of $S_{\mathrm{bh}} \approx 10^{78}k$ is much higher than expected for a star-like object.

Key Point 3.22 *Bekenstein argued that Eq. (3.54) represents missing information. Because of the nature of a black hole, an exterior observer has a deep ignorance about the matter inside a black hole. Black holes are described by only three parameters, their mass M, charge Q and angular momentum L. In principle, no other quantities are measurable by such an observer. Among the quantities about which the observer knows nothing are the composition of elementary particles, temperature, and microstate. This enormous level of ignorance is presumed to be reflected in S_{bh}.*

Key Point 3.23 *Rafael Sorkin conceived a mental picture of a black hole's spherical surface, imagined to be covered with tiny tiles, each of which can be a 0 or 1 (see Fig. 3.18). If there are $\mathcal{M} = A/\ell_{pw}^2$ such Planck-Wheeler sized tiles, the number of possible states is $\Omega = 2^{\mathcal{M}}$, and a Boltzmann-type entropy would imply $S = \mathcal{M}k\ln(2)$. The Beckenstein entropy is similar, being $S_{\mathrm{bh}} = \frac{1}{4}k(A/\ell_{pw}^2)$.*

How is the black hole entropy S_{bh} related to the blackhole's temperature and energy? General relativity leads to the expressions $\mathcal{A} = (16\pi G^2 M^2/c^4)$, and $S_{\text{bh}} = (8\pi^2 kG/hc^4)M^2$. Using the Einstein relation $E = Mc^2$ leads to $S_{bh} \propto E^2$. Thus entropy is *not* a concave downward function of E for this gravity-driven system. The temperature of a black hole is inversely proportional to its mass and thus its energy, $E = Mc^2$: $T = (kGhc^3/16\pi^2)/E$, and $E \propto T^{-1}$. Adding mass and energy to a black hole *lowers* its temperature, so a black hole has negative heat capacity.[17]

Damien Easson and others have proposed a model for the universe that has only one assumption, namely that a horizon has a temperature and entropy associated with it. Their work is based on the concept of information and the holographic principle, and assumes some of the black hole concepts hold for the entire universe.[18] The expansion of the universe is described by Hubble's law $v = H(t)D$ where v is the speed of a distant galaxy a distance D away, and $H(t)$ is the Hubble parameter at time t. Easson, *et al.* view the boundary called the Hubble horizon, which recedes from us at the speed of light, as the boundary of the universe, with radius

$$R_H = \frac{c}{H(t)}. \tag{3.56}$$

As time passes, $H(t)$ gets smaller and R_H gets larger.

Adopting the Bekenstein entropy (3.54) for the universe with boundary radius R_H, the information entropy on the Hubble horizon, at radius R_H is

$$S_H = \frac{k\mathcal{A}}{4\ell_{\text{pw}}^2} = \frac{\pi k R_H^2}{\ell_{\text{pw}}^2}. \tag{3.57}$$

S_H is interpreted as the information entropy of the universe. Had Easson, *et al.* taken the radius to be less than the Hubble radius, the information entropy would also have been smaller. The spherical surface used to calculate the entropic force is viewed as an "information screen." When that screen encloses more of the universe, there is more missing information.

As the Hubble radius increases and accelerates, the entropy in (3.57) increases. Assuming the force that generates the accelerated expansion is an entropic force, as implied by the pressure in the second line of Eq. (3.53). The force here is,

$$F_{\text{ent}} \equiv T_h \frac{dS_H}{dR_H} = \frac{c^4}{G} \approx 1.2 \times 10^{44}\,\text{N.}^{[19]} \tag{3.58}$$

---

[17]Stephen Hawking obtained a negative heat capacity using quantum corrections in S. W. Hawking, "Black holes and thermodynamics," Phys. Rev. D 13, 191–197 (1976).

[18]D. A. Easson, P. H. Frampton & G. F. Smoot, "Entropic Accelerating Universe," Phys. Lett. B **696**, 273–277 (2010).

[19]The definition $F_{\text{ent}} = TdS/dR_H$ for horizon radius r is the negative of that used by Easson, *et al.* Here $F_{\text{ent}} = 4\pi R_H^2 P_{\text{ent}} > 0$ signifies a radially *outward* force.

The last step uses Eqs. 3.55–(3.57) and the horizon temperature definition,

$$T_h \equiv \frac{hH}{4\pi^2 k}.$$ (3.59)

The Hubble constant changes slowly in time, and its current value is,[20]

$$H \approx 70\,\mathrm{km\,s^{-1}Mpc^{-1}}.$$ (3.60)

Using the numerical value in Eq. 3.58 with the estimate for the ordinary mass of the observable universe, $m_{ordinary} \approx 1.5 \times ^{53}$ N implies an acceleration of,

$$a_{ent} = \frac{F_{ent}}{m_{ordinary}} \approx 0.8 \times 10^{-9}\,\mathrm{m\,s^{-2}}.$$ (3.61)

In analogy with the ideal gas, the pressure can be written $P_{ent} = \frac{2}{3}(E/V) = \frac{2}{3}\rho c^2$. Here E/V is the energy density and ρ is the corresponding mass density. This implies $P_{ent} = F_{ent}/(4\pi R_H^2) = \frac{2}{3}(E_H/V) = \frac{2}{3}\rho c^2$. Note that $\rho = (3H^2)/(8\pi G)$ is the mass density of that part of the universe associated with the accelerated expansion.

Key Point 3.24 *The procedure followed here assumes that Eq. (3.58) represents a physical force that produces the accelerated expansion. Neither the origin of that force nor the physical mechanism by which it acts is clear. In this sense, the entropic force approach is as mysterious as the postulate of dark energy. Using the value of the Hubble constant above, the predicted mass density is $\rho = 9 \times 10^{-27}$ kg m^{-3}, which is comparable to the estimated dark energy density, 1.67×10^{-27} kg m^{-3}. I find it particularly interesting that the approaches of dark energy and entropic force lead to similar results, one based on energy and the other on entropy, the focal points of this book.*

---

[20] 1 Mpc=1 million parsecs ≈ 3.26 light-years $= 3.09 \times 10^{-22}$ m.

CHAPTER 4

Gases, Solids, Polymers

CONTENTS

4.1 IDEAL GAS SACKUR-TETRODE ENTROPY

In the early 20th century, it was unclear if only differences in entropy for processes, or entropy's absolute value in a thermodynamic state could be determined. A major question was, "Would entropy always involve an unknown additive constant?" Using the mathematical framework of classical statistical mechanics, the formula for the entropy of a dilute monatomic gas was known to be

$$S(E, V, N) = Nk \left\{ \ln \left(\frac{3}{2} \ln \left(\frac{E}{N} \right) \right) + \ln \left(\frac{V}{N} \right) + s_0 \right\}, \qquad (4.1)$$

where, s_0 was an unknown constant and $E = \frac{3}{2} NkT$.

Classical statistical mechanics considers macroscopic systems with positions and momenta that can be arbitrarily close to one another. In contrast, the Heisenberg uncertainty principle of quantum mechanics, links a molecule's position $\vec{r}$ and momentum $\vec{p} = m\vec{v}$ such that $\Delta r_x \Delta p_x \geq h/(4\pi)$ for the x-components of $\vec{r}$ and $\vec{p}$, with corresponding inequalities for the y- and z-components. Here h is the Planck constant. In 1911, 31-year-old German physical chemist, Otto Sackur, and the 17-year-old Dutch Hugo Tetrode independently sought a way to count the number of possible distinguishable states of motion of the atoms of an ideal monatomic quantum mechanical gas. The objective was to determine Ω in Eq. (3.3), $S = k \ln \Omega$. Their work led to the evaluation of s_0:

$$s_0 = \frac{3}{2} \ln \left(\frac{4\pi m}{3h^2} \right) + \frac{5}{2}. \qquad (4.2)$$

Note that this contains the mass m of each molecule, which is the only parameter that distinguishes one ideal gas from another.

An important point is that in the study of ideal gases using quantum statistical mechanics, which I outline in Sec. 4.1.1, it is helpful to define a quantity called thermal wavelength, λ, defined as

$$\lambda \equiv \sqrt{\frac{h^2}{2\pi mkT}}. \qquad (4.3)$$

Here λ^3 is a volume representing a region in which a molecule will interfere quantum mechanically with other molecules in that region. At at given volume V, at sufficiently low temperatures, λ^3 is large, and quantum effects occur. For high enough temperatures, λ^3 is small and the gas behaves classically. In the classical limit, $v \equiv V/N \gg \lambda^3 \to 0$, the Sackur-Tetrode equation provides a good description. The following form is transparent.

$$S = Nk[\ln(v/\lambda^3) + 5/2] \text{ for } v \gg \lambda^3 \quad \text{(Sackur-Tetrode equation).} \qquad (4.4)$$

The condition $v \gg \lambda^3$ defines a *dilute* gas for a given temperature.

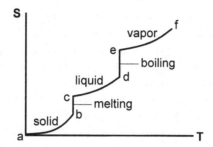

Figure 4.1: Typical entropy *vs.* temperature curve from $T \approx 0$ through the solid, liquid, and vapour phases. Paths bc and de occur at the melting and boiling temperatures, respectively. The "latent heats" of melting and boiling are $T_{\text{melt}} \Delta S_{\text{melt}} = \Delta H_{\text{melt}}$ and $T_{\text{boil}} \Delta S_{\text{boil}} = \Delta H_{\text{boil}}$.

Key Point 4.1 *The straightforward interpretation of the Sackur-Tetrode equation, Eq. (4.4), is that it holds when the volume per particle $v = V/N$ is much greater than the particle's thermal wavelength λ. This defines the classical region. The entropy S in Eq. (4.4) increases with increasing volume, giving energy more space over which to spread, and also with increasing temperature, increasing the amount of energy that spreads.*

The Sackur-Tetrode equation enables calculation of standard entropy values for ideal monatomic gases with given atomic masses. Those values can be compared with entropy values obtained by direct experiment, and tabulated e.g., in the CRC Handbook of Chemistry and Physics.

Suppose we want to determine the entropy of a sample of argon at temperature $T = 298.15$ K. Begin by cooling the sample as much as possible, getting near absolute zero. Then slowly heat the sample, measuring the energy transferred to the argon in a sequence of many small temperature changes. This generates data to calculate the constant-pressure heat capacity (thermal inertia) $C_P(T)$. It is necessary to account for the transitions from solid to liquid and liquid to vapour. The objective is to generate an entropy *vs.* temperature curve similar to that in Fig. 4.1, beginning in state a and ending in state f. Along ab, cd and ef, $\Delta S = \int (C_P/T)\, dT$. and along bc and de, $\Delta S = \Delta H/T$ (recall Eq. (1.12)). Though this is straightforward, it is not the procedure followed by Sackur and Tetrode.

A notable historical footnote is that Sackur and Tetrode tested their equation on mercury vapour, for which the best data were available in 1912. They did not cool the vapour as described above. Their calculation required knowing $C_P(T)$ as a function of T down to the low temperatures. Sackur and Tetrode followed slightly different procedures, each of which entailed using Planck's constant h, which was known from black-body radiation. Sackur directly computed the vapour pressure of mercury and compared his results with the experimental data. In contrast, Tetrode replaced h in the vapour pressure equation by zh and then fit z to existing data. This confirmed the Sackur-Tetrode equation within experimental error.

The element mercury (Hg) has seven stable isotopes with specific nuclear spins, and the mixture of these different isotopes adds to the entropy and causes a nonzero entropy as the temperature approaches $0\,K$. For completeness, one must add to the Sackur-Tetrode entropy these zero-point entropy terms. Although the various mercury isotopes were unknown in 1911, that turned out to be irrelevant because the entropy difference between the vapour and liquid phases is used in the calculation, and both phases have the same zero temperature entropy. Thus, unbeknownst to Sackur and Tetrode, the zero-point entropy terms cancelled out of the result!

4.1.1 Quantum ideal gases

In this section I elaborate on the thermodynamic properties of quantum ideal gases. A quantum mechanical gas of N identical particles is described by a wave function $\psi_k(q_1, q_2, \ldots q_N)$, where $q_1, q_2, \ldots q_N$ are the N 3-dimensional particle coordinates, and k represents the quantum state. Each q_i contains information for the three independent directions (x_i, y_i, z_i). The quantum wave function ψ is a solution to the Schrödinger equation. For a single particle, $|\psi(q)|^2 dx$ represents the probability of finding the particle in a volume element $d^3q = dxdydz$.[1] For example, in a system with two independent particles, the probability of finding the particles in the volume element $d^3q_1 d^3q_2 = (dx_1 dy_1 dz_1)(dx_2 dy_2 dz_2)$ is $|\psi(q_1, q_2)|^2 d^3q_1 d^3q_2$.

Key Point 4.2 *The quantum mechanical description of a system is probabilistic, and is described by a wave function that contains information about the interactions between its constituents. Somewhat mysteriously, a system's behaviour depends on whether the quantum spins of the constituent molecules are integers or half integers.*

Fermions and bosons. Numerous laboratory experiments show that particles with half-integral spin are describable by *antisymmetric* wave functions. Named after Enrico Fermi and Paul Dirac, these particle are called fermions, which obey Fermi Dirac (F-D) statistics. Particles called bosons have integer spin and *symmetric* wave functions. They exhibit very different behaviours. These are called bosons and they obey Bose-Einstein (B-E) statistics.

If two particles are in states i and j respectively, their wave function is, within a multiplicative constant,

$$\psi_{i,j}(q_1, q_2) = \psi_i(q_1)\psi_j(q_2) \pm \psi_j(q_1)\psi_i(q_2) \begin{cases} \text{- for fermions} \\ \text{+ for bosons} \end{cases} \quad (4.5)$$

---

[1] $|\psi|^2$ is defined as $\psi^*\psi$, where the asterisk * connotes the complex conjugate.

Key Point 4.3 *Key points that emerge are:*

1. *For half integer fermions, the minus sign in (4.5) assures that $\psi_{i,j}(q_1, q_2) = -\psi_{i,j}(q_2, q_1)$, i.e., ψ is antisymmetric under an exchange of the two particles.*

2. *That antisymmetry implies that the probability of finding the two fermions in the same state is zero, $|\psi_{k,k}(q_1, q_2)|^2 = 0$. This is the so-called Pauli exclusion principle, named after Wolfgang Pauli, usually stated: No two fermions in a system can occupy the same state.*

3. *Similarly the probability of finding two fermions at the same position is zero, $|\psi_{i,j}(q, q)|^2 = 0$. This has the effect that fermions tend to avoid one another, though there is no mechanical force acting between them to cause this purely quantum effect.[2]*

4. *Examples of fermions are electrons and helium-3 (^{3}He) atoms, with two spin 1/2 protons and one spin 1/2 neutron. There are also 2 electrons, totalling 5 spin 1/2 particles, which guarantees that the total spin is a half integer, i.e., $2 \times 1/2 + 1/2 + 2 \times 1/2 = 5/2$, a half integer.*

5. *Integral spin particles obey Bose-Einstein statistics of quantum mechanics. An example of a "boson" is ^{4}He, whose nucleus contains 2 protons and 2 neutrons, each with spin 1/2. There are also 2 electrons with spin 1/2 each, giving a total of 6 spin 1/2 particles, and ^{4}He has integral spin.*

6. *Bosons can have any number of particles in a given state. This property gives rise to remarkable behaviour of the boson system liquid helium ^{4}He, where at low enough temperatures, the particles condense into the state with lowest energy, the ground state. I describe this Bose-Einstein condensation phenomenon below.[3]*

As a rough model of a fermion system, a small wall with 6 bricks in each of 10 rows is shown in Fig. 4.2. The bricks represent particles. Each row represents a 6-fold degenerate energy and its 6 spin-5/2 particles have spin components (one each) $-5/2, -3/2, -1/2, 1/2, 3/2, 5/2$. The wall height in the figure is analogous to the Fermi energy ϵ_F, the maximum energy for $N = 60$ particles that are fully packed into the lowest possible states. For a fermion gas, such packing occurs in the limit of zero absolute temperature, $T \to 0$. The collection of particles with energy $\epsilon < \epsilon_F$ is sometimes called the "Fermi sea." In Fig. 4.2, several bricks, i.e., particles, have been energised by strong winds, and their added energy takes them from the wall. This is analogous to what happens in

---

[2]Point particles do not exist, and atoms have finite sizes, assuring that no two particles, even bosons, can be at the same position.

[3]^{3}He displays remarkable properties at even lower (mK) temperatures. See Sec. 8.6.4.

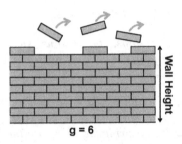

Figure 4.2: Analogy of fermions occupying the lowest energy states. If the highest energy particles are excited, say by a strong wind, they will exit the wall. To make the analogy work, two half bricks at the right and left ends of a row are counted as one whole brick.

a fermion gas at temperatures above absolute zero, when some particles are sufficiently energised to leave the "Fermi sea."

In contrast, bosons have no restrictions on their occupation numbers, and display significant quantum effects when $v \leq \lambda^3$. Specifically, quantum statistical mechanics shows that a sharp departure from normal behaviour occurs at a critical temperature T_c, defined by $\lambda_c^3/v = 2.612$. Cooling below this temperature results in N_0 particles dropping into a zero-energy state, leaving $N_{ex} = N - N_0$ in excited states, with,

$$Fraction\ in\ ground\ state = \frac{N_0}{N} = \begin{cases} 1 - (T/T_c)^{3/2} & for\ T \leq T_c \\ 0 & for\ T > T_c \end{cases}. \qquad (4.6)$$

The fraction of particles in excited states is

$$Fraction\ in\ excited\ states = 1 - \frac{N_0}{N}. \qquad (4.7)$$

These fractions are graphed in Fig. 4.3 and are clarified further below. The main point is that particles in the zero-energy ground state do not contribute to the internal energy; only the N_{ex} particles contribute nonzero energy.

Pressure inequality. Harold Falk found[4] some novel and interesting relationships between the Fermi-Dirac, Bose-Einstein and classical ideal gases, with respective pressures $P_{fermions}(T, V, N), P_{bosons}(T, V, N)$ and $P_{classical}(T, V, N)$.

$$P_{bosons}(T, V, N) < P_{classical}(T, V, N) < P_{fermions}(T, V, N). \qquad (4.8)$$

Equation (4.8) has an interesting interpretation, namely that quantum effects alter the pressure because of quantum correlations. This is very different from a nonideal classical gas, where attractive forces reduce the pressure, which is discussed for the van der Waals model of a gas in Sec. 4.2.3.

---

[4]H. Falk & E. Adler, "On the Pressure of Ideal Quantum Gases," Am. J. Phys. **36**, 454 (1968).

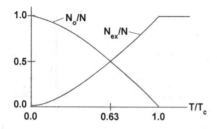

Figure 4.3: Fraction N_0/N of particles in the ground state and N_{ex}/N in excited states, *vs.* T/T_c. The two fractions are equal for $T = 0.63\,T_c$.

Key Point 4.4 *Bosons tend to be close together because of their symmetric wave functions. The quantum correlations reduce the pressure relative to the classical case. In contrast, fermions tend to be farther apart because of their antisymmetric wave functions, increasing the pressure.[5] The inequalities in Eq. (4.8) are solely from quantum correlations that arise from the different symmetry requirements on the wave functions for Bose-Einstein and Fermi-Dirac particles. There are no physical forces in a quantum ideal gas.*

Energy inequality. The Falk inequalities (4.8) have implications for energy. *All* monatomic ideal gases, bosons, fermions, or classical particles satisfy

$$E = \tfrac{3}{2}PV, \tag{4.9}$$

which implies,

$$E_{\text{bosons}}(T, V, N) < E_{\text{classical}}(T, V, N) < E_{\text{fermions}}(T, V, N). \tag{4.10}$$

The fermion energy being largest seems reasonable because the lowest lying particles states are restricted by the Pauli exclusion principle, illustrated in the brick wall analogy in Fig. 4.2.

Energy and entropy for fermions and bosons. The average total energy of a quantum ideal gas with particle energies ϵ_k, and an average number of particles, $\bar{n}_k$, in state k with $k = 0, 1, 2, \ldots$ is,

$$E_{\text{quantum}} = \sum_k \bar{n}_k \epsilon_k, \text{ where } \bar{n}_k = \begin{cases} 0 \text{ or } 1 \text{ for fermions.} \\ 0, 1, 2, \ldots \text{ for bosons.} \end{cases} \tag{4.11}$$

In quantum statistical mechanics, the total energy $\sum_k n_k \epsilon_k$ is placed in the partition function, Eq. (3.18) and some clever mathematics leads to the average number of particles $\bar{n}_k$ for fermions and bosons,

$$\bar{n}_k = \begin{cases} (e^{(\epsilon_k - \mu)/\tau} + 1)^{-1} \text{ for fermions} \\ (e^{(\epsilon_k - \mu)/\tau} - 1)^{-1} \text{ for bosons} \end{cases} \text{ with } \tau = kT. \tag{4.12}$$

---

[5]This is sometimes attributed to *exchange forces*, giving the incorrect impression that a physical force acts.

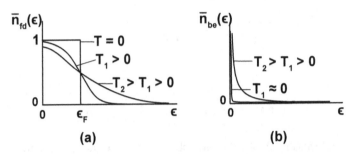

Figure 4.4: Average number of particles in a single-particle state, $\bar{n}(\epsilon)$ for temperatures T_1 and $T_2 > T_1 > 0$. (a) Ideal fermion gas. (b) Ideal boson gas. As $T \to 0$, $\bar{n}(\epsilon) \to N$, i.e., bosons drop into the lowest energy state.

Here, μ is the chemical potential, defined as the energy needed to add one particle to the system at fixed volume and entropy (see Eq. (11.2)). For fermions, when $T \to 0$, if $\epsilon_k < \mu$, $\bar{n}_k \to 1$; and if $\epsilon_k > \mu$, $\bar{n}_k \to 0$. In the limit $T \to 0$, the largest energy in the Fermi sea is called the Fermi energy $E_F = \lim_{T \to 0} \mu(T, P)$. For bosons, as $T \to 0$, $\bar{n}_k \to N$, i.e., all particles drop into the ground state. Similarly, the average energy is,

$$E_{\text{quantum}} = \sum_k \bar{n}_k \epsilon_k = \begin{cases} \sum_k \epsilon_k (e^{(\epsilon_k - \mu)/\tau} + 1)^{-1} \text{ for fermions.} \\ \sum_k \epsilon_k (e^{(\epsilon_k - \mu)\tau} - 1)^{-1} \text{ for bosons.} \end{cases} \quad (4.13)$$

As a consequence of the disparate properties of fermion and boson quantum ideal gases, these two systems display very different behaviours of pressure, energy, entropy and heat capacity. The single-particle energies for an ideal gas in a cubical container of volume $V = L \times L \times L$ are,

$$\epsilon_k = \frac{h^2}{8\pi^2 mL^2}(k_x^2 + k_y^2 + k_z^2), \text{ where } k_x, k_y, k_z = 0, 1, 2 \ldots \quad (4.14)$$

Level spacing between adjacent levels is of order $h^2/(8\pi^2 mL^2)$. For argon gas in a cube with $L = 5$ m, this is about 3.35×10^{-45} J. For comparison, 1 mole of a classical ideal gas at temperature $T = 298$ K has internal energy $E \approx 3716$ J and the average energy per atom is $\approx 6.2 \times 10^{-21}$ J.

Key Point 4.5 *The spacing between adjacent levels of an ideal gas is small compared to the average energy per atom in the gas, so the energy spectrum of the gas can be treated as a continuum, as in Fig. 4.4. The average occupation number per unit energy is $\bar{n}(\epsilon)$ in Eq. (4.15), with $n(\epsilon)d\epsilon$ being the number of states in the interval $(\epsilon, \epsilon + d\epsilon)$.*

$$\bar{n}(\epsilon) = \frac{1}{(e^{(\epsilon - \mu)} \pm 1)} \text{ for (fermions, bosons).} \quad (4.15)$$

Figure 4.4(a) shows $\bar{n}_{\text{fd}}(\epsilon)$ for fermions with temperatures $T \to 0$ and two temperatures, $T_2 > T_1 > 0$. As the temperature gets higher, a smaller fraction

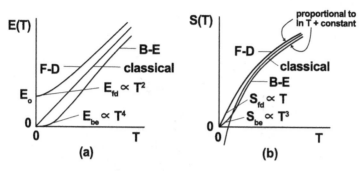

Figure 4.5: (a) Internal energy *vs.* temperature for fermions (FD), bosons (BE) and classical particles. (b) Entropy *vs.* temperature for the same systems as (a). (Copyright 1996 from *Entropy and its Physical Meaning* by J. S. Dugdale, pp. 146–147. Reproduced by permission of Taylor & Francis, a division of Informa plc.)

of the particles is in the Fermi sea, and a greater fraction is excited to have energies $\epsilon > \epsilon_F$. Figure 4.4(b) shows that bosons behave very differently. For $T \to 0$, all the particles would be in the lowest energy state, which is consistent with Fig. 4.3 and Eq. (4.6). At higher temperatures, more particles are excited to higher-energy states.

Statistical mechanics shows that for small T, the total energy for fermions satisfies $(E_{\mathrm{fd}} - E_0) \propto T^2$, and the total energy for bosons satisfies $E_{\mathrm{be}} \propto T^4$. For high temperature, both boson and fermion gases satisfy $(E + \mathrm{constant}) \propto T$. These behaviours are evident in Fig. 4.5(a). Statistical mechanics also shows that at the lowest temperatures, fermions have system entropy $S_{\mathrm{fd}} \propto T$, and bosons have entropy $S_{\mathrm{be}} \propto T^3$. At high temperatures, both cases have entropies that are proportional to $\ln T$ within some constant. This entropy behaviour is shown in Fig. 4.5(b).

Key Point 4.6 *For any temperature T the internal energy for ideal fermions exceeds that of the classical ideal gas, which exceeds that for ideal bosons. Similarly, the fermion entropy exceeds the boson entropy for all temperatures. These behaviours are attributable, at least in part, to the Pauli exclusion principle preventing more than two fermions in a given state. The classical ideal gas entropy at low temperatures is nonphysical (there is no classical gas at those temperature).*

4.2 NONIDEAL GASES & THE VIRIAL EXPANSION

Gases made of molecules are essential to humans. A prime example is air, a mixture of different gas species: approximately 78.09% nitrogen, 20.95% oxygen, 0.93% argon, 0.04% carbon dioxide, plus small amounts of other gases.

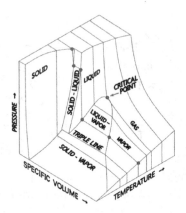

Figure 4.6: Pressure-volume-temperature surface for a real gas. (Source: D. L. Smith, *Thin Film Deposition Principle & Practice*, Addison-Wesley Pub. Co., 1950, 1953 Reprinted with permission by current copyright owner, McGraw-Hill Pub. Co.)

The simplest gases are single-species, dilute gases, i.e., with molecules that are separated sufficiently on average that their interactions with one another are negligible. These are typically well described by the classical ideal equation in Sec. 1.5. All gases take on the shapes of their containers.

A gas that is neither simple nor necessarily dilute is water vapour, which will condense to liquid water when the temperature is low enough at a given pressure. The molecules of real gases, including classical gases, exert forces on one another in accordance with Fig. 1.3. When two molecules get close to one another, they repel and when they are farther apart, they attract.

4.2.1 Liquid-vapour phase transition

Air does *not* always behave like a dilute gas because it contains water vapour, H_2O, whose molecules attract strongly enough to form liquid clusters of molecules. This can cause liquefaction at atmospheric pressures and temperatures, condensing as rain, fog, hail or snow. *Condensation* occurs not only outdoors, but in bathrooms during hot showers when water droplets form on a cold mirror. The hot shower water vaporises, as water molecules become energetic enough to escape from the liquid state. When such molecules get near a cold mirror, their temperature drops and the strong attractive forces between water molecules causes them to cluster into water drops.

Figure 4.6 is a three-dimensional plot of pressure, volume and temperature for a substance, showing regions where it is a vapour, liquid, or solid, or a mix of two or all three of these. In particular, along the so-called *triple line* the substance exists simultaneously in all three phases. The graph of pressure *vs.* volume for a gas that is not ideal deviates from Fig. 1.16b.

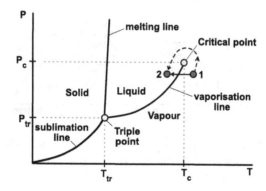

Figure 4.7: Pressure *vs.* temperature phase diagram showing the regions of pure solid, liquid and vapour and the melting, vaporisation and sublimation lines, along which two of the phases coexist. Also shown are the triple point (T_{tr}, P_{tr}) and critical point (T_c, P_c).

A series of constant-temperature pressure-volume curves is shown in Fig. 4.6. For temperatures well above a *critical* temperature T_c, a gas behaves approximately as a dilute gas. However, if the temperature is below T_c, as a gas is compressed isothermally, there is a region in which decreasing the volume ceases to change the pressure.

Key Point 4.7 *Below T_c, compressing a gas isothermally causes molecules to form liquid droplets because of the attractive intermolecular forces. This tends to reduce the pressure and compensates for the pressure increase from compression of the vapour component. The net effect is that the pressure of the vapour-liquid mixture is constant throughout the two-phase gas-liquid region.*

A two-dimensional projection of this on a pressure *vs.* temperature diagram is shown in Fig. 4.7, where the *triple line* appears as a *triple point*. Above the triple point temperature, beginning in the vapour state 1, a sequence of temperature and pressure changes that follow the dashed path takes the former vapour continuously to the liquid state 2. This occurs *without undergoing a discontinuous change in density*. On the other hand, below the critical temperature and pressure, constant-pressure cooling of the vapour brings the vapour to the liquid state with a discontinuous change in density when the vapour-liquid phase line is crossed. Although it appears in Fig. 4.7 that the discontinuity occurs on the vaporisation line, the pressure-volume projection of Fig. 4.6 in Fig. 4.8(a) shows that the change happens along a line of variable volume (dashed horizontal line) at constant pressure.

The *sublimation line* in Fig. 4.6 runs from absolute zero on the temperature axis to the triple point temperature. To the left of that line, the substance is a solid. To the right it is a vapour. Just as a liquid forms a vapour with a temperature-dependent vapour pressure above a liquid, a solid does the same

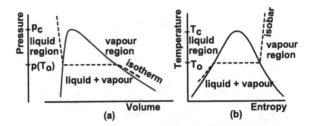

Figure 4.8: Isothermal expansion (dashed line, from left to right) for a liquid-vapour phase transition, shown on (a) pressure-volume and (b) temperature-entropy diagrams.

when the temperature and pressure allow both solid and vapour to coexist. Evidence of this comes from ice cubes in a refrigerator freezer that shrink in size over time. This is *not* a two-step process of melting and then evaporating. Rather, sublimation takes the H_2O from solid to vapour directly, with no liquid phase involved. For water, this happens below the triple point temperature 273.16 K, and triple point pressure of about 6/1000 of standard atmospheric pressure (0.006 atm or < 5 mm mercury).

Key Point 4.8 *The vapour pressure of liquid water at about 300 K is 26 mm mercury, more than 5 times larger than the triple point pressure, and much less than normal atmospheric pressure (760 mm mercury). The ice in a freezer sublimates slowly, just fast enough to achieve the low vapour pressure of water at the freezer temperature. This causes the ice to vanish very slowly.*

4.2.2 Clausius–Clapeyron equation

The slope of each phase line in Fig. 4.7 at a chosen temperature can be expressed in terms of the respective volumes and entropy values on the two sides of each line. This can be seen by writing entropy per particle as a function of the volume per particle $v = V/N$ and the temperature T, i.e., the entropy per particle is $s = S/N = s(v, T)$. Then at a given fixed temperature T, a volume change results in $ds = (\partial s/\partial v)_T dv$. By the second Maxwell equation in Eq. (11.2), $(\partial S/\partial V)_T = (\partial s/\partial v)_T = -(\partial P/\partial T)_v$. Therefore, $ds = (\partial P/\partial T)_v dv$. If we integrate the latter across the phase boundary, from the liquid to vapour, we get the Clausius–Clapeyron equation,

$$\left(\frac{dP}{dT}\right)_{\text{phase boundary}} = \frac{s_{\text{vap}} - s_{\text{liq}}}{v_{\text{vap}} - v_{\text{liq}}} \text{ for given } T, P. \tag{4.16}$$

For fixed temperature and pressure, T, P, $(dP/dT)_{\text{phase boundary}}$ is constant as the liquid becomes vapour. Equation (4.16) generalises to the other phase boundaries in Fig. 4.7. The slope of a phase boundary line equals the entropy change divided by the accompanying volume change across the boundary.

4.2.3 Van der Waals gas

The van der Waals model of a gas accounts for the repulsive part of the force treating the molecules as hard spheres with finite volume b and assuming the spheres reduce the available volume to $(V - Nb)$, which is a rough approximation. Johannes van der Waals introduced this model in his 1873 doctoral dissertation.[6] When the molecules are far apart, they attract one another. Molecules in the interior of the fluid are attracted in all directions but molecules near the walls are attracted mainly inward, reducing the pressure on the walls. Each of the $\frac{1}{2}N(N - 1) \approx \frac{1}{2}N^2$ pairs of molecules is accounted for in a term $-a(N/V)^2$ with $a > 0$. The factor a/V^2 comes from averaging the force between each pair of particles,

$$\overline{F} = \frac{\int d^3r_i \int d^3r_j F(\vec{r}_i, \vec{r}_j)}{\int d^3r_i \int d^3r_j} \equiv \frac{\text{constant}}{V^2}. \tag{4.17}$$

The pressure reduction is assumed to be $-aN^2/V^2$, i.e., $\propto \overline{F}$ for each molecular pair, and to the number of molecular pairs. The result is

$$P = \frac{NkT}{V - Nb} - \frac{aN^2}{V^2} \quad \text{(van der Waals equation)}. \tag{4.18}$$

Three points are worth emphasis:

1. The van der Waals pressure is *intensive*, as it should be. That is, it is unchanged if the volume and number of molecules increase or decrease by the same factor, keeping N/V fixed. The equation is simpler to write and manipulate if we replace the extensive variables V and N by the single *intensive* variable $v \equiv V/N$:[7]

$$P = \frac{kT}{v - b} - \frac{a}{v^2}. \tag{4.19}$$

2. The van der Waals gas is particularly interesting because both its first and second derivatives are zero for a specific temperature called the *critical temperature*. That is, the conditions $(\partial P/\partial v)_T = 0$ and $(\partial^2 P/\partial v^2)_T = 0$ imply the critical values

$$T_c \equiv 8a/(27k), v_c = 3b, \text{ and } P_c = a/(27b^2). \tag{4.20}$$

3. Equations (4.19) and (4.20) imply a material-independent equation with relative variables $T_r \equiv T/T_c, v_r \equiv v/v_c$, and $P_r \equiv P/P_c$.

$$P_r = 8T_r/(3v_r - 1) - 3/v_r^2 \text{ in relative variables } T_r, P_r \text{ and } v_r. \tag{4.21}$$

---

[6] In reality, the excluded volume is more complex, being $> Nb$ for configurations with the spheres close together.

[7] A similar result is obtained for the volume per mole, rather than per molecule, with the number of moles being $n = N/N_A$, where N_A is Avogadro's number (Table 1.4).

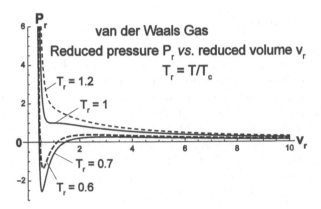

Figure 4.9: van der Waals reduced pressure *vs.* reduced volume for reduced temperatures T_r. Note that this equation gives nonphysical negative pressures.

Key Point 4.9 *The van der Waals gas possesses the critical parameters T_c, v_c and P_c. For temperatures below T_c, the pressure behaves strangely (it wiggles, as described below), suggesting a vapour-to-liquid phase transition. The critical temperature T_c is proportional to the attractive force parameter a, so the larger a is, the higher is the critical temperature, which supports the notion that the attractive forces are responsible for condensation.*

Originally called "the law of corresponding states," Eq. (4.21) was hoped to be a universal, material-independent equation of state. This turned out to not be the case. However, (4.21) allows the simple plot of pressure *vs.* volume for various relative temperatures in Fig. 4.9. The van der Waals gas model displays some properties of real-life gases, and a variety of other properties that bear on energy and entropy:

For temperatures $T > T_c$ ($T_r > 1$), the pressure curve resembles that for an ideal gas. When $T = T_c$ ($T_r = 1$), the pressure curve is flat at reduced volume per particle $v_r = 1$. Below the critical temperature, $T_r < 1$, the pressure has a wiggle in it sometimes called the van der Waals loop. The loop is characterised by P_r decreasing, then increasing, and finally decreasing again. An interpretation is that the van der Waals wiggle, which occurs at temperatures $T < T_c$ signifies that the van der Waals gas is undergoing condensation from gas to liquid.

James Clerk Maxwell proposed a technique, called Maxwell's rule, to get around this deficiency. This replaces the wiggle with a constant pressure segment chosen so the area between the upward part of the loop and the constant pressure line equals the area between the latter line and the downward part of the loop. Strictly speaking, this *ad hoc* "equal-area rule" should not be necessary. Nevertheless, the van der Waals equation is important because it displays a critical temperature and volume, and weird behaviour that suggests a vapour-liquid transition.

The van der Waals equation can be usefully compared with the ideal gas equation of state by expressing it as an infinite series in powers of $1/v$. This is possible for $b/v < 1$, whereupon $(v - b)^{-1} = v^{-1}(1 - b/v)^{-1}$, and

$$\frac{1}{(1 - b/v)} = 1 + \frac{b}{v} - \frac{b^2}{v^2} + \frac{b^3}{v^3} - \dots \tag{4.22}$$

Combining (4.19) and (4.22) leads to

$$\frac{Pv}{kT} = 1 + \frac{(b - a/kT)}{v} + \frac{b^2}{v^2} + \dots \tag{4.23}$$

Key Point 4.10 *Equation (4.23) is a very interesting equation because the second term on the right reveals a competition between the repulsive (b) and attractive (a) parts of the intermolecular force. When repulsion dominates, the pressure exceeds the ideal gas pressure, and when the attraction dominates, the pressure is less than the ideal gas pressure. This is shown in Eq. (4.24).*

$$P \begin{cases} > P_{\text{ideal}} = kT/v \text{ for } b > a/kT \\ < P_{\text{ideal}} = kT/v \text{ for } b \ll a/kT \end{cases} \tag{4.24}$$

Note that for any values of a and b, $P > P_{\text{ideal}}$ for sufficiently high temperature, and $P < P_{\text{ideal}}$ for sufficiently low temperature. In other words, the repulsive force dominates at higher T and the attractive force dominates at lower T. The appearance of the attractive force parameter a in only one term of the infinite series (4.23) suggests that the van der Waals equation is oversimplified. In the limit of infinite dilution, $v = V/n \to \infty$, the van der Waals equation reduces to the ideal gas equation; i.e.,

$$\lim_{v \to \infty} pv/RT = 1. \tag{4.25}$$

The van der Waals equation has energy implications evident in the differential expression $dE = TdS - PdV$ (see Eq. (11.1), with a fixed number of particles so $dN = 0$.) Taking S to be a function of T and V, $TdS = C_v dT + T(\partial S/\partial V)_T dV = C_v dT + T(\partial P/\partial T)_V dV$,

$$\begin{aligned} dE &= C_v dT + [T(\partial P/\partial T)_V - P]dV \\ &= C_v dT + \left[\frac{NkT}{V - Nb} - \frac{NkT}{V - Nb} + \frac{aN^2}{V^2} \right] dV \\ &= \frac{aN^2}{V^2} dV. \end{aligned} \tag{4.26}$$

The internal energy can be found by integrating (4.26), which leads to

$$E(T, V) = -\frac{aN^2}{V} + \chi(T) = \tfrac{3}{2} NkT - \frac{aN^2}{V}. \tag{4.27}$$

The last step assumes that there is no a or b dependence in χ, and when $a = 0$, the internal energy must be $\frac{3}{2}NkT$, the ideal gas internal energy.[8] In Eq. 4.27, $E(T,V)$ is a function of volume as well as temperature, albeit these two dependences occur in separate terms. The volume dependence comes from the attractive part of the intermolecular force through a. There is zero contribution to the energy from the repulsive force parameter b, suggesting the van der Waals pressure is oversimplified. Because repulsion causes a high potential energy when two particles get close together, the lack of dependence of energy on b for a van der Waals gas shows the model's inadequacy.

Pressure and entropy are related through the Maxwell relation (see Eq. (11.2)), $(\partial P/\partial T)_v = (\partial S/\partial V)_T$. That does not imply wiggles in $S(T,V,N)$ because the pressure wiggles involve both parameters a and b, but the van der Waals gas entropy is independent of a.

$$S(T,V,N) = Nk[\tfrac{3}{2}\ln T + \ln(v-b) + \text{constant}]. \qquad (4.28)$$

Therefore S has no van der Waals wiggles.

Baierlein inequality. Comparing (4.28) with the ideal gas entropy, it is clear that $\ln(v-b) < \ln(v)$, so

$$S_{\text{vdw}} < S_{\text{ideal}}. \qquad (4.29)$$

The inequality (4.29) is a special case of Baierlein's theorem,[9]

$$S_{\text{nonideal}} < S_{\text{ideal}} \qquad \text{Baierlein theorem for classical gases.} \qquad (4.30)$$

For a macroscopic gas, we often speak of the number density N/V, which is the average number of particles per unit volume. However forces between molecules induce correlations that cause variations in density over a small number of molecular diameters. Such variations are seen in the statistical mechanics radial (or pair) distribution function $g(r)$, where $g(r)dr$ is the probability of finding a particle between distance r and $r + dr$ from a chosen reference particle. The pair distribution function is an indicator of how the local density varies relative to the ideal gas average, N/V, as a function of distance from a reference particle. This interesting function can be calculated for a potential energy function that models a typical gas. A common example is the Lennard-Jones potential.[10] A graph of $g(r)$ vs. r based on the Lennard-Jones potential is in Fig. 4.10.

Key Point 4.11 *In the discussion of Fig. 3.1, I explained that when samples of the same gas with different densities are able to mix, the result is a gas with an intermediate equilibrium density. Later, in Sec. 4.3, I will show that such a*

---

[8]Equation 4.27 was obtained rigorously by R. H. Swendsen, "Using computation to teach the properties of the van der Waals fluid," Am. J. Phys. **81**, 776–781 (2013).

[9]R. Baierlein, "Forces, Uncertainty, and the Gibbs Entropy," Am. J. Phys. **36**, 625–629 (1968).

[10]Introduced by John Lennard-Jones, 1924, $\Phi(r) = 4\epsilon[(\sigma/r)^{12} - \sigma/r)^6]$. Parameters ϵ and σ can be fitted to best match a particular gas.

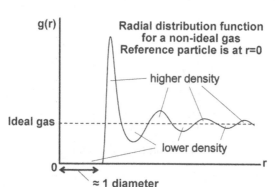

Figure 4.10: Radial distribution function for a Lennard-Jones potential. Wiggles are shown relative to the constant value of $g(r)$ (dashed line) for an ideal gas. This graph shows values of r over $\sim 4 - 5$ molecular diameters. (Adapted from Wikipedia, with permission by Grigory Sarnitskiy.)

new equilibrium state has a higher entropy. With this as a guide, it is natural to expect a nonideal gas with density variations like those in Fig. 4.10 will reduce the gas entropy relative to a classical ideal gas with the same (T, V, N). This is precisely what the Baierlein theorem (4.30) tells us. Intermolecular forces introduce correlations between molecules that lead to, $S_{nonideal} < S_{ideal}$.

4.2.4 Virial expansion

Equation (4.23) is a special case of a useful expansion scheme that expresses the pressure as a series, called the virial series, whose successive terms account for interactions between the particles,

$$\frac{Pv}{kT} = 1 + \frac{B(T)}{v} + \frac{C(T)}{v^2} + \frac{D(T)}{v^3+} \cdots \qquad (4.31)$$

The temperature-dependent quantities $B(T), C(T), D(T) \ldots$ are called the second, third, fourth... virial coefficients. These can be viewed as accounting for clusters of $2, 3, 4 \ldots$ molecules, respectively.

Dilute gases can be well approximated by cutting off the virial series after the second virial coefficient $B(T)$. Statistical mechanics enables us to relate the virial coefficients to the intermolecular potential energy function. Computer calculations and experimental evidence both show that the second virial coefficient has the general behaviour in Fig. 4.11. The series (4.23) shows that the second virial coefficient of the van der Waals gas, $B(T) = (b - a/kT)$ exhibits the same general behaviour. For $T < a/bk$, $B(T) < 0$, and $P < P_{ideal}$. For $T > a/bk$, $B(T) > 0$, and $P > P_{ideal}$.[11] When the virial series first

---

[11] The second virial coefficients of various gases is calculated and graphed in A. Hutem & S. Boonchui, "Numerical evaluation of second and third virial coefficients of some inert gases via classical cluster expansion," J. Math. Chem. **50**, 1262–1276 (2012).

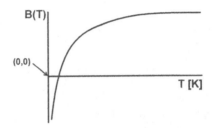

Figure 4.11: Typical second virial coefficient $B(T)$ *vs.* T.

appeared, some people hoped that it would diverge at the critical temperature, which would be a dramatic signal of a phase transition. However, that turned out to *not* be the case.

Key Point 4.12 *Successive terms in the virial series show effects of interaction energies between two, three, ... molecules. A virial series also exists for the entropy, with similar interpretation, showing a connection between entropy and energy interactions of successively larger clusters of molecules.*

4.3 MIXING ENTROPY FUNCTION

4.3.1 Mixing or expansion?

A problem of both historical and ongoing interest is the mixing of two gases. Consider a process that mixes two *distinguishably different* classical ideal gases, A and B, with particle numbers N_1 and N_2, with $N_1 + N_2 = N$. The gases are initially in separate, chambers of volume V, and the system is assumed to be in thermal contact with a constant-temperature reservoir at temperature T. If the partition between the equal-pressure chambers is suddenly removed, the two gases mix together irreversibly, doubling the volume of each. The temperature is unchanged and the final pressure of the mixture equals one half the sum of the initial pressures of the gases. This process is illustrated in Fig. 4.12a.

The entropy increase during this process is

$$\Delta S_{\text{dist}} = Nk\ln 2, \quad \text{for } N_1 = N_2 \text{ or } N_1 \neq N_2, \text{ where } N = N_1 + N_2. \quad (4.32)$$

The subscript "dist" connotes *distinguishable* species, i.e., the gases are physically different from one another.

Key Point 4.13 *Notably $\Delta S_{\text{dist}} = Nk\ln 2$ for the entropy change when two distinguishable gases are mixed holds when the particle numbers of the species are equal or unequal. This is a very general result for dilute gases.*

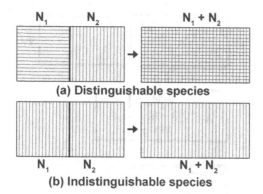

Figure 4.12: (a) Mixing of distinguishably different gases, e.g., N_1 argon atoms and N_2 neon atoms, connoted by horizontal and vertical parallel lines, respectively. (b) Mixing of identical gases, for example, both being neon, each connoted by vertical parallel lines. In either case, $N_1 + N_2 = N$.

Traditionally ΔS_{dist} has been called the *entropy of mixing*. This is unfortunate because the entropy change in Eq. (4.32) comes from the *expansion* of each species, doubling its volume, at constant energy. This can be seen by expanding argon's chamber leftward and neon's rightward in Fig. 4.12(a), and writing the entropy change as the final minus initial entropy,

$$\Delta S_{\text{dist}} = [S_A(T, 2V, N_1) - S_A(T, V, N_1)]$$
$$+ [S_B(T, 2V, N_2) - S_B(T, V, N_2)]. \quad (4.33)$$

The envisaged expansions lead to the top configuration in Fig. 4.13. The gases can be mixed *reversibly* in principle using semipermeable membranes, as shown in Fig. 4.13, without modifying the volume, energy, or entropy of either species. Semipermeable membrane A, SPM$_A$, allows only molecules of species A (argon) to pass through it and SPM$_B$ passes only species B (neon) molecules. The Gibbs theorem states that the entropy change for the process is zero, because the reversible inward telescoping requires zero external work and no other energy exchange with the environment. The sum of the entropy values of the separated species (top pane) equals the entropy of the mixture (bottom pane). Despite mixing in the final configuration, there is no entropy change because each gas maintains its initial temperature T and volume $2V$ throughout.

In Fig. 4.12, both species (or identical particles in separate chambers) expand to twice their initial volumes. The entropy change Eq. (4.32) comes solely from that expansion. This is strengthened with the observation that the latter equation holds when $N_1 \neq N_2$. One might expect *more* mixing and a higher entropy of mixing than for the equal particle case. But, alas, Eq. (4.32) applies in both cases.

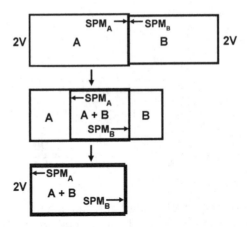

Figure 4.13: Distinguishable gas species A and B, are each in volume $2V$ initially. They are mixed reversibly using semipermeable membranes, with the right cylinder telescoping into the fixed left cylinder. The final mixed configuration has volume $2V$.

What if two *indistinguishable* gases are mixed by central partition removal? Assume $N_1 = N_2 = N/2$. When the partition is removed, no thermodynamic process takes place, i.e., there is no distribution of energy.

$$\Delta S_{\text{ident}} = 0 \text{ when } N_1 = N_2 \qquad (4.34)$$

Key Point 4.14 *The zero value in Eq. (4.34) is understandable because when the partition is removed, nothing changes. The partition can be replaced (slowly to not disturb the molecules), again with zero entropy change.*

When $N_1 \neq N_2$, the situation is different. For example, if $N_1 > N_2$, there is more energy initially in the left chamber, and partition removal leads to a spreading of energy to achieve equity. The entropy change can be calculated using any reversible path that leads from the initial to final state. For example, if $N_1 > N_2$ we can first expand the right chamber reversibly and isothermally to volume $V' = (N_2/N_1)V$, which gives the right chamber gas the same density as the left. This can be done by turning its right wall into a frictionless piston. The entropy change of the gas during the expansion is $\Delta S' = N_1 k \ln(N_1/N_2)$.

Now that the gases in the chambers have the same density, the partition can be removed with entropy change $\Delta S'' = 0$ (as before), and the gas now fills volume $V + V' = V + (N_2/N_1)V$. We can now do a reversible isothermal expansion from $V + V'$ to $2V$. A messy calculation leads to the entropy change $\Delta S''' = (N_1 + N_2)k \ln(2N_1/N)$. Adding the three entropy changes, we find

$$\Delta S_{\text{ident}} = Nk[f_1 \ln f_1 + f_2 \ln f_2 + \ln 2], \text{ with } f_i \equiv N_i/N, \ i = 1, 2. \qquad (4.35)$$

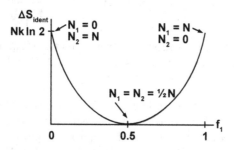

Figure 4.14: Entropy change for the mixing of equal volumes of identical dilute gases as a function of fraction $f_1 = N_1/N$.

I assumed $N_1 > N_2$, but the entire analysis holds also for $N_1 \leq N_2$. In accord with Fig. 4.14 and Eq. (4.35), if $N_1 = N_2$, then $f_1 = f_2 = 1/2$, and $\Delta S_{ident} = 0$. If $N_1 \neq N_2$, then ΔS_{ident} increases monotonically with $|N_2 - N_1|$.

Key Point 4.15 *When two identical dilute gases, initially in separate chambers of equal volume mix, there is an increase in entropy for $N_1 \neq N_2$ because initially the gases have unequal densities and internal energies. The magnitude of the entropy increase increases with increasing density difference, and is maximal when one chamber has all the particles and the other one has none, as is evident in Fig. 4.14.*

I now turn to the mixing of two *distinguishable* gases. The arrangement is the same as in Fig. 4.12(a). Again, to calculate the entropy change for mixing, I use reversible isothermal processes. Imagine converting the far left and far right walls into pistons, and expand each species to double its volume isothermally. The entropy change is $\Delta S^* = N_1 k \ln 2 + N_2 k \ln 2 = N k \ln 2$. Each species has the final volume $2V$, but the gases are not mixed. They can be mixed reversibly using the semipermeable membrane process in Fig. 4.13. This takes zero work and brings zero entropy change, so

$$\Delta S_{\text{dist}} = N k \ln 2 \text{ for all } N_1, N_2, \text{ with } N_1 + N_2 = N. \qquad (4.36)$$

Unlike the indistinguishable species case, for distinguishable species, the entropy change is the same whether $N_1 = N_2$ or $N_1 \neq N_2$.

4.3.2 Mixing entropy function

Here I define a state function S_{mix} called *mixing entropy*, namely, that part of a mixture's total entropy that specifically reflects mixing, as opposed to being

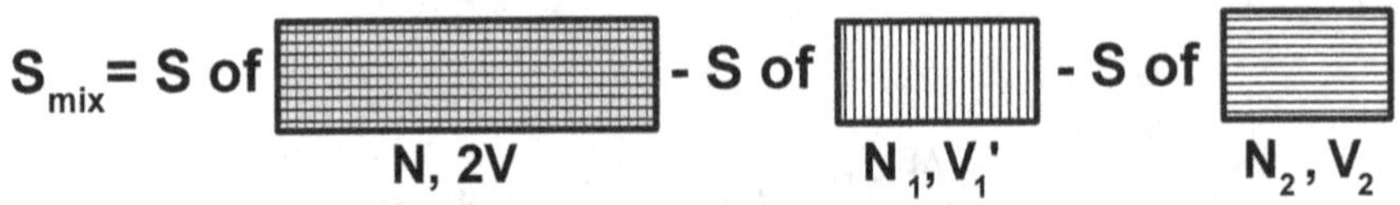

$$S_{mix} = S \text{ of } \boxed{} - S \text{ of } \boxed{} - S \text{ of } \boxed{}$$

$$N, 2V \qquad\qquad N_1, V_1' \qquad\qquad N_2, V_2$$

Figure 4.15: Graphic of mixing entropy function S_{mix} in Eq. (4.37). Volumes V_1' and V_2' assure equal densities: $N/2V = N_1/V_1' = N_2/V_2'$.

related to density-driven diffusion.[12] As before the total number of molecules is $N = N_1 + N_2$. The Gibbs theorem assures that the entropy of the mixture in volume $2V$ equals the sum of the separated species, each in volume V, i.e., has entropy $S_{A+B}(T, 2V, N) = S_A(T, V, N_1 + S_B(T, V, N_2)$.

Key Point 4.16 *The mixing entropy function is defined as the entropy, $S_{A+B}(T, 2V, 2N)$, of the mixture in volume $2V$ minus the entropy of N_1 type A molecules in volume V_1', chosen so the gas has the same density as the mixed system $(N_1/V_1' = N/2V)$ minus a similar entropy for the type B molecules in volume V_2', such that its density equals that of the mixture. The mixture and single species are shown in Fig. 4.15, and S_{mix} can be written*

$$S_{mix} \equiv S_{A+B}(T, 2V, N_1, N_2) - [S_A(T, V_1', N_1) + S_B(T, V_2', N_2)]. \qquad (4.37)$$

If we substitute the Sackur-Tetrode entropy, Eq. (4.4), into each term of Eq. (4.37), the resulting equation is

$$S_{mix} = -Nk[f_1 \ln f_1 + f_2 \ln f_2], \text{ with } f_i \equiv N_i/N. \qquad (4.38)$$

Key Point 4.17 *Equation (4.38) shows that the mixing entropy function has the form of Shannon's missing information function (in Key Point 2.28). The mixing entropy function S_{mix} is largest for $N_1 = N_2 = N$ and $f_1 = f_2 = 1/2$, when it's equally likely that a randomly picked molecule will be from either species. This corresponds to maximal uncertainty. In contrast, if species A had relatively few molecules and species B had relatively many, S_{mix} would be relatively low, because it is very likely that a randomly picked molecule is from species B. In this case, there is less uncertainty and smaller mixing entropy function.*

During mixing, the distinguishable species has the entropy increase $\Delta S_{dist} = Nk \ln 2$. How does this compare with S_{mix}, i.e., can we parse how much of the change comes from mixing and how much comes from diffusion associated with density differences? I'll define the diffusion entropy as the function that when added to S_{mix} gives ΔS_{dist}, i.e., $S_{mix} + S_{diffusion} = \Delta S_{diffusion} = Nkln2$. Then

$$S_{diffusion} = Nk \ln 2 - S_{mix} = Nk[f_1 \ln f_1 + f_2 \ln f_2 + \ln 2]. \qquad (4.39)$$

---

[12]At a specified temperature T, equal density $(= N/V)$ implies equal pressure $P = (N/V)kT$ for two classical ideal gases.

Key Point 4.18 *When two distinguishable species are mixed, there are two entropy contributions: (1) a mixing entropy function, which exists whether the two species have equal or unequal densities, and (2) a diffusion entropy, which is nonzero only if $N_1 \neq N_2$. It is satisfying that the diffusion entropy is identical to ΔS_{ident} for identical species (see Eq. (4.35)), which only have a nonzero entropy change if they have unequal densities.*

4.3.3   Gibbs paradox & information

The Paradox. Josiah Willard Gibbs, one of the key developers of thermodynamics and statistical mechanics, first pointed out the following paradox, which bears his name: *If species A and B have equal initial densities and pressures, two cases arise: (i) the gases are of identical type, e.g., both argon; or (ii) the gases are of different types, say, one is argon and the other neon. For (i), there is no diffusion and thus no entropy change when the partition is removed. The paradox that arises is that if gases A and B differ at all, then $\Delta S = 2Nk \ln 2$, but when they are identical, ΔS vanishes.*

One might think that as two gases become more alike the entropy gain during mixing would approach zero continuously; thus, the paradox. Some thought shows that distinct gases are different in discontinuous ways. For example if A is the isotope helium-3 and B is the isotope helium-4, the atomic weights of the species differ by a finite amount and cannot approach one another. The same is true for the model of a quantum ideal gas, where the particles have either integral or half-integral spin, i.e., they are either bosons or fermions. These species differ in a non-continuous way.

Physicist and historian of physics, Martin Klein examined an interesting problem that illustrates a novel aspect of the Gibbs paradox. Suppose species A and B are identical monatomic ideal gases, except the nuclei of B molecules are in an excited metastable state called an *isomeric* state with energy E_{exc}. Assume that the latter energy is large relative to kT, so $E_{\text{exc}}/kT >> 1$, and the lifetime of this isomer is long compared to the time it takes for diffusion to occur when the partition separating A and B is removed.

Immediately after mixing, the entropy of the mixture of distinguishable gases has increased by an amount $\Delta S_{A,B} = 2Nk \ln 2$. After a long time relative to the isomer's lifetime, each nucleus of species B will have decayed to its ground state. All of the gas atoms will have become type A atoms, and there is no longer a mixture of distinguishable gases, so the mixing entropy function $S_{\text{mix}} = 0$. Most notably, the entropy increase from the initial partition removal and mixing will have vanished, and therefore the entropy change of the gas is $-\Delta S'_{A,B} = -2Nk \ln 2$. If there were no other effect, this would violate the second law of thermodynamics, but the story goes on.

As each excited nucleus decays, it releases an energy E_{exc}, the excitation energy of the metastable state. This energy is transmitted to the constant-temperature reservoir at temperature T. After all the species B atoms have decayed, the reservoir gains entropy $\Delta S_{\text{res}} = N_B E_{\text{exc}}/T$. For this irreversible process, the second law of thermodynamics demands that the entropy decrease from the decay of species B to A atoms is such that,

$$\Delta S_{\text{tot}} = \Delta S'_{A,B} + \Delta S_{\text{res}} = -2Nk\ln 2 + N_B E_{\text{exc}}/T > 0. \qquad (4.40)$$

The assumed condition $E_{\text{exc}} \gg kT$ assures the inequality in Eq. (4.40).[13]

Key Point 4.19 *The Klein example of mixing two gases, one of which decays into the other, illustrates a linkage between energy and entropy. When the partition is removed, each species spreads its energy to a larger volume, with the resulting entropy increase $2Nk\ln 2$. As the N_B species B nuclei decay, the energy added to the gas increases the temperature. Thermal contact with the reservoir keeps the gas temperature constant by transferring energy to the reservoir, increasing its entropy enough that $\Delta S_{total} > 0$.*

4.3.4 The role of information

The example of mixing different isotopes A and B of the same element, e.g., uranium-235 and uranium-238 is amenable to an information-based assessment. The latter two uranium isotopes are distinguishable gases whose atomic masses differ from one another by $\sim 1.3\%$. An experimenter might think the two gases are identical without further information. Only with knowledge, i.e., information, that there were actually two different isotopes present would the nonzero entropy increase upon mixing be known.

To have such knowledge, there would need to be a way to distinguish the isotopes. The uranium isotope example is interesting historically, because uranium-235 was needed to facilitate self-sustaining nuclear fission that made nuclear reactors possible. Nearly 99.3% of naturally-occurring uranium is ^{238}U, and the remainder is mainly ^{235}U, with a negligibly (for our purposes) tiny amount of the third isotope, ^{234}U. The development of nuclear reactors required *enriched* uranium that was $\sim 3\%$ ^{235}U, and nuclear weapons use *highly enriched* uranium at levels of $\sim 80\%$ ^{235}U.

Part of the enrichment was done in a gaseous diffusion plant, which forces gaseous uranium hexafluoride through a semipermeable membrane with small pores, through which molecules with ^{235}U pass slightly faster than those with ^{238}U. This effect is consistent with Graham's law, which states that the rate of effusion through a small opening is inversely proportional to the molecular mass. The semipermeable membrane has many such small pores. Each pass-through brings some isotopic separation, and many pass-throughs can lead to the desired level of ^{235}U enrichment.

---

[13]Klein showed that $\Delta S_{\text{tot}} \geq 0$ holds for *any* value $E_{\text{exc}} > 0$. See M. J. Klein, "Note on a problem concerning the Gibbs paradox," Am. J. Phys. **26**, 80–81 (1958).

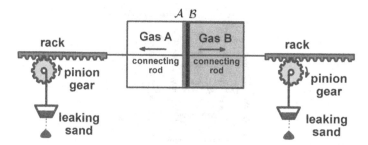

Figure 4.16: A way to effect reversible expansion that accompanies mixing. Membranes $\mathcal{A}$ and $\mathcal{B}$ move frictionlessly leftward and rightward, being pushed by gas species B and A respectively. Sand leaks slowly from the pans attached to each fixed pinion gear and pulley, keeping each rod's force on its connected rack slightly less than the gas force on its semipermeable membrane piston.

In thermodynamics, we use thought experiments (often called Gedanken experiments) with ideal semipermeable membranes (recall Fig. 4.13) that flawlessly separate two species of gas labeled A and B. Suppose two experimenters independently mix species A and B. One does *not* know there are two species present; the other does have that knowledge. The knowledgeable experimenter has access to membranes $\mathcal{A}$ and $\mathcal{B}$, which pass only species A and B, respectively. Both experimenters begin with A and B in the left and right chambers, separated by a fixed partition. The knowledgeable experimenter places membranes $\mathcal{A}$ and $\mathcal{B}$ touching the left and right sides of the central partition.

Both experimenters remove their partitions. For the no-knowledge person, the two gases mix as in Fig. 4.12, with the resulting entropy change of the universe, $\Delta S_{\text{no-knowledge}} = 2Nk \ln 2$. For the knowledgeable one, species B passes through membrane $\mathcal{B}$ and exerts pressure on membrane $\mathcal{A}$. Similarly, membrane $\mathcal{A}$ is *invisible* to species A molecules, which exerts pressure on membrane $\mathcal{B}$. If the membranes are frictionless pistons, each species does positive work on one of the membranes: species A does work on membrane $\mathcal{B}$ and species B does work on membrane $\mathcal{A}$. To assure that this work is done reversibly, the membrane pistons must be linked to work sources that maintain forces on them that are infinitesimally smaller than the gas force, as illustrated in Fig. 4.16.

For the knowledgeable experimenter, the net work lifting the sand-containing pans is $W = 2NkT \ln 2$. The process is reversible, and as the gases experience a net entropy gain of $\Delta S_{\text{dist}} = 2Nk \ln 2$, the constant-temperature reservoir maintaining the system temperature T loses an identical entropy; i.e., $\Delta E = Q - W = 0$, and $\Delta S_{\text{res}} = Q_{\text{res}}/T = -Q/T = -2Nk \ln 2$.

Key Point 4.20 *An experimenter who has knowledge that two gases are distinguishable and separable using semipermeable membranes can use the*

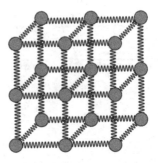

Figure 4.17: Model of a crystalline solid as atoms connected by springs. Each atom stays near its fixed lattice site, in this case, a square lattice.

expansion of each gas as the two species mix to do reversible work, with zero entropy gain of the universe. This is a good example of available energy being converted to work, in accord with Eq. (3.41). Here, $\Delta A = \Delta E - T_a \Delta S$, with $\Delta E = 0$ for an ideal gas at constant temperature, so $\Delta A = -T_a \Delta S = -NkT_a \ln 2$. In contrast, when the no-knowledge experimenter mixes the same two gases, the entropy of the universe increases, and zero work is done. Without information, the available energy is lost.

4.4 MODELS OF SOLIDS

4.4.1 Einstein model

At the very lowest possible temperatures, nearly all materials are solids. It is helpful to begin considering a solid at ultra-low temperature, and then to ponder what happens to it as it is heated. Our mental picture of a solid is a collection of atoms coupled to one another such that each atom stays close to a fixed position, as illustrated in Fig. 4.17. Albert Einstein simplified this picture, introducing a model of a solid that bears his name. Each atom is linked to a fixed lattice site and oscillates with frequency ν. Because the atoms are not coupled and oscillate with a single frequency, the Einstein model gives incorrect low temperature behaviour. However, its simplicity and tractability make it a good starting point.

Consider a simple one-dimensional oscillating atom of mass m with potential energy $\frac{1}{2} k_s x^2$, where k_s is the spring constant and x the displacement from the fixed site. Quantum theory predicts that this oscillator will have the discrete energies, $e_1 = h\nu(n + \frac{1}{2})$. The oscillator frequency $\nu = (2\pi)^{-1}\sqrt{k_s/m}$, and n is a quantum number with possible values $n = 0, 1, 2 \ldots$ An *isotropic* 3-dimensional oscillator, i.e., with the same spring constant in the x-, y-, and z-directions ($k_s = k_{sx} = k_{sy} = k_{sz}$) has energies $e_3 = h\nu(n_x + n_y + n_z + \frac{3}{2})$. A collection of N such oscillators will have allowed energies,

$$E(\{n_{x,j}, n_{y,j}, n_{z,j}\}) = \frac{3}{2}Nh\nu + h\nu \sum_{j}^{N} (n_{x,j} + n_{y,j} + n_{z,j}). \qquad (4.41)$$

This energy is a function of the $3N$ quantum numbers, namely, three for each oscillator. The lowest energy state, the ground state, has energy $3/2\,Nh\nu$.

Key Point 4.21 *The nonzero ground state energy (zero-point energy) is attributed to the quantum uncertainty of each atom's position and momentum. This quantum-mechanical energy is independent of temperature.*

Statistical mechanics enables a straightforward evaluation of the Einstein solid's partition function, which enables evaluation of the internal energy and entropy using Eqs. (3.20) and (3.21):

$$E_E(\tau, N) \equiv \langle E \rangle = 3Nh\nu \left[\frac{1}{2} + \frac{1}{e^{h\nu/\tau} - 1} \right], \text{ with } \tau = kT. \qquad (4.42)$$

In the high temperature (classical) limit, $e^{h\nu/\tau} - 1 \approx (1 + h\nu/\tau) - 1 = h\nu/\tau$, and Eq. (4.42) becomes

$$E_E(\tau, N) \to 3/2\,Nh\nu + 3N\tau = 3/2\,Nh\nu + 3NkT \to 3NkT. \qquad (4.43)$$

In the low temperature limit,

$$E(\tau, N) \to 3/2\,Nh\nu + 3Nh\nu e^{-h\nu/\tau} \to 3/2\,Nh\nu \text{ (zero-point energy)}. \qquad (4.44)$$

It is helpful to define the Einstein temperature,

$$T_E \equiv h\nu/k \qquad \text{Einstein temperature,} \qquad (4.45)$$

and a plot of E vs. (T/T_E) is shown in Fig. 4.18(a). The heat capacity, C_E of the Einstein model is of interest because heat capacity is typically measured experimentally. A plot is in Fig. 4.18(b), and the mathematical formula is

$$C_E = (T_E/T)^2 \frac{e^{(T_E/T)}}{(e^{(T_E/T)} - 1)^2}. \qquad (4.46)$$

Key Point 4.22 *It is clear from Fig. 4.18 that,*

1. *The Einstein temperature T_E (where $T/T_E = 1$) in Fig. 4.18 is a rough indicator of the transition from quantum to classical behaviour as T increases from zero. For $T < T_E$, both the energy E_E and heat capacity C_E approach zero exponentially as $T \to 0$. For $T > T_E$, the Einstein energy E_E becomes linear in T, which corresponds to the heat capacity C_E approaching $3Nk$, the classical value. For example, when $T/T_E = 2$, $C_E = .9794 \times 3Nk$, and when $T/T_E = 10$, $C_E = 0.9992 \times 3Nk$.*

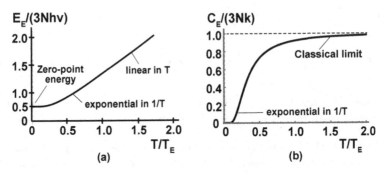

Figure 4.18: (a) Einstein solid's energy $E_E/(3Nh\nu)$ vs. T/T_E. (b) Heat capacity $C_E/(3Nk)$ vs. T/T_E.

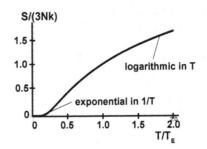

Figure 4.19: Entropy $S_E(T/T_E)$ vs. T/T_E.

2. *The approach to zero of C_E for $T \to 0$ shows that the Einstein model satisfies the third law of thermodynamics.*

The Einstein model's entropy, plotted in Fig. 4.19 is,

$$S_E = 3Nk\left[-\ln(1 - e^{-h\nu/\tau}) + (h\nu/\tau)(e^{h\nu/\tau} - 1)^{-1}\right]. \qquad (4.47)$$

In the low temperature limit, the entropy becomes,

$$S_E \to 3Nke^{-h\nu/\tau}, \text{ which} \to 0 \text{ exponentially as } \tau \text{ and } T \to 0. \qquad (4.48)$$

For high temperatures, $e^{\pm h\nu/\tau} - 1 \approx \pm h\nu/\tau$, so the second term in Eq. (4.47) approaches one, and

$$S_E \to 3Nk\left[\ln(T/T_E) + 1\right] \to 3Nk\ln T. \qquad (4.49)$$

Key Point 4.23 *Some notable points regarding the Einstein solid are:*

1. *In the high temperature limit, $E(\tau, N)$ is the zero-point energy plus the linear term $3NkT$.*

2. In the high temperature limit, the heat capacity (thermal inertia) $C_E = (\partial E_E / \partial T)$ approaches the constant $3Nk$.[14]

3. The latter result is consistent with the classical equipartition theorem: A quadratic energy term in either the position or momentum contributes $1/2\,kT$ to the internal energy. Each oscillator has a kinetic plus potential energy $(p_x^2 + p_y^2 + p_z^2)/2m + \frac{1}{2}k(x^2 + y^2 + z^2)$. The 6 position and momentum variables per particle is commonly referred to as 6 "degrees of freedom." The equipartition theorem tells us that for the N particles, there are $6N$ degrees of freedom, and $E \to 6N \times \frac{1}{2}kT = 3NkT$.

4. The result $C_E \to 3Nk$ for $T \to \infty$ agrees with the law of Dulong and Petit, who discovered in 1819 that the heat capacity of one mole of many solid elements is about $3R \approx 25\,\mathrm{JK}^{-1}\mathrm{mol}^{-1}$, where $R = N_A k$ is the universal gas constant (see Eq. (6.13)) and N_A is the Avogadro number. It follows that $3R$ per mole is equivalent to $3k$ per atom.[15]

5. The entropy behaviour for high temperatures is logarithmic in the temperature, similar to that in the Sackur-Tetrode entropy expression, Eq. (4.4), for a classical ideal gas.

Key Point 4.24 The Einstein solid's energy and entropy display interesting non-classical behaviour at low temperatures.

1. The energy E_E approaches the zero-point energy, in the zero-temperature limit. That approach follows the expression, $+3Nh\nu e^{-h\nu/\tau}$, and as τ and $T \to 0$, $h\nu/\tau \to \infty$, and the exponential is driven to zero. This exponential approach differs from the behaviours of actual solids, as discussed below, which are better captured by the Debye solid of a solid explained in the next section.

2. At lower temperatures, there is not equipartition of energy generally because of the discrete quantum-mechanical energy spectrum. For example, the Einstein solid's internal energy at low temperatures per degree of freedom is $E/6N \approx 1/2\,(h\nu)e^{-h\nu/kT} \neq 1/2\,kT$.

3. For $T \to 0$, the entropy $S_E \to 0$ exponentially and $S \to 3Nk(T_E/T)^2 e^{-(T_E/T)}$, in accord with the third law of thermodynamics.[16]

4. Real-world solids melt and also sublimate under certain conditions. The Einstein model is far too primitive to show such phenomena.

---

[14]The Einstein model does not contain V or P. We can imagine temperature variations to occur at constant pressure, but the model is pressure independent.

[15]Dulong and Petit's work was done prior to the definition of R within the context of kinetic theory of gases.

[16]A common statement of the third law (see Sec. 6.3) is: For $T \to 0$, the entropy of pure crystalline substances approaches a constant (sometimes taken as zero)

4.4.2 Debye solid

In 1912, Peter Debye generalised Einstein's model of a solid. The Debye solid has improved low temperature behaviour, namely, a heat capacity $\propto T^3$. Debye related the oscillations of atoms to sound waves propagating through the solid. Such oscillations, though complex, can be represented as a superposition of the system's *normal modes* of vibration, with each normal mode having a well-defined frequency at which all the atoms oscillate. An example is a vibrating viola string, whose sound is determined by a mixture of its normal modes, which are better known as harmonics or overtones. It is the particular mix of harmonics that make a vibrating viola string sound like a viola. A trumpet playing the same note has a very different mix of the same harmonics.

For the solid, the Debye solid is mathematically equivalent to a mixture of Einstein oscillators, with the ith atom having frequency ν_i, contributing energy $E_i(\{n_{x,j}, n_{y,j}, n_{z,j}\})$ to the internal energy:

$$E_i(\{n_{x,i}, n_{y,i}, n_{z,i}\}) = \tfrac{1}{2}Nh\nu_i + h\nu_i(n_{x,i} + n_{y,i} + n_{z,i}), \tag{4.50}$$

which results in the internal energy,

$$E_D(\tau, N, \{\nu_i\}) \equiv \sum_{i=1}^{N} \left[\tfrac{3}{2}h\nu_i + \frac{h\nu_i}{e^{h\nu_i/\tau} - 1} \right], \text{ with } \tau = kT. \tag{4.51}$$

We do not know the set of frequencies $\{\nu_i\}$, and getting over this hurdle required Debye's genius. He did two things that led to success:

1. Debye related the normal mode frequencies of sound in the solid with the discrete frequencies $\{\nu_i\}$ in Eq. (4.51). The number of normal modes is $3 \times$ the number of atoms, typically of order 10^{20} or greater. Debye treated the oscillation frequency as a continuous variable.

2. He converted the sum in Eq. (4.51) to an integral, using the density of states $g(\nu)$ for sound waves, known to be quadratic, $g(\nu) = B\nu^2$, where B is a constant. The quantity $g(\nu)d\nu$ is the number of normal mode frequencies in the interval $(\nu, \nu + d\nu)$, and a required condition is $\int_0^{\nu_D} g(\nu)d\nu = 3N$, the total number of vibrational modes. Debye assumed a maximum frequency, ν_D, the Debye frequency. Imposing the latter condition determines the constant B, with the results,

$$g(\nu) = (9N/\nu_D^3)\nu^2, \text{ and} \tag{4.52}$$

$$E_D(\tau, N) = \frac{9N}{\nu_D^3} \left[\int_0^{\nu_D} \tfrac{1}{2}(h\nu)\nu^2 d\nu + \int_0^{\nu_D} \frac{(h\nu)\nu^2}{(e^{h\nu/\tau} - 1)} d\nu \right]. \tag{4.53}$$

In analogy with the Einstein temperature T_E, in Eq. (4.45), it is helpful to define the Debye temperature,

$$T_D \equiv \frac{h\nu_D}{k}. \tag{4.54}$$

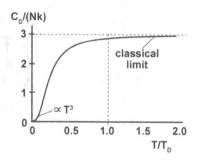

Figure 4.20: Heat capacity $C_D(T)$ for the Debye solid.

The heat capacity is

$$
\begin{aligned}
C_D(T) = (\partial E_D/\partial T) &= \frac{9N}{\nu_D^3}\frac{h^2}{kT^2}\int_0^{\nu_D}\frac{\nu^4 e^{h\nu/\tau}}{(e^{h\nu/\tau}-1)^2}d\nu \\
&= 9Nk\left(\frac{T}{T_D}\right)^3\int_0^{T_D/T}\frac{y^4 e^y}{(e^y-1)^2}dy.
\end{aligned}
\tag{4.55}
$$

In the second line of Eq. (4.55), I made the change of variables $y = h\nu/\tau = h\nu/kT$. This makes the integrand independent of temperature and puts the temperature dependence in the integral's upper limit. The low and high temperature limits evident in Fig. 4.20, are of interest. (1) For $T \to 0$, the upper limit of the integral in (4.55) becomes infinitely large and the integral approaches a constant. Thus $C_D(T) \approx$ constant $\times (T/T_D)^3$. (2) In the high temperature limit, the integral's upper limit becomes arbitrarily small, and for $y \ll 1$, the integrand is well approximated by y^2, and the integral approaches $1/3(T_D/T)^3$. Therefore $C_D \to 3Nk$, the value found experimentally by Dulong and Petit.

Key Point 4.25 *The Debye solid is a more realistic model than the simplistic Einstein model, which has only a single vibration frequency for all atoms.*

1. *The Debye heat capacity resembles the Einstein heat capacity in Fig. 4.18(b), except at the smallest temperatures, where $C_D \propto T^3$, in agreement with the behaviour of metals.*

2. *The Debye temperature T_D serves as a rough indicator of the transition from quantum to classical behaviour as temperature is increased.*

3. *The Debye solid satisfies the third law of thermodynamics.*

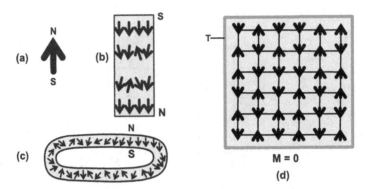

Figure 4.21: (a) Symbolic depiction of a magnetic moment. (b) A ferromagnet ("*permanent*" magnet) that retains its magnetism over time. (c) A paramagnet that becomes magnetised near a ferromagnet or an electromagnet, indicated with N and S for north and south. (d) Two-dimensional lattice with 36 noninteracting spins that can point only up or down, in zero external magnetic field with randomised spins. This is an Ising model, which is explored in Sec. 4.5.3.

4.5 PARAMAGNETS & FERROMAGNETS

Various types of magnetism are known, including paramagnetism, ferromagnetism, and diamagnetism. Here I will examine aspects of paramagnetism and ferromagnetism. Paramagnetism is familiar to kids who take unmagnetised metal paper clips and pick them up with a magnet. When the paper clips are pulled off the magnet and some time has elapsed, they again exhibit unmagnetised behaviour.

Key Point 4.26 *A paramagnet becomes magnetic only in the presence of an external magnetic field. A rough depiction of this is in Fig. 4.21(a)–(c). Part (a) represents a magnetic moment, namely, a microscopic dipole with north (N) and south (S) magnetic poles. Part (b) depicts a permanent magnet, i.e., a ferromagnet.[17] Part (c) shows a paramagnet near a ferromagnet. The magnetic moments closest to the ferromagnet become aligned, magnetising the paramagnet. Paramagnetic materials are common, with examples being aluminium, iron oxide, oxygen, and titanium.*

4.5.1 Ideal paramagnet

A simple, yet elegant model of a paramagnet is a set of N noninteracting spins on fixed sites on a regular lattice. Assume the spins and an associated magnetic

---

[17]The term *permanent* is a misnomer. A ferromagnet can be demagnetised in multiple ways, the simplest being to heat it to a high temperature in a magnetic field-free region.

moment can orient either up or down. In the absence of an external magnetic field, the spin directions are assumed to be randomised by a thermal radiation field produced by container walls at temperature T (radiation is discussed in Ch. 5). This two-dimensional system is shown in Fig. 4.21(b). Although the spins are assumed to be randomised thermally, the model does not account for any kinetic energy; e.g., in the absence of an external field, $E_{\text{ideal par}} = 0$.

Suppose a uniform upward-pointing external magnetic field H is turned on. Then the system of spins experiences a *competition* between randomisation by the thermal radiation field and alignment with the magnetic field. A reasonable expectation is that randomisation is stronger at very high temperatures, and alignment with the field dominates at very low temperatures. This is precisely what happens. The spin-external field interaction energy of the ith spin with magnetic moment μ is

$$e_i = -\mu\sigma_i H, \text{ where } \sigma = \pm 1 \text{ for (up, down) spin}$$
$$= \begin{cases} -\mu H \text{ for spin up } (\sigma_i = 1) \\ +\mu H \text{ for spin down } (\sigma_i = -1) \end{cases} . \qquad (4.56)$$

Alignment of a spin with the H field gives the lowest interaction energy. For the entire N-spin lattice, the energy for a specific configuration $\{\sigma_i\}$ is,

$$E(\{\sigma_i\}) = \sum_{i=1}^{N} -\mu\sigma_i H. \qquad (4.57)$$

The units involved here can be intimidating. Here are some clarifying points.

1. Students are usually introduced to the magnetic induction field B first, through the force exerted on a moving charge q with speed v, moving perpendicular to the direction of $\vec{v}$. The force magnitude $F = qvB$ shows the units of B to be $[\text{NsC}^{-1}\text{m}^{-1}]$, where C is the unit Coulomb. The unit Gauss (G) is defined as $1\,\text{G} \equiv 1\,[\text{Ns C}^{-1}\text{m}^{-1}]$.

2. The H-field is introduced by Ampere's circuital law, where a current I generates a magnetic field. Taking an integral around a loop enclosing the current, $\oint \vec{H} \cdot d\vec{\ell} = I$. This implies that the units of H are $[\text{Am}^{-1}]$.

3. The B and H fields are related to each other in free space (not in a magnetic medium) by $B = \mu_0 H$, where $\mu_0 = 4\pi \times 10^{-7}$ with units $[\text{V s A}^{-1}\,\text{m}^{-1}]$, and $[\text{V}]$ is volts.

4. The magnetic energy of interaction of a magnetic moment μ_p of a particle with an external B-field is $e = -\mu_p B = \mu_0 \mu_p H$. The units for the latter are $[(V\,s\,A^{-1}\,m^{-1})(A\,m^2)(A\,m^{-1}) = V\,A\,s = W\,s = J]$. I have used the fact that current × voltage = power in watts. In Eqs. (4.56) and (4.57) and in what follows, I use the symbol $\mu \equiv \mu_0\mu_p$.

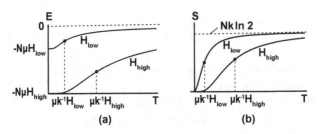

Figure 4.22: (a) Energy and (b) entropy *vs.* temperature for the ideal paramagnet. External magnetic field intensities H_{low} and $H_{high} > H_{low}$ are shown. Circles denote points with magnetic energy per spin $\mu H = kT$.

In a constant-temperature environment, the probability of finding a specific configuration $\{\sigma_i\}$ is the Boltzmann probability, Eq. (3.16),

$$P(\{\sigma_i\}) = Z^{-1} e^{-E(\{\sigma_i\})/\tau}. \tag{4.58}$$

The partition function Z is,

$$
\begin{aligned}
Z &= \left[e^{\mu H/\tau} + e^{-\mu H/\tau}\right]^N \\
&= \left[2\cosh(\mu H/\tau)\right]^N,
\end{aligned} \tag{4.59}
$$

where $\cosh(x) \equiv 1/2(e^x + e^{-x})$, called the hyperbolic cosine. From this, we can find the average energy, E,

$$E = \sum P(\{\sigma_i\}) E_s(\{\sigma_i\} = -N\mu H \tanh(\mu H/\tau), \tag{4.60}$$

where $\tanh(x) \equiv (e^x - e^{-x})/(e^x + e^{-x})$, the hyperbolic tangent. We can also write the total *magnetisation*, namely,

$$M = \sum_i P(\{\sigma_i\}) \mu \sigma_i = -E/H = N\mu \tanh(\mu H/\tau). \tag{4.61}$$

The entropy can be obtained using Eq. (3.21), together with (4.59) and (4.60),

$$
\begin{aligned}
S(T, V, N) &= k \ln Z + \frac{E}{T} \\
&= Nk \left[\ln(2\cosh(\mu H/\tau)) - (\mu H/\tau) \tanh(\mu H/\tau)\right].
\end{aligned} \tag{4.62}
$$

Energy E and entropy S are plotted as functions of temperature in Fig. 4.22(a) and (b), which is rich in interpretive value.

Key Point 4.27 *The energy E and entropy S show a transition from low temperatures, where energy dominates, to high temperatures, where entropy has more influence. Curves for two different values of the magnetic field intensity lead to telling behaviour patterns of energy and entropy and shows an interesting competition between them, as described in the following.*

1. The average energy E is negative for all finite values of temperature T, and approaches zero for $T \to \infty$. This is consistent with Eq. (4.56) when H is positive, namely, $E = \sigma\mu H < 0$ for spins pointing upward ($\sigma = +1$), in alignment with the field.

2. Minimum energy occurs in the zero-temperature limit when all spins align with the field, i.e., $E \to -N\mu H$.

3. The energy curve for the high field value lies below the low field curve because in a stronger magnetic field, more spins align with the upward-pointing external field. This is an energy effect.

4. As temperature is increased from zero, the thermal agitation causes more spins to point randomly up or down. This is an entropy effect, which opposes the energy effect of the field tending to align spins.

5. In Figs. (4.22)(a) and (b), temperatures $\mu k^{-1} H_{\text{low}}$ and $\mu k^{-1} H_{\text{high}}$ indicate transitions between low and high temperature regions where energy or entropy dominates.

6. The entropy begins at zero value for $T \to 0$, and rises to a maximum value of $Nk \ln 2$, which is the value at which half the spins are pointing upward and half downward.

7. The zero-temperature limit of entropy, regardless of the value of the magnetic field is consistent with the third law of thermodynamics.

8. Entropy is lower for higher values of the magnetic field intensity, where a larger fraction of spin alignment exists on average.

9. This model of paramagnetism is one for which the interpretation of entropy in terms of disorder makes sense. For higher temperatures, the spins are more disorganised, pointing more randomly either up or down. For low temperatures, energy and entropy approach their respective minimum values, as spins align and have lowest entropy and least disorder.

A plot of magnetisation M *vs.* T is shown in Fig. 4.23 for high (H_{high}) and low (H_{low}) values of the magnetic field.

4.5.2 Negative temperature

One of the interesting features of the ideal paramagnet is that it reveals the possibility of a negative absolute temperature, albeit only for short time intervals in temporary, slowly varying non-equilibrium states that relax slowly to equilibrium states. To see this, we can express entropy in terms of the system energy and examine the temperature, defined by $(\partial S/\partial E) = 1/T$.

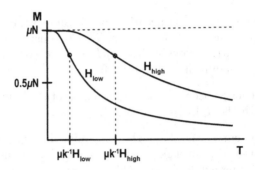

Figure 4.23: Magnetisation M vs. temperature T. Circles show points with $\mu H = kT$.

Consider an ideal paramagnet in a magnetic field H as before, with n_+ spins pointing upward, each with energy $e = -\mu H$. The other $N - n_+ \equiv n_-$ spins point downward, with higher energy $e_- = \mu H$. Therefore, the total energy for the envisaged spin assignment is,

$$E = (n_- - n_+)\mu H. \tag{4.63}$$

Also, n_- and n_+ satisfy,

$$n_+ + n_- = N, \tag{4.64}$$

and combining (4.63) and (4.64) we obtain,

$$n_- = \frac{N + E/\mu H}{2} \quad \text{and} \quad n_+ = \frac{N - E/\mu H}{2}. \tag{4.65}$$

It follows that $n_- = n_+ = N/2$ corresponds to $E = 0$, and $n_- = 0, n_+ = N$ implies $E = -N\mu H$.

The number of distinct spin configurations, i.e., the number of microstates, with total energy E is,

$$\Omega(E) = \Omega(n_+) = \frac{N}{n_+! (N - n_+)}. \tag{4.66}$$

This is the number of ways one can pick n_+ objects from N identical objects. There are N ways to pick the first one, $(N - 1)$ ways to pick the second, $\ldots$, $(N - n_+ + 1)$ ways to pick the n_+^{th}. This gives $N(N - 1)\ldots(N - n_+ + 1) = N!/(N - n_+)!$ configurations, each of which is a microstate of the system. We have overcounted, because there are $(n_+)!$ permutations of the n_+ identical up spins. Dividing by $n_+!$ removes the overcounting and results in Eq. (4.66).

For a specified value of energy E, n_+ is determined by (4.65), and the number of states of the system can be written $\Omega(n_+)$. Recalling Eq. (3.3), the entropy is $S(E) = k \ln \Omega(E) = k \ln \Omega(n_+)$. To use Eq. (4.66) in the entropy expression, we can use the Stirling approximation for the logarithm of the

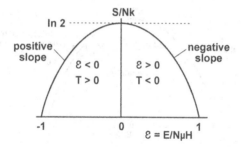

Figure 4.24: Entropy *vs.* $\mathcal{E} == E/N\mu H$ for an ideal paramagnet.

factorials: $\ln N! \approx N \ln N - N$. This approximation is good within 0.9% for $N = 100$ and 0.07% for $N = 1000$. Because we are working with systems that have at least 10^{20} particles, the Stirling approximation suffices, and implies

$$S = -n_+ k \ln(n_+/N) - n_- k \ln(n_-/N) \qquad (4.67)$$

It is convenient to define

$$\mathcal{E} \equiv \frac{E}{N\mu H}, \qquad (4.68)$$

whereupon Eqs. (4.66) and (4.67) combine to give,

$$\frac{S(\mathcal{E})}{Nk} = -\tfrac{1}{2}(1 + \mathcal{E})\ln((1 + \mathcal{E})) - \tfrac{1}{2}(1 - \mathcal{E})\ln(1 - \mathcal{E})) + \ln 2. \qquad (4.69)$$

It is easy to see that if $\mathcal{E} \to 0$, then $S \to Nk \ln 2$, and when $\mathcal{E} \to \pm 1$, then $S \to 0$. It is also clear that $S(\mathcal{E}) = S(-\mathcal{E})$; i.e., the entropy is a symmetric function of the energy. The plot of entropy *vs.* $\mathcal{E}$ in Fig. 4.24 illustrates these findings.

Key Point 4.28 *The ideal paramagnet is notable for several reasons.*

1. *It is a model that approximates reality qualitatively and enables straightforward interpretations of energy and entropy at very low and very high temperatures.*

2. *It has the entropy function shown in Fig. (4.24), which is unlike any we have seen. The entropy increases from zero to $Nk \ln 2$ over a negative energy region, and then decreases to zero over a positive energy region. This is attributable to the energy spectrum having a maximum value. For $\mathcal{E} < 0$, where normal equilibrium states exist, $(\partial S/\partial \mathcal{E}) > 0$, which is consistent with positive temperatures because $(\partial S/\partial \mathcal{E}) = Nk\mu H/\tau$. For $\mathcal{E} > 0$, entropy decreases with increasing energy, $(\partial S/\partial \mathcal{E}) < 0$, which implies a negative absolute temperature. The states for $\mathcal{E} > 0$ are unstable, transitory non-equilibrium states.*

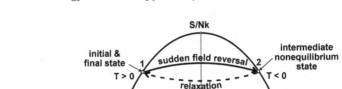

Figure 4.25: A paramagnet, initially in the positive-temperature state 1 is forced into a negative-temperature, nonequilibrium state 2 by a quick field reversal. Subsequently, it relaxes back to the initial state, i.e., $2 \to 1$.

3. *In 1951 Edwin Purcell and Robert Pound reported on experiments with initially magnetised LiF crystals that showed surprisingly long relaxation times when placed in a strong external magnetic field. The sudden reversal of the field produced a negative temperature for the spin system that persisted over the measured spin-lattice relaxation time of about 5 minutes. This negative temperature spin system loses internal energy as it gains entropy and, notably, was not in thermal equilibrium with the LiF lattice until the relaxation was complete.*

4. *When both negative and positive kelvin temperatures occur, an interesting measure of hotness is $1/T$. In Fig. 4.25, $1/T \to +\infty$ as the minimum energy (zero entropy) is approached; i.e., $\mathcal{E} \to -1$. As maximum energy (also zero entropy) is approached, $\mathcal{E} \to 1$, $1/T \to -\infty$. For $\mathcal{E} = 0$, S is maximum, and $1/T = 0$. Increasing the system energy from minimum to maximum, the inverse temperature changes continuously from positive infinity through zero to negative infinity. The spin temperature is coldest at minimum energy and hottest at maximum energy. Negative kelvin temperatures are hotter than positive ones!*

4.5.3 Ferromagnets

Just as an ideal gas does not exhibit a vapour-liquid phase transition, an ideal paramagnet does not exhibit permanent magnetism. These phenomena arise because of interactions between particles. Perhaps the most familiar ferromagnets are the elements iron and nickel and the rare earth compound of neodymium, boron and iron, called a neodymium magnet. What is meant by *permanent* magnet is the existence of a nonzero magnetisation when there is zero external field.

Key Point 4.29 *A ferromagnet is a substance that has nonzero "spontaneous" magnetisation M_0 when it is in zero external magnetic field at*

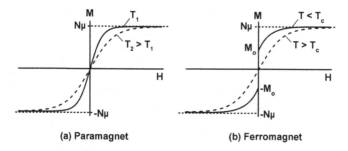

(a) Paramagnet   (b) Ferromagnet

Figure 4.26: Magnetisation *vs.* magnetic field H at two temperatures for (a) a paramagnet and (b) a ferromagnet.

temperature $T < T_c$, the Curie temperature. Above T_c, there is no nonzero spontaneous magnetisation in zero external field. The magnetisation curves at two temperatures shown in Fig. 4.26(b), indicate that a phase transition between the paramagnetic and ferromagnetic phase occur for $T < T_c$. The magnetisation $M_0 > 0$ is obtained when $H \to 0$ through positive values, and $M_0 < 0$ when $H \to 0$ through negative values. The absolute value $|M_0|$ depends on the temperature at which the field is reduced to zero. At temperature T_c, $|M_0| = 0$, and is largest ($|M_0| = N\mu$) for $T \to 0$.

In mathematical language, Fig. 4.26(b) shows that,

$$\lim_{H \to 0} M(H) = \begin{cases} 0 \leq M_0 \leq N\mu & \text{for } H \to 0+ \\ -N\mu \leq M_0 \leq 0 & \text{for } H \to 0- \end{cases} \quad \text{when } T \leq T_c. \quad (4.70)$$

The Curie temperature is the magnetic analog to the critical point for gases and liquids, and there is an analogy between the pressure-temperature phase diagram for fluids and the magnetic field-temperature phase diagram for ferromagnets. This is illustrated in Fig. 4.27, where in part (a), phase lines separate liquid and vapour (evaporation/condensation line), solid and liquid (melting/freezing line), and solid and vapour (sublimation/deposition line). In part (b) a phase line separates regions of positive and negative spontaneous magnetisation $\pm M_0$.

Unlike fluids, the phase boundary for ferromagnets is not a region where separated phases coexist. Rather as $T > T_c$ is lowered toward T_c with $H = 0$, i.e., along the temperature axis in Fig. 4.27, the tendency for alignment of microscopic magnetic moments leads to clusters of atoms with nonzero net magnetisations. This is an energy effect, minimising the energy of interaction between atomic/molecular magnetic moments. The non-macroscopic clusters interact weakly with one another, and their net magnetic moments point in random directions. The clusters are uncorrelated statistically, and the magnetisation for the entire system is zero.

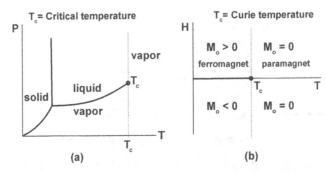

Figure 4.27: (a) P-T phase diagram for a fluid. (b) H-T phase diagram for a ferromagnet.

On the other hand, if a nonzero external magnetic field exists, the net magnetic moments of the clusters will align with the field, at least partially, leading to a nonzero magnetisation M. Suppose T is lowered below T_c and subsequently the field is reduced to zero. As H decreases (or increases, if the field is negative) toward zero, the total magnetisation magnitude is also reduced. When $H = 0$, the interparticle forces of alignment are sufficiently strong to "freeze" in a configuration with spontaneous magnetisation, $M_0 \neq 0$.

Key Point 4.30 *The ferromagnet is a particularly clear example of a competition between energy and entropy. Interacting magnetic moments tend to align parallel with one another. This alignment tendency is offset by the thermal motion of atoms, which tends to randomise spin orientations. This is an entropic effect that depends on how high the temperature is. The Curie temperature T_c is that temperature where these two phenomena have nearly equal effects. It is a demarcation point between the paramagnetic and ferromagnetic domains. This is illustrated by the vertical line going through the point $(T = T_c, H = 0)$ in Fig. 4.27(b).*

Ingenious theories have been developed to explain critical phenomena, based upon the idea that near a critical point, there are long-range correlations between particles, which are largely independent of the details of interparticle forces. These play an important role in determining critical behaviour. The long-range correlations can be understood, at least in part, by the following qualitative argument for ferromagnets. Above the Curie temperature, magnetised clusters of particles form, as described earlier. Within any cluster there are strong correlations among the particle spins in the sense that knowing the direction of one particle's spin makes it highly probable that another particle in the cluster has its spin pointing in the same direction. The probability is not unity because the spins in a cluster are not all perfectly aligned. In the absence of a magnetic field, i.e., when H = 0, the clusters are not strongly correlated with one another, and the net magnetisation is zero. Below the Curie temperature, there might or might not be a nonzero magnetisation, depending upon the material's history. The degree to which particles in the material are correlated with one another depends upon this history too.

However at the Curie temperature, the situation is special, with the entire material seeming to *know* this is a special transition point. Mathematically, using statistical mechanics, this *knowledge* can be linked to the fact that the range over which the particles are correlated becomes unbounded. The long-range correlations do not depend upon the details of the interparticle forces, and are present at both, the liquid-vapour critical point and the magnetic Curie point. These correlations give rise to universal thermodynamic behaviour in the vicinity of T_c.

The most successful theory of critical points, renormalisation group theory, was developed by Kenneth Wilson as an outgrowth of theories of Leo Kadanoff and Benjamin Widom that incorporated the scaling concept. Scaling means that near a critical point, because of long-range correlations, groups of particles interact with one another in a way that is similar (in a specific mathematical sense) to individual particle interactions. Looking at groups of particles rather than individual particles corresponds to a change of spatial scale. The details are beyond the scope here but the ideas are fascinating and lead to new understanding.[18]

Fluids and magnets display weird behaviour close to their critical points. An example is the constant-pressure specific heat as a function of temperature. Data for iron are plotted in Fig. 4.28(a), and the specific heat has a singularity at the Curie temperature $T_c = 1041.3\,\text{K}$. The derivative of C_p at $T = T_c$ depends on whether the approach is from below or above T_c.

Key Point 4.31 *As T_c is approached, a ferromagnet fluctuates between being paramagnetic and ferromagnetic, with neither entropy nor energy dominating. The heat capacity is singular at $T = T_c$, behaving differently for $T < T_c$ (where energy dominates) and $T > T_c$ (where entropy dominates).*

In 1924, Ernst Ising examined a one-dimensional model intended to capture the essence of ferromagnets. That model consists of N lattice sites, each with a magnetic moment that can point up or down. Only nearest neighbour magnetic moments, e.g., spins i and $(i+1)$, interact with one another via the potential energy $-J\sigma_i\sigma_{i+1}$, with $J > 0$. Note that

$$-J\sigma_i\sigma_{i+1} = \begin{cases} -J \text{ for parallel spins, both up or both down} \\ +J \text{ for antiparallel spins, one up and one down} \end{cases} \quad (4.71)$$

Thus, parallel spin configurations have lower energies than antiparallel ones and are energetically favoured. If an external field is present, this adds a term $\mu H\sigma_i$ for the ith spin.

---

[18]H. J. Maris & L. P. Kadanoff, "Teaching the renormalization group," Am. J. Phys. 46, 652–657 (1978).

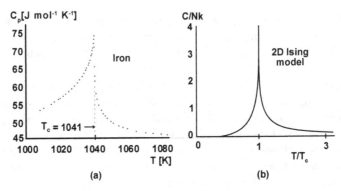

Figure 4.28: (a) Specific heat of iron near its Curie temperature. (Data plotted with with permission from F. L. Lederman, M. B. Salamon & L. W. Shacklette, "Experimental verification of scaling and test of the universality hypothesis from specific-heat data," Phys. Rev. B9, 2981–2988 (1974). Copyright (1974) by the American Physical Society. (b) Specific heat of 2D Ising model at the Curie temperature. Reprinted with permission from L. Onsager, "Crystal Statistics. I. A Two-Dimensional Model with an Order-Disorder Transition," Phys. Rev. **65**, 117–149 (1944). Copyright (1944) by the American Physical Society.)

Ising found that his model does not have a para-ferromagnetic phase transition for $T > 0$. He argued (incorrectly) that there was not a phase transition in higher dimensions either.[19] Other work using a "mean field" approximation ensued over the years. In 1941, Hendrik Kramers and Gregory Wannier used a symmetry argument to relate the 2D Ising model's high and low temperature behaviour, and proved that there was indeed a phase transition. Kramers and Wannier introduced the seminal idea of using a "transfer-matrix" method of analysis for treating statistical mechanics problems, a method used subsequently by Lars Onsager.

At a meeting of the New York academy of Sciences February 28, 1942, Wannier presented a paper, and at the end of his talk, Lars Onsager made a jaw-dropping announcement: He had obtained the exact solution for the 2-dimensional Ising model with a square lattice. His result, which was for zero external magnetic field, was not published for two years, leaving physicists puzzled about the details of Onsager's work. In two dimensions, the Ising model on a square lattice produces a transition from paramagnetic to ferromagnetic behaviour for temperatures below the Curie temperature

$$T_c = \frac{2J}{k \ln(1 + \sqrt{2})}. \tag{4.72}$$

---

[19] Frustrated with his inability to make progress with his model ferromagnet he gave up physics research. Ising, a Jew, was prosecuted in Germany and dismissed from his job. Only years later did he learn of the fame of his model, which had borne fruit through the work of others.

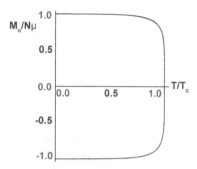

Figure 4.29: Spontaneous magnetisation, $M_0/N\mu$ vs. T/T_c for a 2D Ising model on a square lattice. (Reprinted with permission from C. N. Yang, "The Spontaneous magnetisation of a Two-Dimensional Ising Model," Phys. Rev. 85, 808–816 (1952). Copyright (1952) by the American Physical Society.)

Key Point 4.32 *The two-dimensional Ising model has a phase transition at the temperature in Eq. (4.72), which is a linearly increasing function of the coupling constant J. For larger J values, T_c is higher, i.e., stronger coupling between nearest-neighbour spins enables the transition to ferromagnetic behaviour at higher temperature, offsetting the stronger entropic tendency for spins to randomise. The specific heat for the 2D Ising model shows a logarithmic singularity at the Curie temperature, shown in Fig. 4.28(b). The similarity to the experimental result for iron is striking.*

On August 23, 1948 on a blackboard at Cornell University, which was hosting a conference on phase transitions, Lars Onsager once again unexpectedly announced a new result. He wrote on a chalkboard that the spontaneous magnetisation of the 2D Ising model has the form,

$$M_0 = N\mu\{1 - [\sinh(\tfrac{2J}{\tau})]^{-4}\}^{1/8} \text{ with } \tau = kT. \qquad (4.73)$$

Onsager gave no indication of how he derived this expression, and as it happened, he never published his result, which turned out to be entirely correct. Finally, in 1952, Chen-Ning Yang published a derivation of the Onsager spontaneous magnetisation result in what can only be called a *tour de force*. Figure 4.29 shows a plot of Eq. 4.73. Using Eq. (4.72), the quantity $2J/kT = \ln(1 + \sqrt{2})/(T/T_c)$, so when $T = T_c$, $M_0 = 0$ in Eq. (4.73) because $[\sinh(2J/\tau)]^{-4} = [\sinh(\ln(1 + \sqrt{2}))]^{-4} = 1$.

Key Point 4.33 *Figure 4.29 shows that the spontaneous magnetisation $M_0 \rightarrow \pm\mu N$ as $T \rightarrow 0$. As the temperature increases toward the Curie temperature, the entropic effect increases driving $|M_0|$ to zero at $T = T_c$. For $T < T_c$, energy tends to be minimised, with $|M_0| \neq 0$, and for $T > T_c$, entropy tends to be maximised, with $M_0 = 0$.*

As might be expected, clever people have discovered ways to use the Curie point for technological applications. One such application is to write to a small area on magnetic media when it is just above the Curie temperature, and to then immediately cool it, preserving the information in the ferromagnetic phase. Laser-heating is used to effect erasure in such magneto-optical systems.

4.6   RUBBER BANDS

If one heats a gas, it tends to expand. Molecules zoom around faster and push harder on their containers. This is consistent with the observation that when gases are heated, they tend to spread out. The same happens with liquids like water. When heated, a liquid's vapour pressure increases and some of the liquid molecules escape as gas molecules. Most solids behave similarly; they expand, pushing outward when heated.

Rubber is different, as illustrated in Fig. 4.30. When you heat a rubber band, it shortens, thereby lifting a weight F attached to its bottom. Lifting is by the rubber's entropic force, described in Sec. 3.5.4. The left pane shows a rubber band hanging from a support, with a weight F attached to its lower end. The rubber band temperature is T, and the length of the stretched rubber band is L. In the right pane, a heat lamp is pointed at the rubber band, raising its temperature to $T = T + \Delta T$. As it is heated, the rubber band contracts, lifting the weight.

4.6.1   Rubber band experiment

An activity that shows the properties of rubber, is to quickly stretch and contract a rubber band, sensing the temperature just after the stretch, and just after the rapid contraction.

• Hold the rubber band firmly at each end, keeping it taut, but not stretched.

• Touch it against your forehead, sensing its temperature.

• Remove it from your forehead and quickly stretch it to several times its relaxed length, and quickly touch it again against your forehead, sensing its temperature.

• Remove the rubber band from your forehead and quickly let it contract to its unstretched length, and again, quickly touch it to your forehead, sensing its temperature.

What is invariably found is that the expansion leads to a temperature increase for the rubber band. The forehead is quite sensitive and typically detects that the rubber band is warmer than initially. The relaxation of the rubber band is accompanied by a temperature decrease, again, easily detectable by the forehead. Some people do this experiment using one lip as the temperature sensor, but that is a bit messy and unsanitary, so I prefer using the forehead.

These properties are attributable to the fact that rubber is a polymer. It consists of long-chain molecules that can coil and uncoil with relative ease,

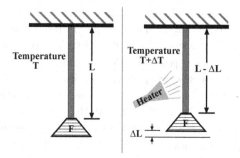

Figure 4.30: A rubber band, hanging from a support, with a weight F attached to it. The stretched length is L at temperature T. Heating the rubber band, contracts it, lifting the weight via an entropic force, as described in the text.

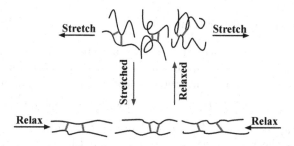

Figure 4.31: Two coiled layers of long-chain molecules. (a) Relaxed state. (b) Stretched state.

making rubber flexible and elastic. We may usefully envisage these long chains as weakly interacting, so that the internal energy depends only weakly on the rubber band length, L. Recall Fig. 3.16 and the discussion of entropic force in rubber.

Figure 4.31 depicts a small region of the rubber band's interior. In the top configuration, the rubber band is relaxed and the two shown strands (polymer chains) are linked by forces at a small number of regions. Because of this linkage, when the rubber band is stretched horizontally, as shown, the lower pane chains are set into vigorous vibration. This raises the kinetic energy of the chains, and is detectable by an increased temperature. Energetically, a quick stretch of the rubber band requires work by your muscles, and that work increases the intermolecular potential energy *and* temperature. The quickness of the stretch means that there is insufficient time for the temperature increase to correct itself by energy transfer to the rubber band's environment. This is an approximately an *adiabatic* process.

If the rubber band is held in the stretched configuration for ten seconds, and is allowed to relax, returning to its original, length, that length contraction is caused by the internal forces between chains. The potential energy decreases as the chains coil up again. As the chains return to being coiled, the forces between chains lessen, the internal vibrations diminish, and the temperature decreases. As for the expansion, the contraction is also *adiabatic*.

If the same steps were carried out slowly, namely, with a slow expansion, there would be time for the warmer rubber band to cool, sending energy to the surroundings. If the process were slow enough, there would be no perceptible temperature change throughout the expansion, and similarly for the contraction. In this case the processes would be approximately *isothermal*.

The expansion of a rubber band can be likened to the quick, adiabatic compression of a gas, which also results in a temperature increase. The quick relaxation of the rubber band can be likened to adiabatically decreasing an external magnetic field surrounding a magnetised paramagnet. In both cases, there is an adiabatic temperature drop.

4.6.2 Model of a rubber band

The results of the simple experimental activities above can be further understood and interpreted using an equation that approximates a rubber band under tension F, with unstretched length L_0 and temperature T,

$$F = bT[L/L_0 - (L/L_0)^{-2}] \text{ with } b = \text{constant.} \qquad (4.74)$$

When $L = L_0$, the tension $F = 0$, as it must. If we stretch the rubber band slowly by dL, the work done on it is FdL, and the internal energy change is $dE = TdS + FdL$. The corresponding Helmholtz free energy change is $dA = FdL - SdT$. With the technique in Appendix 11.1, we can obtain the Maxwell relation, $(\partial S/\partial L)_T = (\partial F/\partial T)_L$. Combining this with Eq. (4.74) and integrating leads to

$$S(T, L) = -b\left(\frac{L^2}{2L_0^2} + \frac{L_0^2}{L}\right) + \text{a function of } T. \qquad (4.75)$$

The entropy is expected to increase with temperature for each value of the length L. Thus, the isotherms $F(T, L)$ *vs.* L/L_0 and $S(T, L)$ *vs.* L/L_0 look as shown in Fig. 4.32, and the isotherms are for temperatures $T_1 < T_2 < T_3$.

In the rubber band experiment described above, suppose the rubber band is initially relaxed, in state a with temperature T_2. The sequence of actions and results are visibly evident in Fig. 4.32(b).

1. Upon quick stretching, the (approximately) reversible, adiabatic expansion leads to state b with higher temperature $T_3 > T_2$, and $L/L_0 \approx 2.7$.

2. If the rubber band is allowed to quickly relax, it will (approximately) reverse its path, returning to state a with temperature T_2 and $L = L_0$.

3. If instead the rubber band is held stretched for say 5 seconds, it will cool back to temperature T_2 in at its stretched length in state c.

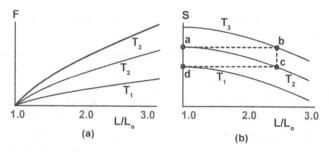

Figure 4.32: (a) Rubber band tension *vs.* reduced length L/L_0. (b) Rubber band entropy *vs.* reduced length.

4. If the rubber band then contracts quickly and (approximately) adiabatically and reversibly, it will end in state d, at temperature $T_1 < T_2$.

5. If no further action is taken, the rubber band will heat up back to temperature T_2 in the original state a.

Key Point 4.34 *The rubber band experiment and graphs is an excellent example of the interplay of energy and entropy.*

1. *For path ab, positive work is done on the rubber band at constant entropy. The temperature increase keeps the entropy from dropping as the length increases.*

2. *Path bc entails a transfer of energy from the rubber band to surroundings, decreasing the entropy of the rubber band and increasing that for the environment. Path da is a similar process in reverse, heating the rubber band and cooling the surroundings.*

3. *Notably, the rubber band's entropy is a decreasing function of the length L, at constant temperature, supporting the interpretation that the number of accessible states drops as L increases and it vibrations of the long chain molecules become constrained.*

4.7 NUCLEAR BINDING ENERGY, FISSION, FUSION

We saw that electrons are bound to nuclei in atoms and the mutual energy of interaction is negative. Binding energies are negative because it takes positive external work to pull atoms apart, i.e, to disassemble them into individual, separated electrons and nuclei. The single electron in the ground state of a hydrogen atom has energy -13.6 electron volts (eV).[20] This means that 13.6 eV

---

[20] 1 eV= 1.6×10^{-19} J, the energy gained by an electron that is accelerated through a 1 *volt* electric potential difference.

of energy is required to remove the electron from being bound to the single proton nucleus.

Binding energies are about 10^5 times larger for nuclei than for atoms. Protons attract other protons and neutrons, and neutrons attract one another, all by the strong nuclear force. Typically the binding energy per nucleon is $1-10$ MeV. Nuclei are classified in terms of their number of protons z (atomic number) and total number of nucleons A.[21] The number of neutrons is $(A-z)$.

In the nucleus of He^4, $z = A - z = 2$. The negative binding energy of helium-4 means that its mass differs from the masses of two protons and two neutrons outside the nucleus by $\Delta m < 0$. The Einstein mass-energy relationship enables a calculation of the energy equivalent of any mass m, namely, with energy $= mc^2$, where $c = 2.9979246 \times 10^8$ ms^{-1} is the speed of light in a vacuum.

- For an isolated proton, the mass is $m_p = 1.6726219 \times 10^{-27}$ kg and the corresponding energy is $m_p c^2 = 938.27208$ MeV.

- The mass of an isolated neutron is $m_n = 1.674927410^{-27}$ kg and the corresponding energy is $m_n c^2 = 939.56541$ MeV.

- Adding the masses of two isolated neutrons and two isolated protons and calculating its energy equivalent gives total energy $= 3755.675$ MeV.

- The mass of a ^{4}He nucleus is 6.644657×10^{-27} kg and the corresponding energy is 3737.379 MeV.

- The difference is $3727.379 - 3755.675 = 28.3$ MeV $= BE$ the binding energy of ^{4}He.

The nuclear force has interesting characteristics. (1) It is a very short range force, acting only over distances of about 10^{-15} m For larger distances it is effectively zero. (2) The nuclear force *saturates*; i.e., as more nucleons are packed into a nucleus, the electrostatic repulsive force between protons grows but the binding energy per nucleon diminishes. Thus, high A nuclei are unstable. An informative graph of binding energy, $|B|/A$, per nucleon is in Fig. 4.33. The lightest elements have the smallest $|B|$, and $|B| = 0$ (by definition) for hydrogen which has $z = A = 1$. As A increases, with a few exceptions, $|B|/A$ increases until nickel-62 (^{62}Ni), which is near the labeled ^{56}Fe, is reached. For $A > 62$, the envelope of the curve $|B|/A$ *vs.* A decreases from its maximum value of nearly 8.8 MeV per nucleon.

Key Point 4.35 *Notably, for low atomic mass nuclei, fusion leads to higher atomic mass and larger binding energy magnitude. For high atomic mass nuclei, fission leads to smaller atomic masses and again, a larger binding energy magnitude. Total nuclear energy decreases in both fission and fusion.*

---

[21]This should be distinguished by context from the symbol A for the Helmholtz free energy.

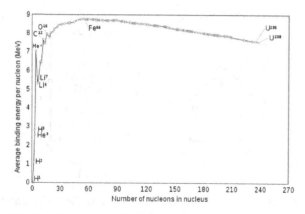

Figure 4.33: The absolute value of the negative energy of binding per nucleon, $|B|/A$, *vs.* A. Graphing the algebraic counterpart BE/A would turn the graph upside down. (Reprinted with permission by Fastfission.)

Entropy is produced in both fission and fusion, which are irreversible processes. The tendency to associate entropy with an increase with *disorder* associated with production of more particles in fission is misleading, because nuclear fusion *decreases* the number of particles and still generates entropy.

4.8   JARZYNSKI FREE ENERGY EQUALITY

NOTE: This section is more advanced mathematically than most of this monograph. To close this chapter, I return to the topic of free energy, discussed in Sec. 3.5. In 1997, Christopher Jarzynski proved a mathematical identity that has been influential in the field of non-equilibrium statistical mechanics.[22] His result has become known as the Jarzynski equality, which relates a system's Helmholtz free energy change between two given states to an average $\langle e^{-W/\tau} \rangle$, where W is the work done *on* the system to effect the change.

$$\langle e^{-W/\tau} \rangle = e^{-\Delta A/\tau} \text{ (Jarzynski equality)}, \qquad (4.76)$$

with $\tau = kT$ (see Eq. (3.14)). No single *specific* work path, e.g., fast or slow, is implied, but rather, *all* possible paths are contained in the averaging process, as explained here.

1. The angled brackets $\langle\,\rangle$ connote an average over all possible experiments that can connect the two equilibrium states.

---

[22]C. Jarzynski, "Nonequilibrium Equality for Free Energy Differences," Phys. Rev. Lett. **78**, 2690–2693 (1997); C. Jarzynski, "Equilibrium free-energy differences from nonequilibrium measurements: A master-equation approach," Phys. Rev. E **556**, 5018–5035 (1997).

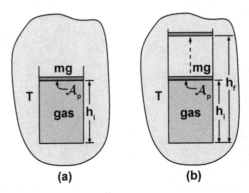

Figure 4.34: Expansion of an ideal gas, which can be fast, slow, and everything in between. All such processes are captured in the Jarzynski equality for an expansion from $h_i \rightarrow h_f$.

2. The paths are characterised by a parameter $\lambda(t)$, which at time $t = 0$ is $\lambda(0)$, and at the final time $t = t_f > 0$ is $\lambda(t_f)$. In the example in Fig. 4.34, $\lambda(0) = h_i$, the height of the gas at time zero and $\lambda(t_f) = h_f$, the final height at time t_f. We envisage carrying out the same experiment numerous times, each time following a different path $\lambda(t)$. This results in a set of work values, $\{W_1, W_2, \ldots\}$. The continuous probability distribution $\mathcal{P}(W)$ reflects the distribution of the large set of discrete work values, i.e., $\mathcal{P}(W)dW$ represents the fraction of experiments for which the measured work was in the interval $(W, W + dW)$.[23]

3. Jarzynski used the probability distribution $\mathcal{P}(W)$ to define the left side of Eq. 4.76,

$$\langle e^{-W/\tau} \rangle = \int \mathcal{P}(W)e^{-W/\tau}dW, \qquad (4.77)$$

where the integral includes all possible work values.

The function $f = e^{-W/\tau}$ is a convex function, for which all chords lie above (or on) f, all tangents lie below (or on) f, and the second derivative is nonnegative, as in Fig. 4.35. The average of f must lie above f itself, i.e., in the shaded area, while $f(\langle W \rangle/\tau)$ obviously lies on f. Jensen's inequality guarantees that[24]

$$\langle e^{-W/\tau} \rangle \geq e^{-\langle W \rangle/\tau}. \qquad (4.78)$$

---

[23] Theoretically, for a classical system, $\mathcal{P}(W)$ can be found in principle from the distribution of trajectories in the $6N$-dimensional phase space (3 momentum and 2 position components for each of N molecules), beginning with an ensemble of systems in a given equilibrium *macroscopic* state. Each classical microstate will have its own initial state and will evolve differently, in accord with Hamiltonian's equations of motion.

[24] Jensen's inequality states that if X is a random variable and $f(X)$ is a convex function of X, then denoting by $\langle \rangle$ an average over X, then $\langle f(X) \rangle \geq f(\langle X \rangle)$.

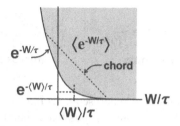

Figure 4.35: The average $\langle e^{-W/\tau} \rangle$ is in the shaded area, above $e^{-\langle W \rangle/\tau}$, in accord with Jensen's inequality, Eq. (4.78).

That this is reasonable is suggested by examples of the left and right sides of Jensen's inequality. By Jarzynski's equality, the left side of Eq. (4.78) equals $e^{-\Delta A/\tau}$, and (4.78) becomes $e^{-\Delta A/\tau} \geq e^{-\langle W \rangle/\tau}$, implying $\langle W \rangle \geq \Delta A$. This and Eq. (4.76) generalise and extend Eq. 3.41, and are particularly useful for small systems, where thermal fluctuations make individual work values for a single experiment unreliable. To get a better sense of what this all means, consider two examples that are inspired by the work of Jarzynski.[25]

Macroscopic example. First we revisit the rubber band example introduced in Sec. 4.6. Here, let the rubber band be fixed to a left wall with a blob of glue, and subjected to an external force at its right end, as in Fig. 4.36(a). Initially, the rubber band is not stretched, but is taut, with length x_i. During the stretching process, the external force does work W on the rubber band, increasing its length from x_i to x_f (not shown). The process is carried out by choosing a function $\lambda(t) = x(t)$ and a positive time t_f, with $0 \leq t \leq t_f$. These determine the path for given initial and final values $x(0) = A$ and $x(t_f) = B$ that governs the speed of the stretching process.

1. If the rubber band is stretched very slowly, i.e., with a slowly varying $\lambda(t)$ and large t_f, the process is approximately isothermal because there is sufficient time for energy exchanges with the environment. It is also nearly reversible, with approximately the same resulting free energy difference $\Delta A = A_f - A_i$ for each repeated experiment. In this case, $\langle W \rangle \approx W_{\text{rev}} = \Delta A$, where the average includes only very slow expansions. Minimal work W_{rev} is done and minimal energy is delivered to the surroundings.

2. For a relatively fast stretching process, more work is done, i.e., $W > W_{\text{rev}}$, and there is insufficient time for energy transfer to the environment *during* the process, and the rubber band's temperature increases. There is a larger energy transfer to the surroundings *after* the fast stretch than for the slow stretch.

[25]C. Jarzynski, "Nonequilibrium work relations: foundations and applications," The European Physical Journal B**64**, 331–340 (2008).

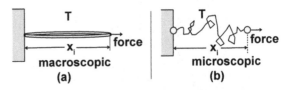

Figure 4.36: (a) Stretching of a macroscopic rubber band. (b) Stretching of a long-chain molecule. The environment has fixed temperature T.

3. The Jarzynski equality, $\langle e^{-W/\tau} \rangle = e^{-\Delta A/\tau}$, entails an average over *all possible* stretches that connect states A and B. It turns out that for some problems, it is possible to evaluate the left side of the Jarzynski equality and thus obtain a value for the process-independent free energy change $\Delta A = A_B - A_A$. We shall see with the next example, this averaging process is essential for sufficiently small systems, where thermal fluctuations are dominant.

Microscopic example. Figure 4.36(b) shows a similar example, with the rubber band replaced by a polymeric molecule that is visibly coiled, and can be stretched in principle, by attaching its ends to tiny beads. The left bead is held at the left end by a micropipette, and the right bead is in a laser trap that enables it to be moved left or right. Thermal fluctuations occur at both ends, making the initial length x_i uncertain.

For this *microscopic* system, use of the second law becomes questionable. Some quantities hold in an average sense, albeit with relatively large fluctuations about the average. As Jarzynski has described it, the system is immersed in aqueous solution at room temperature and pressure. The distance $\ell \equiv x_f - x_i$ is varied by moving the centre of the optical trap. Initially, the system is in equilibrium with the environment. Work W is done *on* the molecule as the optical trap is moved rightward, following a prescribed path $\lambda(t)$. After completion of the stretch, the system is allowed to equilibrate with the environment, with the pipette-to-trap distance held fixed at ℓ.

If the experiment is repeated, thermal fluctuations of the long-chain molecule and of the molecules of the aqueous solution cause variations in the work required to stretch the molecule. Repetition gives a set of work values, $\{W_1, W_2 \ldots\}$. The distribution of work values might look like Fig. 4.37(a). Typically, $W > \Delta A$, but with a small probability that $W < \Delta A$. The latter behaviour violates the second law of thermodynamics, occurs only rarely, and cannot be repeated at will. It is an example of Boltzmann's observation that the entropy of an isolated system (a target system plus environment) increases with a probability near, *but less than*, one.

Because of the substantial fluctuations in this microscopic system, the equation $W \geq \Delta A$ fails to have meaning and must be replaced by

$$\langle W \rangle \geq \Delta A, \text{ which is evident in Fig. 4.37(a).} \tag{4.79}$$

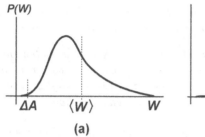 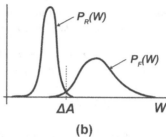

Figure 4.37: (a) Work distribution for: (a) a *forward* process, e.g., stretching a long-chain molecule. (b) Asymmetric *forward* and *reversed* processes, e.g., relaxation of a stretched molecule. The processes follow the protocol $\lambda(t)$ and $\lambda(-t)$ (time-reversed path), respectively.

This ends the discussion of the microscopic example. For the latter two examples, the Jarzynski equality is a strong statement that provides a strict equality (rather than an inequality) between average work behaviour and Helmholtz free energy change.

In addition to the Jarzynski equality, there are two other equalities that deserve mention. One is a work fluctuation theorem, proven by Gavin Crooks. It compares the probability distribution, $\mathcal{P}_F(W)$, that external work W is done *on* a system for the path $i \to f$ using the protocol $\lambda(t)$, with the probability distribution $\mathcal{P}_R(W)$ that the *conjugate* process $f \to i$ entails external work $-W$ (*on* the system). The conjugate process follows the time-reversed protocol $\lambda(-t)$. The qualitative behaviour of the work distributions for both paths is shown in Fig. 4.37(b).

$$\frac{\mathcal{P}_F(W)}{\mathcal{P}_R(-W)} = e^{(W-\Delta A)/\tau}. \tag{4.80}$$

Key Point 4.36 *The following points are notable.*

1. *Equation (3.24) is $W - \Delta A \geq 0$, which implies $\mathcal{P}_F(W) \geq \mathcal{P}_R(W)$. However because of the statistical nature of the second law of thermodynamics, that is not always true, but holds almost all the time. This is clear in Fig. 4.37(b). Such violations of the second law are rare, unrepeatable, and are fluctuations rather than average behaviour.*

2. *The asymmetry between the forward and reversed processes is evident in Fig. 4.37(b). It reflects the second law of thermodynamics, which is evident from the fact that $\langle W_F \rangle + \langle W_R \rangle > 0$. If the latter inequality were reversed that would imply that the rubber band did positive work in a cyclic process, using energy from a single reservoir, which violates the Kelvin-Planck statement of the second law.*

3. *Figure 4.37(b) shows that when the work $W = \Delta A$, the right side of (4.80) is one, and the forward and reverse distributions are equal.*

4. *It might seem that the inequality $\langle W \rangle \geq \Delta A$ is a statement of the second law of thermodynamics. However, we have assumed processes that connect equilibrium states i and f. This implicitly assumes the assumption that the second law of thermodynamics holds.*

A related expression for the probabilities of achieving *total* entropy change of the system plus environment, ΔS_{tot}, for forward and reverse processes was derived by Crooks,[26]

$$\frac{\mathcal{P}_F(\Delta S_{tot}/k)}{\mathcal{P}_R(-\Delta S_{tot}/k)} = e^{\Delta S_{tot}/k} \text{ Crooks equality.} \qquad (4.81)$$

When $\Delta S_{tot} > 0$, $\mathcal{P}_F(\Delta S_{tot}/k) > \mathcal{P}_R(-\Delta S_{tot}/k)$, and if $\Delta S_{tot} < 0$, the reverse is true. A connection between (4.81) and (4.76) can be made using the fact that for process $i \to f$, the system's internal energy change is $\Delta E = Q + W$, the entropy change is ΔS, and the environment's entropy change is $-Q/T$. The system's change in the Helmholtz free energy is $\Delta A = \Delta E - T\Delta S = Q + W - T\Delta S$. This implies that the total entropy production is

$$\Delta S_{tot}/k = (W - \Delta A)/kT + Q/kT - Q/kT = (W - \Delta A)/\tau. \qquad (4.82)$$

Let $\omega \equiv \Delta S_{tot}/k$ temporarily, which leads to $\mathcal{P}_F(\omega)e^{-W/\tau} = e^{-\Delta A/\tau}\mathcal{P}_R(-\omega)$. Integrating both sides of the latter over $(-\infty < \omega < \infty)$, we have

$$\int_{-\infty}^{\infty} P_F(\omega)e^{-W/\tau}\,d\omega = e^{-\Delta A/\tau}\int_{-\infty}^{\infty}\mathcal{P}_R(-\omega)\,d\omega. \qquad (4.83)$$

Note that $\omega = W - \Delta A/\tau$. Each of the experiments in the ensemble of measurements has the same ΔA because the initial and final equilibrium states are the same. We can replace the variable ω by W/τ with the same limits of integration. Thus, the left side of (4.83) is the average $\langle e^{-W/\tau} \rangle$. The right side reduces to $e^{-\Delta A/\tau}$, so Eq. (4.81) for entropy fluctuations reduces to the Jarzynski equality (4.76) for work fluctuations.

Key Point 4.37 *The mathematical arguments above illustrate that by considering ensembles of measurements, we can develop a framework for dealing with processes between equilibrium states that entail non-equilibrium paths. This formalism is suitable for small (non-macroscopic, and even microscopic) systems and irreversible processes. The energy-based Jarzynski equality (4.76) can be derived from the entropy-based Crooks equality (4.81), revealing once again an intimate relationship between energy and entropy.*

---

[26]G. E. Crooks, "Entropy production fluctuation theorem and the nonequilibrium work relation for free energy differences," Phys. Rev. E **60**, 2721–2726 (1999). Crooks used different notation. The ratio S_{tot}/k is used here because it is dimensionless.

The fluctuation equations of Jarzynski and Crooks have caused a considerable stir in the scientific community. Numerous articles have been published covering applications of these equations. This is because, they represent a new way to treat and understand non-equilibrium phenomena, which extends the applicability of traditional statistical mechanics.

4.8.1 Examples of the Jarzynski equality

Ideal gas expansion. Among the simplest examples that reveal the essence of the Jarzynski equality is a single-molecule gas in an ideal gas in a container with a movable piston. A classical ideal gas can be viewed as a collection of one-molecule gases, as depicted in Fig. 4.38, where the horizontal component of the molecule's velocity is v and the piston is moving rightward with velocity v_p.

Suppose the speed $v_p \ll 1$, i.e., the expansion is approximately quasistatic. In a collection of say 10^{21} one-molecule gases, a full range of speeds would exist in accord with the Maxwell-Boltzmann distribution, Eq. (3.1). In principle the collisions of each molecule with the piston can be followed mathematically in order to demonstrate that the Jarzynski equality is satisfied. This has been done by Lua and Grosberg,[27] who assume each molecule bounces off the walls and piston elastically. With each collision with the assumed fixed-speed piston, the molecule's velocity change is $\Delta v = -2(v - v_p)$. The piston transfers the momentum transferred to it to a load, and is assumed to maintain the constant velocity v_p. In this quasistatic case, $W = \Delta A$, and the left side of the Jarzynski equality reduces immediately to the right side, $e^{-\Delta A/\tau}$.

Key Point 4.38 *For a quasistatic volume change, the Jarzynski equality (4.76) is automatically satisfied because $W = \Delta A$. More generally, it was found that for an ideal gas, with initial and final state partition functions, Z_i and Z_f,[27]*

$$\langle e^{-W/\tau} \rangle = \frac{L + v_p \Delta t}{L} = \frac{Z_f}{Z_i} = e^{-\Delta A/\tau} \quad (see\ Eq.\ (3.26)). \qquad (4.84)$$

Now consider the following puzzle. Suppose the process entails a rapidly, highly non-equilibrium piston motion, $v_p > v$ for *every* molecule in the gas. Although this might seem quite reasonable at first glance, it turns out to be impossible.[27] This is because as the piston moves from say L_i to $L + v_p t$ in Δt seconds, the gas does zero work, and the left side of the Jarzynski equality (4.76) is $\langle e^{-W/\tau} \rangle = 1$. This implies $\Delta A = 0$, which is obviously false, because $\Delta A = \Delta E - T\Delta S$, and although $\Delta E = 0$, $\Delta S \neq 0$. So what has gone wrong? The problem is that our assumption that no molecules can catch up with the fast, receding piston is inconsistent with the Maxwell-Boltzmann velocity distribution for an ideal gas in (3.1).

---

[27] R. C. Lua & A. Y. Grosberg, "Practical applicability of the Jarzynski relation in statistical mechanics: a pedagogical example.," J Phys. Chem. B **109**, 6805–6811 (2005).

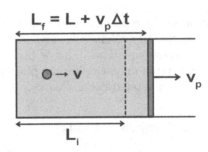

Figure 4.38: Single-molecule gas in cylinder with moving piston, starting with cylinder length L_i and ending Δt seconds later with length L_f.

Key Point 4.39 *Regardless of the piston's speed, there is a finite probability that some molecules move fast enough to catch up with and collide with it, thereby doing nonzero work. One must maintain consistency with the Maxwell-Boltzmann distribution when calculating the left side of the Jarzynski equality, (4.76). For rapid piston motion, it was found[27] that the probability that the piston does positive work during Δt is well approximated by*

$$P(W > 0) \propto \frac{1}{Lv_p\Delta t}e^{-\frac{1}{2}mv_p^2/\tau}. \tag{4.85}$$

Thus, all molecular velocities must be accounted for in any attempt to recover the Jarzynski equality.

The gas expansion considered here is instructive because:

1. The details can be worked out collision by collision. With a laborious calculation,[27] one can verify the validity of the Jarzynski equality.

2. The latter calculation requires careful inclusion of the far tail of the Maxwell distribution because the fastest molecules do much of the work when the piston moves very rapidly.

3. It is possible to estimate how many independent computer-experiments are needed to recover the equilibrium free energy in the Jarzynski equality accurately. Indications are that the number of needed trials grows exponentially with the system size L.[27]

Calculation of free energy. The accurate computer calculation of Helmholtz free energy differences is desirable in order to better understand the interactions responsible for the thermodynamic behaviour of materials. A variety of computational techniques have been used to accomplish this, and an entire book has been published on the topic.[28] The equations of Jarzynski

[28] Chipot, Ch., Pohorille, A., Eds. *Free Energy Calculations: Theory and Applications in Chemistry and Biology;* Springer Series in Chemical Physics; Springer: Berlin, 2007.

and Crooks offer one such approach. The Jarzynski equality can be used by first writing it in the form, $\Delta A = -\tau \ln \langle e^{-W/\tau} \rangle$. Then using N experimental computational trials, each giving a calculated work value W_i generated by the protocol $\lambda(t)$, one can calculate an approximate free energy difference,

$$\Delta A_{approx} = -\tau \ln \left(\frac{1}{N} \sum_{i=1}^{N} e^{-W/\tau_i} \right). \tag{4.86}$$

Recall that the protocol $\lambda(t)$ determines the speed of the work process that connects the initial and final equilibrium states. Typically, the approximation will improve for a larger number of trials N. Notice that Eq. (4.86) enables an estimate of the work distribution function. Let us renumber the trials so $W_1 \leq W_2 \leq W_i \leq W_{i+1} \cdots \leq W_N$, and divide the interval $(W_N - W_1)$ into M equal intervals, each with an average work $\overline{W}_j$. If the number of trials in interval j is $\mathcal{N}_j$, the probability of finding a work value in interval j is approximately $\mathcal{P}_j = \mathcal{N}_j/N$. In the limit $N, M \to \infty$, the work values approach a continuum, so $\mathcal{P}_j \to \mathcal{P}(W)$ for $N, M \to \infty$; i.e.,

$$\Delta A_{approx} = -\tau \ln \left(\sum_{i=1}^{M} \mathcal{P}_j e^{-W/\tau_j} \right) \to -\tau \ln \left(\int_0^{\infty} \mathcal{P}(W) e^{-W/\tau} \, dW \right).$$
$$\tag{4.87}$$

Equation (4.87) shows that for small values of W, when the probability density $\mathcal{P}(W)$ is near zero, the factor $e^{-W/\tau}$ has its largest value. This enhances the small-W contributions to the right side of the Jarzynski equality. However, for large W, both $\mathcal{P}(W)$ and $e^{-W/\tau}$ are small and it is only by including contributions over the long tail of the distribution that one can obtain a substantial contribution to $\langle e^{-W/\tau} \rangle$. In 2002, Liphardt and co-workers tested the Jarzynski equality experimentally.[29] Specifically, their test consisted of mechanically stretching a single molecule of ribonucleic acid,[30] both reversibly and irreversibly between chosen initial and final molecular states. This test of Jarzynski's equality provides a bridge between the traditional statistical mechanics of equilibrium and that of nonequilibrium.

The measurements entail tiny forces in the pico-Newton ($1 \, \text{pN} \equiv 10^{-12} \, \text{N}$) range, and the stretching and folding of single molecules over distances of $30 - 50 \, \text{nm}$. All experiments were done at approximately room temperature, $298 - 301 \, \text{K}$. One end of the RNA molecule was attached to $2 - 3 \, \mu\text{m}$ polystyrene beads. One of the beads was held in an optical trap that enabled the measurement of forces. The other bead linked to the RNA molecule via a micropipette tip, which was driven by a piezoelectric actuator that expanded and contracted under voltage changes. This generated forces to unfold and

---

[29] J. Liphardt, S. Dumont, S. B. Smith, I. Tinoco & C. Bustamante, "Equilibrium Information from Nonequilibrium Measurements in an Experimental Test of Jarzynski's Equality," Science **296**, 1832–1835 (2002).

[30] Also known as messenger RNA (mRNA), a polymeric molecule that conveys genetic information

fold the RNA molecule. Molecules were manipulated by moving the tip bead with the actuator, and the force acting on the RNA molecule was found by measuring the deflection of the trapping laser beams with position-sensitive photodetectors. These were difficult and complex high tech experiments.

Key Point 4.40 *Notably, the application of Jarzynski's equality to work trajectories obtained at two different nonequilibrium rates yields free energy difference estimates that are within* $0.3\,\tau$ *of one another. The difference between the nonequilibrium and reversible work values is less than* $0.6\,\tau$, *independent of the switching rate. For slow folding and unfolding, the measured free energy difference*[31] *was* $60.2\,\tau \pm 1.6\,\tau$. *For rapid, nonequilibrium processes, the corresponding free energy change was* $59.6\,\tau \pm 0.2\,\tau$, *in excellent agreement with theoretical predictions and independent single molecule measurements. These results show that the Jarzynski equality can be used to accurately measure free energy differences.*

---

[31]Liphardt, *et al.*[29] use the Gibbs free energy difference, the appropriate thermodynamic variable for their measurements.

CHAPTER 5

Radiation & Photons

CONTENTS

5.1 EM RADIATION & TEMPERATURE

When I took high school chemistry, I learned to adjust a Bunsen burner, varying the air supply to get more or less complete combustion. When combustion took place at higher temperatures, the flame was very blue. With complete combustion, a methane, natural gas flame temperature is about 2240 K, and a propane flame burns at about 2260 K. In contrast, when I decreased the air supply to the Bunsen burner, the cooler flame became red or yellow. I learned later that the yellow colour is from the incandescence of fine soot particles, produced because of incomplete combustion. The temperature of the yellowish flame is about 1270 K.

One of our children had a teacher who taught the colour spectrum using the mnemonic acronym, "ROY G BIV," which stands for the order of colours, **R**ed, **O**range, **Y**ellow, Green, **B**lue, **I**ndigo, and **V**iolet.[1] This is the order in which colours appear when an object is heated enough to glow as its

---

[1]Indigo is the dark blue dye obtained from the indigo plant, with a colour between blue and violet.

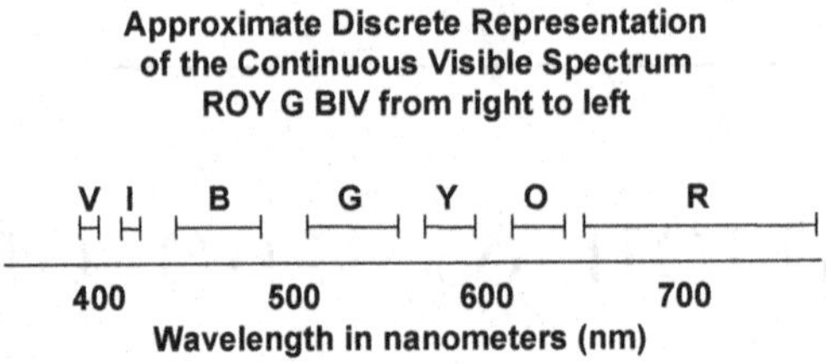

**Approximate Discrete Representation
of the Continuous Visible Spectrum
ROY G BIV from right to left**

V I   B   G   Y   O   R

400   500   600   700

Wavelength in nanometers (nm)

Figure 5.1: Black and white representation of the continuous colour spectrum of visible light. Colours vary continuously with increasing wavelength, from violet to blue, green, yellow, orange and red. The reverse order, from left to right, for increasing frequency, reads ROY G BIV.

temperature increases. I never forgot ROY G BIV and used it in my teaching many times. That representation is a simplification of reality, because all colours are superpositions of a continuum of frequencies; i.e., there is no single wavelength for red, blue or any colour.[2]

As explained in Ch. 1, Albert Einstein won the Nobel Prize for his work on the photoelectric effect. To explain it, he hypothesised the existence of "light quanta," discrete packets of energy that are now called photons, each with a specific frequency ν, wavelength λ, and

$$\text{Photon energy} = h\nu = h\left(\frac{c}{\lambda}\right). \tag{5.1}$$

The constants in Eq. (5.1) are Planck's constant h and the speed of light c.

Key Point 5.1 *In a plot of colour vs. wavelength, the ROY G BIV order is from right to left, the direction of increasing photon energy, in Fig. 5.1. Using an incandescent light bulb attached to a dimmer, as the brightness is increased from the bulb barely glowing, the filament initially glows red, then yellow, and then whiteish. At higher filament temperature, more higher-frequency light is emitted. Whiteish light results from an additive mixture of the entire visible spectrum.*

When an incandescent lamp is OFF, the filament temperature is that of the room, and the mixture of emitted frequencies is primarily in the infrared part of the spectrum. Our human eyes cannot detect these frequencies. The number of photons emitted in the visible spectrum, though not zero, is too small for human eyes to detect. In Sec. 5.3, we consider a gas of particles that are photons, in an enclosure whose walls are at temperature T. The walls emit photons just as a gas flame does, with a range of photon energies, consistent with the temperature. They also continually absorb photons. If a hole is punched in a wall, photons can escape and that escaping radiation is called blackbody radiation.

---

[2]Additionally, colour is subjective and can be affected by colour blindness.

5.2 BLACKBODY RADIATION

Blackbody radiation has a long history. In a notable experiment in 1800, astronomer William Herschel used a prism to decompose a beam of sunlight into a spectrum of colours onto a thermometer, and found that the temperature varied across that spectrum. Temperatures increased from the violet end to the red end of the spectrum. The highest temperature he found was beyond red; i.e., as he moved his thermometer through the dark area *beyond* red, the temperature continued rising. With those observations, Herschel discovered what is now known as *infrared radiation*. At the time, it was remarkable to learn that there was a form of light that we could not see! We have since discovered a substantial *invisible* electromagnetic spectrum including infrared, radio waves, microwaves, x-rays and gamma rays.

Herschel's findings showed a relation between light waves and thermal phenomena, and called the light rays "calorific rays." We now know that what he measured came from absorption of energy from electromagnetic radiation. The reason that Herschel found the highest temperature in the infrared region is that at Earth's surface, more than half the energy of sunlight is in the infrared, with 40–45 percent in the visible region and about 5 percent in the ultraviolet. Herschel's accidental discovery has led to the development of infrared technology including non-invasive measurements of body organs and temperature measurements of ocean temperatures and dust galaxies.

In the mid 1850s, Balfour Stewart compared the intensity of radiation of lamp-black surfaces with radiation from other sources at the same temperature. He found that a material that is a good absorber is also a good emitter, and established that radiation takes place throughout a radiating body and not simply at its surface.[3] Lamp-black surfaces had the greatest absorption of radiation, and also emitted the largest radiation intensity.

A modern example of the connection between emission and absorption is a lightweight thermal blanket with a highly reflective metallic coating that absorbs little, i.e., it mainly reflects radiation. Such a blanket also emits relatively little radiation and is definitely not a blackbody. In 1859, Gustav Kirchhoff defined a blackbody as an object that is a *perfect* emitter and absorber of radiation. By the 1890s, experimentalists were attempting to measure the spectral energy distribution of radiating objects. Modern radiation detectors fall into three main categories: disappearing filament optical pyrometers, thermal detectors, and photon or quantum detectors.

The mental construction of a model that leads to an example of blackbody radiation is straightforward. Imagine a solid with a cavity that opens through a small hole in the surface. It has temperature T, and the cavity emits blackbody radiation through the opening with an energy distribution matching that

---

[3]Stewart's work generalised Pierre Prevost's 1791 *law of exchanges* and the work of Count Rumford . A historical account is in H. Chang, "Rumford and the Reflection of Radiant Cold: Historical Reflections and Metaphysical Reflexes," Phys. Perspect. **4**, 127–169 (2002).

temperature. A suitable radiation-measuring device pointed at the opening can detect radiation with the blackbody radiation spectrum.

5.3   THE PHOTON GAS

5.3.1   What is a photon gas?

In order to generate radiation in a cavity, I introduce what is perhaps the simplest, yet scientifically richest gas: the photon gas. Its richness comes from at least four things:

1. Photons are observed throughout the universe and their number varies.

2. It is through photons, the quanta of electromagnet radiation, that we receive most of the known information about planets, stars, black holes, galaxies, the expanding universe, and the like.

3. Three fundamental constants from quantum mechanics, relativity, and thermal physics are central to the behaviour of a photon gas and appear in the equations that describe it: Planck's constant h (quantum mechanics), the speed of light c in a vacuum (relativity), and the Boltzmann constant k (thermal physics).

4. The thermal behaviour of photons in blackbody radiation played a pivotal role in the development of quantum mechanics. The number of photons per unit volume in an enclosed cavity, and the distribution of their energies depends *solely* on the temperature of the cavity's walls.

How can we generate a photon gas? Photon gases are made automatically, and the room you are in contains a photon gas along with the gas of air molecules. All macroscopic objects radiate photons with a frequency mixture consistent with their temperature. The cavity we envisaged earlier that emits the blackbody spectrum contains a photon gas.

The nature of the electromagnetic radiation depends on the temperature of the emitting matter. A room without furniture or other objects still contains photons, because its walls always radiate, as indicated symbolically in Fig. 5.2. The term *empty* is misleading, really meaning no *material* objects are present. Unlike a *material* gas of molecules, for which there are the three independent variables, the number of molecules N, temperature T and volume V, the photon gas has only two independent, controllable variables, T and V, which determine the average number N of photons.

Key Point 5.2 *A closed container that is empty of any molecules will be filled with a gas of photons. The average number of photons increases with the container walls' temperature and the container's volume (for a given temperature).*

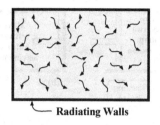

Radiating Walls

Figure 5.2: Depiction of an empty room except of photons. Only a small number are shown here for clarity.

Photon gases are dynamic and are constantly changing because new photons are continually emitted *and* old ones are absorbed by the room's walls. In thermodynamic equilibrium, the average numbers of photons absorbed each second equals the average number emitted per second. In a room with volume $V = 3\,\text{m} \times 4\,\text{m} \times 2.5\,\text{m} = 60\,\text{m}^3$, with a typical room temperature of 300 K, the average number density of photons is $\langle N \rangle / V = N(T,V)/V = 5.5 \times 10^{14}\,\text{m}^{-3}$, and the total average photon number $\langle N \rangle = 3.3 \times 10^{16}$ photons. For comparison, the number density of air molecules at the same temperature and atmospheric pressure: $N_{air}/V = 2.5 \times 10^{25}\,\text{m}^{-3}$.

At any instant of time, the number of photons N generally differs from the average value $N(T,V)$. The relative rms (root-mean-square[4]) fluctuations for the photons are

$$\text{relative rms fluctuations} \equiv \frac{\sqrt{\langle (N - N(T,V))^2 \rangle}}{\langle N \rangle} \tag{5.2}$$

$$= 5.5 \times 10^{-9} \text{ at room temperature.}$$

Key Point 5.3 *The fluctuations in the non-constant number of photons in the volume V are about one billion times smaller than the average number of photons, $N(T,V)$.*

The photons of blackbody radiation span a wide range of energies. As expected, if the wall temperature is increased, a larger fraction of photons is emitted at higher energies. The average energy per photon in a photon gas is $\sim 2.7\,\tau$, where k = the Boltzmann constant when the wall temperature is $T = \tau/k$. For comparison, the average energy per particle is $1.5\,\tau$ for the classical ideal gas of material particles, e.g., a dilute helium gas at room temperature.

Because photons have wave as well as particle properties, under certain circumstances, they can interfere constructively or destructively with one another. Yet photons in a photon gas have sufficiently randomised directions,

---

[4]The rms fluctuations are defined as the square root of the mean (average) of the nonnegative square $(N - \langle N \rangle)^2$ of the fluctuation from the average $\langle N \rangle$. The relative rms fluctuations are the rms fluctuations divided by the average $\langle N \rangle$.

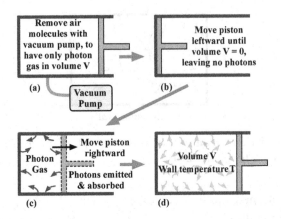

Figure 5.3: Idealised thought experiment: (a) Air molecules are pumped out of the cylinder, leaving only photons. (b) The piston is moved leftward to the left cylinder wall, until volume and photon number are zero. (c) The walls are held at temperature T and the piston moves rightward, bringing the volume to $V > 0$. Photons pour out of the walls and some photon absorption occurs. (d) The photon volume is V.

wavelengths, and phases that they do not ordinarily affect one another's energies or directions of motion. This lack of interactions simplifies photon thermodynamics. Unlike the classical ideal gas, which is discussed later in this chapter, the internal energy function for the photon gas depends on the container volume as well as the temperature, and the equilibrium pressure of the photon gas depends *only* on temperature, independent of the volume.

We can usefully imagine building a photon gas from energy stored in the container walls. Consider the cylinder and movable piston in Fig. 5.3. Suppose further that the air in the cylinder is pumped out of the cylinder using an idealised *perfect* vacuum pump (Fig. 5.3(a)). This leaves the volume V empty of material particles, but still with a photon gas. If we move the piston leftward, reducing V, the number of photons adjusts automatically, decreasing to a value that is appropriate for the existing wall temperature and volume. If we continue to move the piston leftward as far as it can go, the volume $V \to 0$ and the photon number $N \to 0$ as well (Fig. 5.3(b)). Then the piston is moved rightward (Fig. 5.3(c)), keeping the wall temperature constant, *building* the photon gas to that in Fig. 5.3(d).

Key Point 5.4 *When the process of building a photon gas is complete in Fig. 5.3, there will be a time-averaged value of $\langle N \rangle = N(T, V)$ photons. This average photon number is fully determined by V and T.*

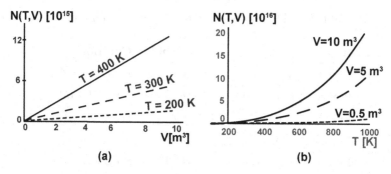

Figure 5.4: Average number of photons *vs.* (a) volume, for temperatures $200\,K, 300\,K, 400\,K$, and (b) temperature, for volumes $0.5, 5.0, 10\,m^3$.

5.3.2 Photon gas equations & graphs

The equations of state of the photon gas provide information on how N depends on T and V, and similarly how the internal energy, pressure, and entropy depend on these variables. I will begin with the equation for N. Using the framework of statistical mechanics, the equation of state for N shows how it behaves as a function of volume at fixed temperature (Fig. 5.4(a)), and as a function of temperature at fixed volume (Fig. 5.4(b)). Notably, these graphs indicate that

Key Point 5.5 *As you read this, the number of photons in your room is likely to be $10^{16} - 10^{17}$, compared with $10^{25} - 10^{26}$ air molecules.*

Statistical mechanics enables a determination of the average number, $N(T, V)$, of photons is,

$$N = rVT^3, \tag{5.3}$$

where

$$r = 60.4\left(\frac{k}{hc}\right)^3 = 2.03 \times 10^7 \mathrm{m}^{-3}\,\mathrm{K}^{-3}. \tag{5.4}$$

Similar expressions exist for energy, pressure, Helmholtz and Gibbs free energies, and entropy of the photon gas, each of which also depends on the three fundamental constants (h, c, k). A summary of relevant thermodynamic functions is in Table 5.1, where

$$b = 7.56 \times 10^{-16}\mathrm{J}\,\mathrm{K}^{-4}\mathrm{m}^{-3}. \tag{5.5}$$

Key Point 5.6 *Figure 5.4 shows that the average number $N(T, V)$ of photons increases in direct proportion with the volume; i.e., the photon number density $N(T, V)/V$ is a function only of the wall temperature. The number of photons adjusts automatically because the emission rate depends on the temperature and the absorption rate depends on the number of photons impinging on the*

Table 5.1: Thermodynamic functions for a photon gas

Item	Formula	Constant
#photons $N(T,V)$	rVT^3	$r = 60.4k/hc^3$
Pressure $P(T)$	$\frac{1}{3}bT^4$	$b = 163k^4/15hc^3$
Energy $E(T,V)$	bVT^4	See above
Entropy $S(T,V)$	$\frac{4}{3}bVT^3$	See above
Enthalpy $H(T,V)$	$\frac{4}{3}bVT^4$	See above
Helmholtz function $A(T,V)$	$-\frac{1}{3}bVT^4$	See above
Gibbs function $G(T,V)$	0	

walls each second. If V is increased to V', the rate of photons impinging on the walls is (temporarily) too low for equilibrium, so more photons are emitted each second than are absorbed, and N increases until the two rates are again equal and a new steady-state equilibrium number $N(T,V)$ is achieved.

Key Point 5.7 *The photon gas is a quantum mechanical, relativistic, and thermal system, and its equations contain the fundamental constants h, c and k. The internal energy, pressure, enthalpy and Helmholtz function are each proportional to T^4, while the average number of photons and entropy are proportional to T^3. The extensive functions N, E, S, H, A, G are all proportional to V, the extensive controllable variable.*

Fig. 5.5 shows how the internal energy depends on temperature for three values of V. Conventionally, the internal energy is measured in joules (J) and the temperature is in kelvins (K). For larger V, there are more photons for a given temperature, which leads to higher internal energy for *any* value of that temperature. Notably, at the same temperature and volume, the internal energy of a photon gas is lower by a factor of about 10^{-10} than the internal energy for a typical dilute gas of molecules like room temperature air.

Key Point 5.8 *The average photon energy has the same order of magnitude as a molecule's energy, but because the number of photons is so much smaller than the number of air molecules in a room, the internal energy of the photon gas is much smaller than the air's. Similarly, the pressure P of a photon gas is much lower than that for a gas such as air at the same temperature and volume, primarily because of the smaller number of photons.*

As expected, the entropy increases with the volume V because the number of photons increases when there is more space for them to fill. It also increases with temperature, as the average energy per photon increases. A

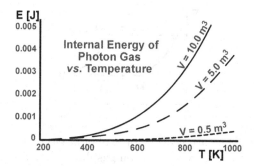

Figure 5.5: Internal energy of a photon gas as a function of the temperature T for three different values of volume: $V = 0.5$ m^3 (small-dash line), 5.0 m^3 (large-dash line), and 10 m^3 (solid line).

very interesting observation is that both the entropy and average number of photons are proportional to VT^3. As a consequence, S is proportional to N, and combining them leads to the interesting result,[5]

$$S(T, V) = 3.6\, kN(T, V). \qquad (5.6)$$

Entropy is proportional to Boltzmann's constant k, which has the same units as entropy, and Eq. (5.6) shows something notable:

Key Point 5.9 *The average entropy per photon is 3.6k for any T and V. A physical reason for this is not known.*

Table 5.1 shows that the Gibbs function $G = 0$. For material particles, where N is an independent variable, the derivative $\mu \equiv (\partial G/\partial N)_{T,P} = $ chemical potential.[6]

Key Point 5.10 *Because G is identically zero, it cannot change when photons are added or removed. This implies that the chemical potential of a photon gas is zero.*

5.3.3 Photon gas processes

It is easy to plot how isothermal and adiabatic processes look on a pressure-volume graph. From Table 5.1, because pressure depends *only* on temperature, it is clear that for an isothermal process, the pressure *vs.* volume curve is a horizontal line segment. Similarly, for a reversible adiabatic

---

[5]The factor 3.6 is an approximation. A better one is 3.60157. In either case, this number and the simple proportionality between S and N has no known explanation.

[6]The chemical potential can also be written as $\mu = (\partial A/\partial N)_{T,V}$, $\mu = (\partial E/\partial N)_{S,V}$, and $\mu = -T(\partial S/\partial N)_{E,V}$.

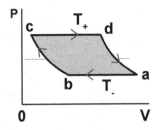

Figure 5.6: Reversible Carnot cycle for a photon gas. The shaded area $abcda$ represents the work done by the photon gas during one cycle.

process, i.e., with constant entropy, Table 5.1 shows that $T^3V = constant$ and $PV^{4/3} = constant$. This has immediate implications for a Carnot cycle with a photon gas working fluid.

Key Point 5.11 *A Carnot cycle with a photon gas working fluid, operating with temperatures T_- and $T_+ > T_-$ is illustrated on a PV plot in Fig. 5.6.*

Along the upper horizontal (isothermal) path, the entropy change is $\Delta S_+ = \frac{4}{3}b(V_d - V_c)T_+^3 > 0$ and along the lower isothermal path, $\Delta S_- = -\frac{4}{3}b(V_a - V_b)T_-^3 < 0$. For a reversible Carnot cycle, the *magnitudes* of the heating input and output are $Q_+ = T_+\Delta S_+$ and $Q_- = -T_-\Delta S_-$. Along the adiabatic expansion, the entropy is constant, so $VT^3 = constant$, the photon gas cools, and

$$V_cT_+^3 = V_bT_-^3 \text{ and } V_dT_+^3 = V_aT_-^3 \text{ , or } \frac{(V_a - V_b)}{(V_d - V_c)} = \frac{T_+^3}{T_-^3}. \qquad (5.7)$$

Using the latter result, it follows that the thermal efficiency is,

$$\eta_{\text{photon car}} = 1 - \frac{Q_-}{Q_+} = 1 - \frac{T_-^4(V_a - V_b)}{T_+^4(V_d - V_c)} = 1 - \frac{T_-}{T_+}, \qquad (5.8)$$

the well known Carnot efficiency, Eq. (2.16).

Key Point 5.12 *Notably, the entropy of a photon gas is linear in the volume V, very different from the Sackur-Tetrode equation (4.1) and (4.4) for the entropy of an ideal gas, which is has a logarithmic volume dependence.*

Now suppose we build the photon gas by choosing the initial volume to be zero and then allowing the piston to move to the right slowly and isothermally to volume V, creating the photon gas. At zero volume, the photon number $N = 0$, so $S = 0$; that is, there can be no entropy if there are no photons, in agreement with Eq. (5.6). As shown earlier, the heating input from the walls is $Q = T\Delta S = \frac{4}{3}bVT^4 = \Delta H$, as expected for a constant pressure process.

Key Point 5.13 *The enthalpy H is the energy needed to establish the photon gas and to do the work needed to make available the volume V it occupies.*

Next, consider a slow adiabatic volume change.

Key Point 5.14 *For a photon gas, adiabatic means no photons are emitted or absorbed by the container walls. In this case the* only *possible type of heat process is via such photons. Therefore to achieve an adiabatic volume change, the container walls must be perfectly reflecting mirrors that can neither emit nor absorb photons.*

Emission and absorption go hand in hand: if a surface cannot absorb, then it cannot emit because the average number of photons $N(T, V) =$ constant, which is consistent with Eq. (5.6). Note that an isentropic process satisfies (5.7), $T^3 V =$ constant, which can also be recast as $PV^{4/3} =$ constant.

Key Point 5.15 *The result* $PV^{4/3} =$ *constant is similar to the isentropic condition for a monatomic ideal gas,* $PV^\gamma =$ *constant, where* $\gamma = C_p/C_V = (5/2)/(3/2) = 5/3$. *This similarity is fortuitous because* C_p *does not exist for a photon gas. This is because it is impossible to vary the temperature at constant pressure. The photon gas pressure* $P(T)$ *is solely a function of temperature, so any change in* T *must change* P *as well. In contrast, the constant-volume heat capacity is well defined for a photon gas.*

5.4 KIRCHHOFF'S & PLANCK'S LAWS

I discussed the absorption and emission of radiation from matter, but have not examined the relationship between these two phenomena. I argued that a perfectly reflecting mirror has zero absorption *and* emission, and here I generalise that fact. To do so, envisage a body in thermodynamic equilibrium within a cavity whose walls have temperature T. There is a photon gas in the cavity, and the body and the walls continually absorb and emit photons. Let $u(T)$ be the photon energy density, i.e., the energy per unit volume at temperature T. Similarly, let $u(\lambda, T)$ be the spectral energy density per unit wavelength.[7] The photon energy in a volume element dV and in the wavelength interval $(\lambda, \lambda + d\lambda)$ is $u(\lambda, T)d\lambda dV$. The objective is to pin down the function $u(\lambda, T)$.

The analysis that follows is one that is common in *kinetic theory*, a theoretical framework that preceded statistical mechanics. First I'll find the fraction of photons moving within a specified narrow range of angles. To do this, I use the spherical coordinate system in Fig. 5.7. Assume the photon directions are distributed uniformly over 4π steradians, so the fraction of photons moving toward the origin, and passing through surface area $d\sigma$ is,

$$df(\theta, \phi) = d\sigma/(4\pi r^2) = \left(\frac{1}{4\pi}\right)\sin(\phi)d\theta d\phi. \tag{5.9}$$

---

[7]I will show that $u(T)$ and $u(T, \lambda)$ depend *only* on T and (T, λ), respectively.

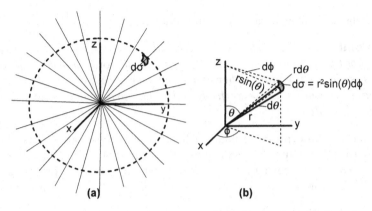

Figure 5.7: Fraction of photons moving radially toward $(0,0,0)$ and passing through the grey area $d\sigma = r^2 \sin(\theta)d\theta d\phi$, is $d\sigma/(4\pi r^2)$.

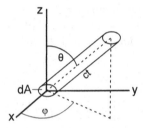

Figure 5.8: Collision cylinder, defined by angles (θ, ϕ), for photons moving along the cylinder axis toward area dA during time dt.

The second part of the argument uses a "collision cylinder," shown in Fig. 5.8 The axis of this imagined cylinder is defined by angles θ and ϕ, as in Fig. 5.7. At any moment of time, there are

$$dN_{cyl}(\theta, \phi) = (N/V)df(\theta, \phi)dV_{cyl} \qquad (5.10)$$

photons in the cylinder, traveling parallel to the cylinder's axis toward the origin. The length of the cylinder is chosen to be cdt, the distance travelled by a photon in a small time dt. Here

$$dV_{cyl} = cdt(dA \cos(\theta)). \qquad (5.11)$$

Because the photons move with speed c, all photons in the collision cylinder traveling along (θ, ϕ) reach dA in time interval dt. The area of the cylinder perpendicular to its axis, $dA \cos(\theta)$, decreases from dA to 0 as θ increases from 0 to $\pi/2$ radians. This is reflected in the expression (5.11) and in the following expression, which combines Eqs. (5.9)-(5.11),

$$dN_{cyl}(\theta, \phi) = \frac{Nc}{4\pi V}(dt)dA \sin(\theta) \cos(\theta)d\theta d\phi. \qquad (5.12)$$

Table 5.2: Important radiation quantities and their units

Name	Symbol	Description & Units
Particle flux	Φ_N	# photons per second impinging on a surface, per unit area [s^{-1}m^{-2}]
Energy density	$u(T) = bT^4$	Total photon gas energy per unit volume [Jm^{-3}]
Spectral energy density	$u(T, \lambda)$	Total photon gas energy per unit volume, per unit wavelength for the photon gas [Jm^{-3}]
Blackbody spectral radiance	$\mathcal{E}_{\text{bbrad}}(T, \lambda)$	Photon energy emitted by a blackbody in the wavelength $(\lambda, \lambda + d\lambda)$ interval, per unit area, per unit time [Js^{-1}m^{-2} or Wm^{-2}]
Spectral radiance	$\mathcal{E}_{\text{rad}}(T, \lambda)$	Photon energy emitted (and absorbed in equilibrium) by a non-blackbody in the wavelength interval $(\lambda, \lambda + d\lambda)$, per unit area, per unit time [Js^{-1}m^{-2} or Wm^{-2}]
Spectral absorptivity and emissivity	$\alpha(T, \lambda), \epsilon(T, \lambda)$	Fraction of radiation absorbed ($\alpha(T, \lambda)$ or emitted ($\epsilon(T, \lambda)$ in wavelength interval $(\lambda, \lambda + d\lambda)$ [dimensionless]

We are interested in these photons, which travel along the collision cylinder and through the area of its base. Integrating Eq. (5.12) over θ over $(0, \pi, 2)$ and ϕ over $(0, 2\pi)$, and dividing both sides by $(dA\,dt)$ leads to the desired photon flux,

$$\text{Photon flux} = \Phi_N = \frac{dN}{dA\,dt} = \frac{1}{4}\left(\frac{N}{V}\right)c. \tag{5.13}$$

Table 5.2 gives a preview of the functions that are introduced in the remainder of this section.

Inside a cavity whose walls are at temperature T, the radiation is called *blackbody radiation spectral radiance* $\mathcal{E}_{bbrad}(T)$, which is the photon flux (5.13), times the average energy per photon $\bar{e}$,

$$\mathcal{E}_{bbrad}(T) = \frac{1}{4}\left(\frac{N}{V}\right)c\bar{e} = \frac{1}{4}\left(\frac{N}{V}\right)c\left(\frac{E}{N}\right) = \frac{1}{4}cu(T). \tag{5.14}$$

In Eq. (5.14), the average photon number N divides out and the photon gas energy $E(T, V) = bVT^4$ from Table 5.1 is used in the last step to show that

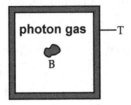

Figure 5.9: Body B in cavity with wall temperature T, in a photon gas.

the energy density $u(T) = E(T, V)/V$ is solely a function of temperature T. It is useful to also introduce the *spectral energy density* $u(T, \lambda)$, the energy per unit volume, per unit wavelength, which turns out to be most useful in relating non-blackbody to blackbody radiators.

If a perfect absorber is placed in a photon gas at temperature T, it absorbs energy at the rate $\mathcal{E}_{bbrad}(T)$ per unit area. In thermal equilibrium, it must emit energy with the same rate or it would either heat or cool spontaneously, in violation of the second law of thermodynamics. This must hold in each small wavelength interval for the given temperature, which can be understood by imagining the body's surface being covered by a filter that allows passage *only* of photons with wavelengths in $(\lambda, \lambda + d\lambda)$. If more photons were emitted through the filter than were received by the body, there would be a spontaneous heat process between regions of the same temperature. Again, this would violate the second law of thermodynamics..

Envisage an opaque non-blackbody B in an enclosure containing a photon gas at temperature T, as in Fig. 5.9. Generally, radiation arriving at B can be absorbed, reflected, or transmitted. Because B is opaque, the fraction transmitted is zero. The fraction absorbed for photons with wavelength λ is called the spectral absorptivity, $\alpha(T, \lambda)$, and the reflected fraction is necessarily $(1 - \alpha(T, \lambda))$. In order for B to remain in equilibrium during any time interval dt, the energy absorbed in $(\lambda, \lambda + d\lambda)$ during dt must equal the energy emitted. The absorption cannot exceed the emission in one wavelength interval and have the reverse true in another interval, such that total absorption equals total emission. This follows from our argument using a filter that passed only wavelengths in $(\lambda, \lambda + d\lambda)$. The Kirchhoff law summarises this:

Key Point 5.16 *Kirchhoff's Law For an opaque body B in a cavity with radiation at temperature T, the energy absorbed per unit time by B equals energy emitted per unit time in each wavelength interval $d\lambda$.*

In a cavity containing thermal photons, the incident radiation energy per unit time, per unit area, per unit wavelength is $\mathcal{E}_{bbrad}(T, \lambda) = \frac{1}{4}cu(T, \lambda)$. Of that amount, $\alpha(T,\lambda)\mathcal{E}_{bbrad}(T,\lambda$ is absorbed by the opaque non-blackbody B. In equilibrium, the corresponding spectral radiance must be,

$$\mathcal{E}_{rad}(T,\lambda) = \epsilon(T,\lambda)\mathcal{E}_{bbrad}(T,\lambda) = \alpha(T,\lambda)\mathcal{E}_{bbrad}(T,\lambda). \qquad (5.15)$$

It follows that

$$\epsilon(T,\lambda) = \alpha(T,\lambda) \quad \text{(corollary to Kirchoff's law).} \qquad (5.16)$$

For a perfect emitter, namely, a blackbody,

$$\epsilon(T,\lambda) = \alpha(T,\lambda) = 1, \text{ and } \mathcal{E}_{rad}(T,\lambda) = \mathcal{E}_{bbrad}(T,\lambda) = \frac{1}{4}cu(T,\lambda). \qquad (5.17)$$

Key Point 5.17 *The energy density $u(T,\lambda)$ is a universal function that is applicable for the thermal radiation in a cavity at temperature T, namely, blackbody radiation. An opaque object will absorb fraction $\alpha(T,\lambda)$ and emit fraction $\epsilon(T,\lambda)$ of the blackbody radiation flux that impinges on it.*

Key Point 5.18 *Equation (5.16) assures that a good absorber is a good emitter and a bad absorber is a bad emitter. The arguments that led to this result involve energy and the second law of thermodynamics in essential ways.*

Statistical mechanics provides an explicit expression for the spectral radiance of a blackbody,

$$\mathcal{E}_{bbrad}(T,\lambda) = \frac{2\pi hc^2}{\lambda^5(e^{-hc/\lambda\tau} - 1)} \quad [Wm^{-2}]. \qquad (5.18)$$

Equation (5.18) is the celebrated Planck radiation law, discovered initially experimentally and subsequently derived mathematically by Max Planck in 1900. The integral of this quantity is,

$$\int_0^\infty \mathcal{E}_{bbrad}(T,\lambda)\,d\lambda = \frac{2\pi^5 k^4}{15h^3c^2}T^4 \equiv \sigma T^4. \qquad (5.19)$$

The constant $\sigma = 5.67 \times 10^{-8}\,W\,m^{-2}\,T^{-4}$ carries the name *Stefan-Boltzmann*.

5.4.1 Incandescent lamps

In 1990, I published an article in *The Physics Teacher* entitled, "Illuminating physics with light bulbs."[8] Incandescent lamps are excellent teaching tools, covering aspects of electricity: voltage, current, power, efficiency, and series and parallel circuits. In this section, I use incandescent lamps to illustrate the thermodynamics of electromagnetic radiation.

---

[8]Phys. Teach. **28**, 30–35 (1990).

Worldwide, lighting accounts for $10 - 15\%$ of total electrical energy use. Because they are much less efficient than more modern light-emitting diode (LED) lamps and compact fluorescent lamps, incandescent lamps are being phased out by governments around the world. However, many incandescent lamps are still in operation and special-use incandescent lamps for which no alternatives are available are likely to continue to be used.

Key Point 5.19 *The key points for an incandescent lamp, which were discussed briefly in Example E2 in Ch. 1, are:*

1. *Electric current is sent through a thin metal wire filament, which is coiled tightly to achieve an effectively larger emitting surface at a higher temperature than for a straight wire. The need to coil a thin wire limits workable filament metals to those that are maleable.*

2. *Because incandescent lamp filaments reach typical temperatures of $2500 - 2800\,\mathrm{K}$, the filament metal must have a high melting temperature. Tungsten (the element wolfram), with a melting temperature of $3697\,\mathrm{K}$, is commonly used in these lamps.*

3. *To account for the eye's sensitivity, a different measure of the power radiated in the visible part of the electromagnetic spectrum, is used. That unit of measure is the Lumen,[9] which is adjusted for the spectral luminous efficiency $S(\lambda)$ of the average human eye. The eye is most sensitive at wavelength $\lambda^* = 550\,\mathrm{nm}$ where the maximum of $S(\lambda^*) = 1$ by definition. The shape of $S(\lambda)$ is shown in Fig. 5.10.*

4. *Higher temperatures bring higher efficiencies, i.e., more visible light for a given electrical energy input. The filament spectrum at temperature $\sim 6600\,\mathrm{K}$ would best match the sensitivity of the human eye, but no suitable metals have melting temperatures much above tungsten's. Recall that the Sun emits a spectrum consistent with a temperature of $5800\,\mathrm{K}$*

5. *The light from an incandescent lamp is similar to that from a blackbody at the lamp's filament temperature, namely, covering a wide range of wavelengths, but with much of the energy emitted being infrared radiation. A relatively small part of the spectrum is in the visible spectral region, which means that the lamp is inefficient.*

The main quantities of interest are the spectral radiance $\mathcal{E}_{bbrad}(T,\lambda)$ for a blackbody in Eq. (5.18), and the spectral emissivity $\epsilon(T,\lambda) \equiv \mathcal{E}_{rad}(T,\lambda)/\mathcal{E}_{bbrad}(T,\lambda)$ in Table 5.2. The spectral emissivity $\epsilon(T,\lambda) \leq 1$ is the fraction of the Planck blackbody spectral radiance that is emitted by the filament at temperature T for each wavelength λ. Using $\mathcal{E}_{bbrad}(T,\lambda)$ and $\epsilon(T,\lambda)$, one can estimate (i) an incandescent lamp's efficacy in Lumens/Watt,

---

[9]The luminous flux is the amount of light emitted per second in a unit solid angle of one steradian from a uniform source of one candela.

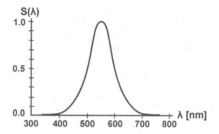

Figure 5.10: Spectral luminous efficiency $S(\lambda)$ for human vision. Maximum efficiency is for a greenish light, roughly the colour of grass. (Plot of data, with IES permission, from *IES Lighting Ready Reference*, edited by J. E. Kaufman & J. F. Christensen (Illuminating Engineering Society (IES), New York, 1985), p. 35.)

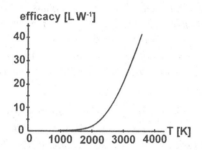

Figure 5.11: Efficacy *vs.* filament temperature. (Reproduced (adapted) from D. C. Agrawal, H. S. Leff & V. J. Menon, "Efficiency and efficacy of incandescent lamps," Am. J. Phys. **64**, 649–654 (1996), with permission of the American Association of Physics Teachers.)

and (ii) the efficiency defined as the power output in the visible part of the spectrum divided by the power input.[10]

A summary of the salient points is as follows:

1. The efficacy in Lumens/Watt is defined by

$$\text{efficacy} \equiv \frac{\text{Adjusted output in}(\lambda_0, \lambda_f)}{Input}$$

$$= \frac{\int_{\lambda_0}^{\lambda_1} 683 S(\lambda)\mathcal{E}_{rad}(T,\lambda)d\lambda}{\int_0^\infty \mathcal{E}_{rad}(T,\lambda)d\lambda}. \qquad (5.20)$$

The factor 683 (Lumens per Watt) of light represents the maximum luminous flux detected by an average human at $\lambda^* = 550$ m, and output

---

[10]D. C. Agrawal, H. S. Leff & V. J. Menon, "Efficiency and efficacy of incandescent lamps," Am. J. Phys. **64**, 649–654 (1996).

is in the visible spectrum with $(\lambda_0, \lambda_f) = (380\,\text{nm}, 760\,\text{nm})$. Using an approximation for $S(\lambda)$, the efficacy can be calculated for various temperatures. Figure 5.11, shows efficacy *vs.* filament temperature.

2. The dimensionless lamp efficiency is,

$$\text{Lamp efficiency} \equiv \frac{\text{Output power in } (\lambda_0, \lambda_f)}{\text{Input}}$$
$$= \frac{\int_{\lambda_0}^{\lambda_f} \mathcal{E}_{rad}(T, \lambda)d\lambda}{\int_0^\infty \mathcal{E}_{rad}(T, \lambda)d\lambda}. \tag{5.21}$$

Key Point 5.20 *Some losses are not accounted for in the above equation, as explained below.*

1. *Equations (5.20) and (5.21) do not show losses due to the absorption of about 10% of the radiated energy by the glass, i.e., the transmittance of the glass lamp envelope is about 0.9. Similarly, the lamp's base absorbs a fraction $f = 0.13 - 0.25$ of the radiated energy. These effects reduce the numerators in the efficacy and efficiency equation by the (approximate) factor, $0.9(1 - f)$, leading to the result, radiated power $= \bar{\epsilon}(T)\sigma AT^4 = 0.9(1 - f_c)P_{in}$.*

2. *With these losses accounted for, the estimated efficacy values are those in Fig. 5.11. For example, a 60 W lamp with filament temperature $\sim 2800\,K$, has efficacy $\sim 15\,L\,W^{-1}$, and a 100 W lamp with filament temperature $\sim 2900\,K$ has efficacy $\sim 18\,L\ W^{-1}$. Higher efficacies are expected at higher temperatures because a larger fraction of the emitted light is in the visible spectrum.*

3. *Efficiency calculations show that typical lamp efficiencies are $2 - 13\,\%$, with the highest values for the largest-power lamps.*

4. *Incandescent lamp lifetimes are limited by the sublimation of tungsten because tungsten's vapour pressure and sublimation rate increase with the filament temperature. This thins the filament and leads to eventual filament breakage. Thus sublimation reduces lamp lifetimes. The sublimation rate can be diminished by the presence of inert argon or another gas, which is done for bulbs with power ratings above $\sim 40\,W$. For lower wattage bulbs, conduction and convection losses in a gas can nearly offset the benefit of increased lifetime, so such lamps are generally evacuated so they operate in a near vacuum.*

5. *The effects of sublimation in common household light bulbs can be seen because convection currents in gas-filled lamps carry the vaporised tungsten to the bulb's highest point–above the filament–which depends on its orientation. This is illustrated in Fig. 5.12. Old gas-filled bulbs can reveal the bulb's orientation during its operating lifetime.*

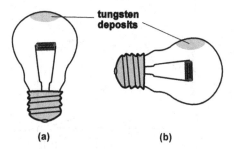

Figure 5.12: Light bulbs in configurations (a) and (b) develop tungsten spots above the filament.

5.4.2 Cosmic microwave background radiation

One of the most inviting examples of a photon gas is the cosmic microwave background radiation (CMBR), which is a gas of *old* photons that were created in the early, hot universe, approximately 13.8 billion years ago. Over time, as the universe expanded, that gas cooled to 2.7 K. If we substitute this temperature and the numerical value of r from Eq. (5.4) into Eq. (5.1) we find,

Key Point 5.21 *On Earth, the photon number density of cosmic background radiation is* ~ 416 *photons* cm^{-3}. *These photons, which make even dinosaur bones seem young, are in our vicinity all the time. Awareness of the photon gas opens the door to an understanding and appreciation of the CMBR.*

Immediately after the Big Bang, photons became trapped in a hot collection of positively charged ions and electrons. The photons that make up the CMBR are believed to have been emitted about 300,000 years after the Big Bang, after the universe cooled sufficiently for electrons and ions to form neutral atoms. Then the photons were believed to be in equilibrium with the hot plasma at about 3000 K. Subsequent expansion of the universe caused continual cooling to the current value, 2.7 K.

Physicists Arno Penzias and Robert Wilson discovered the CMBR serendipitously in 1965. Earlier, in 1941, Andrew McKellar had found evidence of a temperature of 2.3 K and published it in an article, "Molecular Lines from the Lowest States of Diatomic Molecules Composed of Atoms Probably Present in Interstellar Space." Those measurements indicated that the radiation from different directions was nearly uniform.

Despite those measurements, which showed that the distribution of mass in the early universe was reasonably smooth, there must have been fluctuations from uniformity. Otherwise, what would have caused the formation of the galaxies we observe now? The point is that any ripples in the density, would be accompanied by ripples in temperature of photons emitted from distinct regions. Regions of higher density would have caused a larger redshift of released photons, reducing the temperature of the spectrum. Researchers

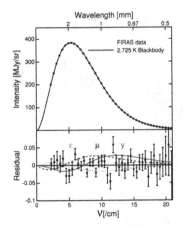

Figure 5.13: Upper graph: The Planck spectrum as measured by the COBE project. Solid circles are data points and the solid line is Eq. 5.18 for temperature $T = 2.725$. Lower graph: Deviations from the Planck spectrum (too small to see in the upper graph). (Data from FIRAS on NASA's COBE satellite. Figure by Ned Wright - used with permission.)

believe that such fluctuations could have resulted in clumping of matter, ultimately forming galaxies.

During 1989–1993, a NASA satellite called COBE (Cosmic Background Explorer) orbited Earth, measuring CMBR. The results supported the Big Bang theory. The spectrum of the radiation was extremely close to the Planck radiation law, Eq. 5.18, albeit with small anisotropies. George Smoot and John Mather received the Nobel Prize in Physics in 2006 for their work on COBE, which was hailed as the starting point of precision-measurement cosmology. It was followed by two more advanced projects: the Wilkinson Microwave Anisotropy Probe (2001–2010), and the Planck spacecraft (2009–2013). The spectrum measured by the COBE mission is shown in Fig. 5.13. Note that the data points are right on the blackbody radiation curve for $T = 2.725$ K.

Key Point 5.22 *The Planck radiation law is not simply an idealistic equation with no relevance to anything real. It accurately describes the CMBR, which goes back to the Big Bang. Additionally the Planck law is used in analyses of light from our sun and other stars, electric lamps, infrared detectors, radiative cooling, and the like.*

5.4.3 Hawking radiation from black holes

What is a black hole?

Imagine shooting a projectile of mass m out of a cannon from a star with mass M. Suppose m's initial speed is v. As the mass leaves the barrel of the cannon, the total energy of the projectile-star system is,

$$E_{\text{tot}} = \text{kinetic} + \text{potential energy} = \frac{1}{2}mv^2 - \frac{mMG}{r}. \tag{5.22}$$

The second term is the mutual gravitational potential energy for the projectile-star system, where r is the separation distance and G is the universal gravitational constant in Eq. (3.55). For sufficiently large v, $E_{\text{tot}} > 0$ and the projectile will escape the star's gravitational field and never return. On the other hand, for smaller v, $E_{\text{tot}} < 0$, and the projectile will slow down, stop instantaneously, and fall back to the star. In this case,

$$v < \sqrt{\frac{2MG}{r}}. \tag{5.23}$$

Of interest here is the limiting case where the projectile has the maximum possible speed, c, the speed of light and M is just large enough that $E_{\text{tot}} = 0$:

$$c < \sqrt{\frac{2MG}{r}} \ \text{or} \ M > \frac{rc^2}{2G}. \tag{5.24}$$

Key Point 5.23 *The mass $rc^2/2G$ is the minimum mass for a spherical object of radius r to be a black hole, namely, an object from which nothing can escape, not even light. For a given mass M, the radius*

$$R_s = \frac{2MG}{c^2} \sim (1.5 \times 10^{-27}\,\text{m}\,\text{kg}^{-1})\,M. \tag{5.25}$$

is called the Schwarzschild radius, discovered in 1916 by Karl Schwarzschild. It is the maximum radius within which the object behaves as a black hole.

Consider the example of Earth's sun, with a mass $M \approx 10^{-30}\,\text{kg}$. Its Schwarzschild radius $R_s \sim 1\,\text{km}$, which can be compared with its measured radius of $7 \times 10^5\,\text{km}$. For an object with Earth's mass, about one million times smaller than our sun's, $R_s \sim 1\,\text{mm}$!

Key Point 5.24 *The Schwarzschild radius, Eq. (5.25), shows that in order for a black hole to have astronomical significance, i.e., to be large enough to have observable behaviour, its mass must be much greater than that of our sun.*

In 1783, John Michel used the Newtonian ideas here to determine that an object with the density of the Sun but 500 times its radius would be a black hole. The rigorous treatment of black holes requires the use of general relativity which, perhaps fortuitously, predicts the Schwarzschild radius Eq. (5.25). In Einstein's theory of general relativity, gravity has the effect of producing a curvature of spacetime. In the centre of a black hole, the curvature approaches infinity, a daunting prospect.

Key Point 5.25 *Black holes are the result of the death of stars, which evolve through birth, maturation, and ultimate death. The end of a star's life occurs as its nuclear fuel, mainly hydrogen, becomes exhausted. As the hydrogen diminishes and the nuclear fusion reactions slow, the inward force of gravity is no longer compensated by outward fusion pressure. With an insufficient outward force the star collapses. For stars with masses of roughly 5 − 20 or more solar masses, as the star volume diminishes, the gravity at its surface and also the speed and energy needed for a particle to escape increase. When the escape speed becomes greater than the speed of light, nothing, including light, can escape the dead star, which has become a black hole. The mathematical surface with the Schwarzschild radius is called the event horizon. Note that matter from the original star need not be at the horizon. However any matter that passes inward through that horizon cannot escape. There are also super-massive black holes, millions or billions of solar masses, near the centres of many galaxies. It is unclear how these formed, but they certainly do not arise from the death of a single star.*

In seminal articles by Stephen Hawking in 1974 and 1975, he emphasised that because of their small size, indicated by a singularity at the centre of the mathematical horizon sphere in general relativity, black hole theories must be based on both quantum mechanics and general relativity. He showed that a black hole emits a type of thermal radiation known as Hawking radiation in all directions. An observer outside a black hole with mass M_{bh}, located a large distance from the horizon, would see a blackbody radiation spectrum of energies that correspond to the black hole temperature,

$$T_{\text{bh}} = \frac{hc^3}{16\pi^2 kGM_{\text{bh}}}. \tag{5.26}$$

Remarkably, Hawking showed that black holes emit exact black-body radiation. This is difficult to believe given that nothing can escape from the event horizon. The radiation released is consistent with a black body whose temperature is given by Eq. (5.26).

How can black holes radiate? The answer is not simple. Hawking's work is highly mathematical and has no obvious direct interpretation. A simplified model, which does not conform exactly to what Hawking did, contains the following elements. The release of radiation is a multi-step process involving virtual particle–antiparticle pairs that form just outside the horizon, under the influence of the black hole's strong gravitational field. Such virtual pairs form constantly, but normally remain *virtual*, disappearing soon after they form, which is viewed as as a fluctuation. They are not observable directly.

However, one can envisage formation of a virtual pair, with the antiparticle falling into the black hole, carrying *negative* energy. Its pair particle then leaves the vicinity of the black hole with positive energy. The energy radiated is made possible by the reduced energy and mass of the black hole from the added negative energy particle. This picture makes Hawking radiation seem

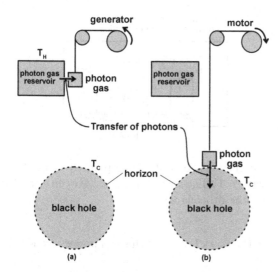

Figure 5.14: Heat engine conceived by Robert Geroch described in the text. (a) Transfer of photons to a box at temperature T_h. The filled box is pulled by a black hole. (b) Transfer of photons to the black hole at lower temperature T_c.

to be particles with mass, like electrons, coming from the horizon. However Hawking's mathematics showed the radiating energy to be primarily from photons, i.e., electromagnetic radiation.

The need for black hole entropy. Why must a black hole have entropy? In the early 1970s, John Archibald Wheeler of Princeton University imagined a wicked demon, who could commit the perfect crime against the second law of thermodynamics. It would accomplish this by dropping a package, say a container of gas with entropy S_{gas} into a black hole. The gas entropy would disappear for any observer outside the black hole, who had no knowledge about what was inside the black hole. The loss of entropy was part of Jacob Bekenstein's motivation to hypothesise that the entropy of a black hole is proportional to the surface area of its horizon, Eq. (3.54). If a black hole acquires positive mass, its horizon's radius increases according to Eq. (5.25). Thus, although the entropy of the gas disappeared, the entropy of the black hole that had absorbed it would increase by at least S_{gas}; i.e., $\Delta S_{\text{bh}} \geq S_{\text{gas}}$. This is a generalised statement of the second law of thermodynamics.

Black hole heat engine. General relativist Robert Geroch imagined a gravitationally-driven heat engine using a black hole as an energy sink for photons. The operation is illustrated in Fig. 5.14, operating as follows:

1. The system is a box that is filled with a photon gas at temperature T_h and is then lowered toward the horizon of a black hole. The box never quite reaches the horizon. During lowering, positive work W_{down} is done

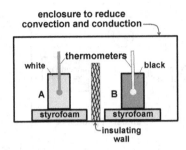

Figure 5.15: Two similar cans of water, one painted white, the other painted black from temperature T. The only difference between A and B is that one's surface is black and the other is white.

on the box by the black hole, and this can power a generator that charges a battery.

2. When the box is just above the horizon, radiation is pulled into the black hole by its strong gravitational field.

3. The less massive box is raised back up using a motor powered by the charged battery to do work $W_{up} < W_{down}$. The box is then refilled with photons from the high temperature reservoir.

This cycle, which can be repeated *ad infinitum*, charging a battery each cycle, is an imagined "black hole heat engine."

5.4.4 What you see is not always what you get

Equation (5.16) tells us that the spectral absorptivity of a substance equals its spectral emissivity at the same temperature for each wavelength. In this sense, "a good absorber is a good emitter." However a poor absorber at one wavelength could be a good absorber at a different wavelength. Wearing white clothing in summer seems like a good idea because it appears to be a relatively good reflector of radiation. But what we see is not always what we get.

Consider two similar cans filled with hot water. One of the cans is painted white and the other is painted black. The objective is to compare the rates of cooling of the cans. This and several other radiation experiments were done by Richard Bartels. The cans were insulated from one another, and the entire setup is covered to minimise conduction and convection energy exchanges, as illustrated in Fig. 5.15. Each run comparing the black and white cans was done twice. The cans' positions were interchanged to verify that there was no asymmetry. The cooling rates of the two cans are plotted in Fig. 5.16(a), showing that the black-painted and white-painted cans cooled at nearly the same rate. This result runs counter to our intuition.

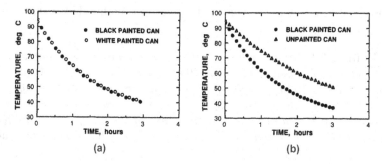

Figure 5.16: Cooling rates for a can painted black and another painted white. (Reproduced from R. A. Bartels, "Do darker objects really cool faster," Am. J. Phys. **58**, 244–248 (1990), with the permission of the American Association of Physics Teachers.)

Table 5.3: Results of experiments with roofing shingles

Object	Max. Day Temp °C	Night Temp Diff*	Comments
Black shingle	78	2	Excellent absorber of visible and near-IR. Hottest in sunlight; absorption exceeds emission. Excellent night IR emitter.
Bare shingle	73	0.5–1	Relatively poor absorber of visible and near IR. Hot in the sunlight because it's a poor IR emitter.
White shingle	46	2	Relatively poor visible absorber. Not hot in the sunlight. Good emitter in the IR.

*Temperature drop below ambient (°C)

Key Point 5.26 *Although the emissivities of the cans were very different in the visible part of the spectrum, the black and white cans cool at nearly the same rate because their emissivities at infrared wavelengths are nearly equal.*

A similar experiment compared a shiny unpainted metal can with a black painted can. The unpainted metal can cooled much more slowly than the painted one, as illustrated in Fig. 5.16(b).

Key Point 5.27 *Unpainted metals are typically poor absorbers and emitters throughout the visible and infrared regions. Here, $\epsilon_{\text{unpainted}}(8 - 10\,\mu m) < \epsilon_{\text{painted}}(8 - 10\,\mu m)$, and cooling is slower for a bare metal can than for a black painted can.*

A third experiment done by Bartels with Dena Russell[11] entails determination of which of three roof shingle types, black, white, or shiny metal (unpainted),

---

[11]D. G. Russell & R. A. Bartels, "The temperature of various surfaces exposed to solar radiation: An experiment," The Physics Teacher **27**, 179-181 (1989).

is hottest during summer days and coolest at night. The results are in Table 5.3.

Key Point 5.28 *These results are consistent with the common lore that black objects get warmer in sunlight because they are "good absorbers." As above, the reason is related also to emission qualities. The bare sheet metal, though a poor absorber in the visible, gets nearly as hot as the black shingle because it is a poor emitter in the infrared (IR). The moral of the story is: For radiation, one cannot draw conclusions based upon properties in the visible region alone.*

CHAPTER 6

Numerical Entropy

CONTENTS

6.1   NUMERICAL ENTROPY

I taught thermodynamics for the first time in 1963 and subsequently taught both statistical mechanics and thermodynamics many times. These subjects fascinated me, with the most fascination centred around entropy, its properties, and conceptual meaning. Many have written about entropy, but the topic is by no means a settled one. One aspect of entropy that has not received much attention is that it is a measurable property for any system in thermodynamic equilibrium, i.e., a system's entropy has a definite numerical value when its temperature, pressure, volume, and the like are unchanging and there is no energy or matter flowing through it. Rarely do discussions of entropy recognise that entropy is a numerical entity.

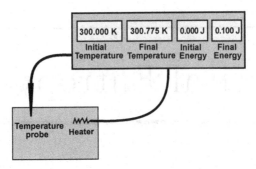

Figure 6.1: To measure the entropy change over a small temperature difference you can in principle use the equipment above.

Furthermore, the entropy of any object, element, or compound approaches zero or a positive number as the lowest possible temperature, absolute zero, is approached. As heating occurs and the temperature rises, so does the entropy. At any subsequent temperature, the corresponding entropy depends upon how much energy was added to the material, and how much of that energy is stored within it. This illustrates a close connection between energy and entropy.

Unfortunately, entropy discussions are commonly divorced from reality, ignoring both the numerical entropy and the close linkage between energy and entropy. Entropy is a measurable thermodynamic state function, i.e., it is a function of the equilibrium state of a thermodynamic system. In principle, the measurement of entropy *changes* is very simple, as illustrated in Fig. 6.1. For the system of interest, one can use a temperature probe to measure an initial temperature, 300 K, then heat the system a small amount, say with a 0.100 W heater for 1.00 s, which supplies energy $\delta Q = 0.100$ J. I use δQ for a small but finite change, while dQ is reserved for infinitesimal changes and Q is a finite change that can be large or small. If this brings the temperature to 300.775 K, then the average temperature is $T_{av} = 300.388$ K and entropy change is estimated to be $\Delta S \approx \delta Q/T_{av} = 0.1/300.388 = .000333$ J K^{-1}.

All needed data can be measured, as shown in the figure, and you can use it to calculate the total entropy change. You can heat a little, make a temperature measurement and repeat the process as many times as you require. You can calculate the ratio $\delta Q/T_{av}$ for each step and add up the contributions to the entropy for all steps. Begin with the system at temperature T_0 and end at T_M, and label the intervals $1 \ldots M$,

$$\Delta S = \delta Q_1/T_{av,1} + \delta Q_2/T_{av,2} + \cdots + \delta Q_M/T_{av,M} \rightarrow \int_{T_0}^{T_M} dQ/T. \quad (6.1)$$

The individual heating steps need to be sufficiently small that the system never goes out of equilibrium perceptibly. This is because the exact Clausius entropy change $\Delta S = \delta Q/T$ is to follow a reversible path.[1] We can bring the system back from 300.775 K to its original temperature, 300.0 K using cooling steps with the intermediate system temperatures encountered for heating reached in reverse order. Each δQ must be as small as possible, with the number of individual steps as large as possible. As described in Eq. (6.1), the experimental measurements give an *approximate* ΔS, though that result can be correct to $4 - 6$ significant digits.

Key Point 6.1 *The procedure in Eq. (6.1) determines the entropy difference ΔS, and not S itself. In order to establish a value for entropy S, a reference entropy is needed. This is provided, at least for pure crystalline solids, by the third law of thermodynamics, discussed in Sec. 6.3.*

One can in principle determine the standard entropy values of elements and compounds (inorganic and organic) this way. See for example the online *Handbook of Chemistry and Physics*, or Standard Thermodynamic Values at 25°C and pressure 1 atm, at URL: http://www.chemistry-reference.com/.

It is common for researchers to measure the *heat capacity* at constant pressure, C_p, and to calculate entropy changes with it. Determining the heat capacity experimentally is similar to determining the entropy change itself. Referring once again to Fig. 6.1, the heat capacity is defined as

$$C_p \approx \frac{\delta Q}{\Delta T}. \tag{6.2}$$

The sign $\approx$ becomes an equal sign when the interval ΔT approaches zero size. If the system has mass m, the *specific heat capacity* c_p is defined as

$$c_p = \frac{C_p}{m} \quad \text{i.e.,} \quad C_p = mc_p. \tag{6.3}$$

Note that for the *specific heat capacity* or simply, *specific heat*, the lower case letter c is commonly used while the total heat capacity of a system is referred to using the upper case C. For the temperature interval in the figure, this implies $C_p \approx (0.100 \, \text{J}/(300.775 - 300) \, \text{K} = 0.129 \, \text{J K}^{-1}$. Then if the system is a 1 g block of gold, the *specific heat capacity* is $0.129 \, \text{J K}^{-1} \text{g}^{-1}$. It is c_p that is tabulated for a variety of elements, compounds and building materials, and is generally available in public databases. A relatively short, but illustrative table of specific heat values, with units $\text{J K}^{-1}\text{g}^{-1}$ is in the Wikipedia entry for Heat Capacity.

When a sufficient set of tabulated values of c_p is available then a temperature region from T_{initial} to T_{final} of interest can be divided up into M intervals, labeled $1, 2, \ldots M$, each of which has a known value of c_p. The Clausius entropy expression can be used to calculate the entropy change for each interval and these can be summed to obtain ΔS for the full interval.

---

[1] If the path is quasistatic, it is typically reversible. However, that is not always the case. See Fig. 2.13 and its accompanying discussion. In Ch. 8, I use idealised quasistatic paths that are irreversible for heat engine and refrigerator cycles.

Key Point 6.2 *In broad brushstrokes, the mathematics of this is as follows. The heat capacity $c_p(T)$ is defined as the ratio of the energy added with a heat process, divided by the corresponding temperature change in the neighbourhood of temperature T, namely,*

$$c_p(T) \equiv \frac{1}{m}\frac{(dQ)}{(dT)} \quad \text{at temperature } T \text{ and pressure } p. \qquad (6.4)$$

Note that dQ divided by dT is the ratio of two infinitesimal numbers, and not a calculus derivative. It cannot be a derivative because Q is not a function of the system state, but rather is an energy transfer for a process.

Key Point 6.3 *Once the definition (6.4) is understood, the Clausius entropy change equation can be usefully rewritten,*

$$dS = \frac{m\, c_p(T)}{T}\, dT. \qquad (6.5)$$

For a sequence of small, but finite, temperature changes, a summation similar to that in Eq. (6.1) gives an approximation for the entropy change,

$$\Delta S = \int_{T_{\text{initial}}}^{T_{\text{final}}} \frac{c_p(T)}{T} dT. \qquad (6.6)$$

If either T_{initial} or T_{final} is chosen to be a reference temperature at which the entropy value is specified, then Eq. (6.6) gives an absolute entropy.

Specific heats over a wide temperature range, from near absolute zero on the kelvin scale to room temperature and beyond, have been measured. Figure 6.2 shows a graph of *molar* specific heat capacity *vs.* temperature for several solids. Two of them, copper (Cu) and lead (Pb), are monatomic solids, and both level off as the temperature increases, seemingly approaching the same value. Sodium chloride (NaCl), table salt) with two atoms per NaCl unit, also levels off, but to a higher value. Finally, iron chloride ($FeCl_2$), with three atoms per unit, levels off to an even higher value as temperature increases. The heat capacity for each of the four materials becomes zero in the limit of zero absolute temperature, consistent with the third law of thermodynamics (see Sec. 6.3).

Using available data for dozens of solids, a plot of the entropy per mole (mol) *vs.* the heating energy (enthalpy change) needed to heat the system from the lowest temperature to room temperature (specifically, $T_{\text{room}} = 298.15\,\text{K}$) is shown in Fig. 6.3. Molar quantities are per mole (rather than per gram) where $1\,\text{mol} = 6.023 \times 10^{23}$ molecules.

Although most of the added energy, which equals the enthalpy change, raises the temperature of the system in question, *some* of it goes to expanding the material, thereby doing work on the atmosphere. The data points represented by triangles are for monatomic solids, while those indicated by

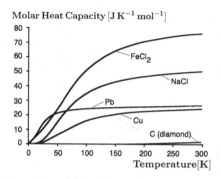

Figure 6.2: Molar heat capacity per mol *vs.* temperature. (Reprinted with permission from F. L. Lambert & H. Leff, "The Correlation of Standard Entropy with Enthalpy Supplied from 0 to 298.15 K," J. Chem. Ed. **86**, 94–98 (2009). Copyright (2009) American Chemical Society.)

circles are for diatomic and polyatomic solids. The last data point (farthest right) is for tripalmitin, whose molecular formula is $C_{51}H_{98}O_6$, i.e., one molecule contains 51 carbon atoms, 98 oxygen atoms, and 6 oxygen atoms. That's one big molecule!

Key Point 6.4 *The importance of Fig. 6.3 is that as the molecules become more complex, the energy needed to bring the system to T_{room} increases. This is because more complex molecules with more atoms can store energy in a greater number of ways. There are more ways the molecules can jiggle, and each way offers a storage mode for internal energy. Thus the energy needed to bring more complex systems to room temperature is greater and so is the system's entropy value. With more microscopic states $\Omega(E)$ accessible for the given system internal energy E, the entropy is higher, in accord with Eq. (3.3).*

6.2 ENTROPY OF ELEMENTS & COMPOUNDS

How can we determine the absolute entropy of a sample as a function of temperature at constant pressure? In 2015, Francisco Paños and Enric Pérez reviewed the validity of the Sackur–Tetrode equation (STE) for the entropy of monatomic ideal gases.[2] They examined the principal theoretical assumptions and discussed how entropy can be measured in the laboratory. Focusing on Krypton, they calculated ΔS from near $T = 0\,\mathrm{K}$ to $T = 119.81\,\mathrm{K}$, removing the contribution to the vapour from nonideal gas contributions. They obtained $S_{Kr}(119.81) = 145.19\,\mathrm{J\,K^{-1}\,mol^{-1}}$, by integrating c_p/T from 0 to T_v, the

---

[2]F. J. Paños & E. Perez, "Sackur–Tetrode equation in the lab," Eur. J. Phys 36, 055033 (2015).

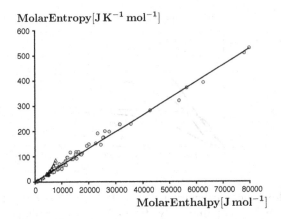

Figure 6.3: Molar entropy *vs.* molar enthalpy. (Adapted with permission from F. L. Lambert & H. Leff, "The Correlation of Standard Entropy with Enthalpy Supplied from 0 to 298.15 K," J. Chem. Ed. **86**, 94–98 (2009). Copyright (2009) American Chemical Society.)

temperature at which krypton liquid vaporises. They do not explicitly display a term $S(0)$, and later in their manuscript, reject the suggestion that $S(0) = 0$, writing that it is possible that $S(0) =$ a constant $\neq 0$. Nevertheless, comparing $S_{Kr}(119.81) = 145.19 \, \text{J K}^{-1} \, \text{mol}^{-1}$ with the Sackur-Tetrode equation value $S_{ste}(119.81) = 145.03 \text{J K}^{-1} \, \text{mol}^{-1}$ shows *rough* consistency with $S(0) = 0$. The value of the work by Paños and Perez is that they examine, step by step, how entropy values can be established experimentally.

6.3 THIRD LAW OF THERMODYNAMICS

6.3.1 Nuances of entropy

As I have emphasised, the Clausius entropy expression deals only with *changes* in entropy, not in its absolute value. Specifically according to the Clausius procedure, only entropy changes $\Delta S \equiv S_{final} - S_{initial}$ can be calculated. To obtain a value for, say, the entropy S_{final} in the final state, we would need to know the numerical value of the entropy $S_{initial}$ for the initial state. But without extending the framework of thermodynamics, there is no obvious reference state for which a specific value is specified. The establishment of an absolute entropy is accomplished, at least in part, by the *third law of thermodynamics.*

The need for this law became evident years after Clausius's work, when newly developed refrigeration technologies became available that revealed interesting behaviour of entropy at extremely low temperatures. Researchers found a previously unknown behaviour pattern: The heat capacities of materials decreased and approached zero as their temperatures were lowered toward absolute zero temperature, 0 K. The low temperature behaviour of heat capacity was predicted for some model physical systems using the then newly developed quantum theory. Happily, experiment and theory agreed well. How does entropy behave in the zero temperature limit? The answer depends on the type of material being studied. The third law of thermodynamics addresses this in the following ways, with three common statements.

6.3.2 Three statements of the third law

Key Point 6.5 *Nernst theorem (third law of thermodynamics). In 1905 Walther Nernst introduced his New Heat Theorem: The entropy difference between two different states of a substance at temperature T approaches zero as $T \to 0$. In mathematical language*

$$\Delta S = S(T, p) - S(T, p') \to 0 \; for \; T \to 0. \tag{6.7}$$

The statement also applies to different allotropes, e.g., carbon's graphite, diamond and graphite allotropes, i.e.

$$S_i - S_j \to 0 \; for \; T \to 0, \; for \; any \; two \; allotropes \; (i, j). \tag{6.8}$$

In Eq. (6.7), p and p' could be two different pressures, but the temperature T is the same at the beginning and end of the process from p' to p.

For example, suppose one begins with samples of sodium (Na) and chlorine (Cl) at ultra-low temperature. These are heated, measuring the added energy in each small temperature interval. The two samples are mixed, forming NaCl, which is then cooled to ultra-low temperature, with energy and temperature measurements along the way. Nernst did this experiment and found that the difference between the entropies of the NaCl and separated Na and Cl was very small. He believed it approached zero as the low temperature approached 0 K, as shown in Fig. 6.4a. In contrast, Fig. 6.4b shows entropy curves that have *different* zero-point entropies.

If the Nernst statement holds, one can in principle cool adiabatically and reversibly along path *ab*, i.e. at constant entropy. In the same way one can reduce the entropy isothermally along path *bc*, and again adiabatically along *dc*. Continuing this way, Fig. 6.4a shows that one cannot reach absolute zero using a *finite* number of adiabatic and isothermal processes. In contrast, if the Nernst theorem does *not* hold, as in Fig. 6.4b, one *could reach absolute zero* in the three steps: *ab*, *bc*, and *cd*. This assumes a third entropy vs. temperature curve, ending at state *d*, exists. A substantial body of experimental evidence shows that many materials do in fact satisfy the Nernst theorem, as illustrated in Fig. 6.4(a).

The third law drew a lot of interest in the early to mid 20th century, and evolved, with a number of people, including Albert Einstein and Max Planck, tweaking the statement. A special case of the Nernst theorem follows.

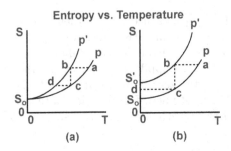

Figure 6.4: (a) Entropy *vs.* temperature along two different paths p' and p for a substance that obeys Nernst's theorem. (b) Entropy *vs.* temperature along the same paths for a substance that violates the Nernst theorem.

Key Point 6.6 *Planck statement of the third law. In the limit of zero temperature, the entropy of any pure crystalline solid tends to zero. This provides a reference level for entropy that is consistent with the entropy of quantum ideal gases and the Debye and Einstein models of solids.*

A third common statement of the third law is this.

Key Point 6.7 *No thermodynamic process can reach the temperature of absolute zero by a finite number of steps and within a finite time.*

None of these statements of the third law of thermodynamics can be proven generally. They do not follow from the first and second laws of thermodynamics. However, each is consistent with empirical evidence. The statement that $S \to 0$ as $T \to 0$ is known to fail for substances that are not *pure crystalline* materials. If a crystal consists of different isotopes, the mixing entropy from the isotopic *mixture* causes a nonzero entropy as $T \to 0$. If the isotopes are separated into two or more *pure* crystals, each satisfies the Planck form of the third law: $S \to 0$, but separation in a solid mixture is too slow to be observed, and $\lim_{T \to 0} S > 0$.

The statement that absolute zero is not an attainable temperature is interesting because if the absolute zero of temperature *could* be reached, then the cold reservoir temperature in a Carnot cycle could be taken to be absolute zero. This would enable a thermal efficiency $= 1$ (i.e., 100%) for a reversible Carnot cycle. The conversion of heat energy entirely to work energy would be accomplished, in contradiction with the Kelvin-Planck statement of the second law of thermodynamics.

Key Point 6.8 *A consequence of the third law of thermodynamics is that the heat capacity (thermal inertia) of a substance approaches zero as the temperature approaches zero. Figure 6.2 illustrates examples of five solids that have this behaviour. Equation (6.6), $S(T_{\text{final}}) - S(0) = \int_0^{T_{\text{final}}} c_p(T)/T \, dT$ will diverge unless $c_p(T) \to 0$ at least as fast as T. Typically, solids have $c_p(T) \propto T^3$ at low temperatures.*

6.3.3 Metastable states and residual entropy

The statement $S \to 0$ as $T \to 0$ can fail for several reasons. As mentioned, when a substance contains a mixture of distinct species, e.g., a substance with two or more isotopes, there can be a nonzero entropy that persists as $T \to 0$. Additionally, if the system does not achieve equilibrium with its surroundings in the low temperature limit, neither the Nernst nor Planck statements will be satisfied.[3] For example, vitreous (glassy) materials, which lack a crystal structure, can exhibit such behaviour. At ultra low temperatures, the relaxation time for equilibration with the surroundings is longer than the typical length of experiments. Such systems can become *frozen into* a metastable state as they are cooled, and cannot achieve equilibrium with the surroundings, whose temperature is decreasing. Once the *frozen-in* state exists, further cooling cannot lower their entropies, which results in a "residual entropy" for $T \to 0$.

For example, carbon monoxide which, upon freezing has its linear CO molecules aligning either head-to-head, CO-OC-CO-OC... or head-to-tail, CO-CO-CO-CO.... These can be viewed as dipole moment configurations $\to \leftarrow \to \leftarrow \; ...$ or $\to \to \to \to \; ....$ For N molecules, there are 2^N possible linear dipole configurations. In the 1930s, William Giauque and co-workers hypothesised that such crystals have randomly oriented head-to-tail orientations at higher temperatures, and that the rate of molecular reorientation becomes vanishingly small as cooling occurs between the CO's freezing point and the unattainable absolute zero. The result is that some randomisation becomes "frozen-in" below the freezing temperature. As the temperature of the environment is lowered, the state of the solid does not change, which gives rise to a positive residual entropy. About 50 years later, Kevin Nary and coworkers confirmed the Giauque hypothesis via dielectric experiments on solid CO that measured the molecular reorientation rate as a function of temperature, confirming that Giauque was correct: As cooling occurs below the freezing point, solid CO's angular orientations are out of thermodynamic equilibrium relative to its lattice vibrations and surroundings. Thermal relaxation times of a few hours at were found at $17\,\mathrm{K}$. These are believed to be from slow head-tail reorientations.

Frozen-in behaviour can be understood in part using an analogy with a single particle in each of the three possible potential energy wells in Fig. 6.5. In (a) the particle is in stable equilibrium at the lowest possible potential energy. If its x position shifts either leftward or rightward, the force on it brings it back toward $x = 0$. In (b) the particle is in a metastable equilibrium in a shallow potential well, such that if it is pushed leftward hard enough it can overcome the potential barrier V_b and fall to the lower, deeper potential

---

[3]This occurs when the relaxation time for equilibration is much larger than the observation time of the experiment.

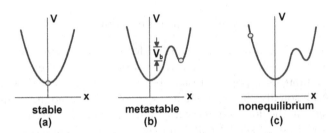

Figure 6.5: A particle in three different potential energy wells, illustrating stable mechanical equilibrium, metastable equilibrium and nonequilibrium.

energy well at $x = 0$. In (c) the particle is in a nonequilibrium state and the force on it moves it rightward toward stable or metastable equilibrium.

A positive residual entropy happens when glasses and solids like CO are *frozen into* metastable thermodynamic states. These systems cannot transfer energy to the colder environment, just as the particle in Fig. 6.5(b) could not get to a lower energy well because of insufficient energy. When a system is in a frozen-in state, its free energy is higher than that of the equilibrium state that cannot be reached.

6.3.4 Comparison of the third and other laws of thermodynamics

The third law is fundamentally different from the first and second laws, and it is worth listing some of those differences.

1. The first and second laws are universal, but the third law is not. It is primarily an ultra-low temperature law, albeit with implications for higher temperature behaviour. The Planck form of the third law holds only for single-phase crystals, not mixtures of phases, solutions, isotopes or *any* mixed species systems. Different forms of the law are needed for different systems, and some substances do not achieve equilibrium with their environment at ultra low temperatures and satisfy only the unattainability (of $T = 0$) statement of the third law.

2. The first law of thermodynamics gives rise to the fundamental thermodynamic function called internal energy, and with the first law, the second law leads to the fundamental entropy function. In contrast, the third law has no new function associated with it.

3. For these reasons the third law of thermodynamics does not have the same status as a fundamental law as the first and second laws of thermodynamics.

6.3.5 The third law and model systems

Various mathematical models for gases and solids exist and these can be studied to see if common forms of the third law of thermodynamics hold for them. Some such models and their results are listed here.

1. The classical monatomic ideal gas entropy is given by the Sackur-Tetrode equation (4.4), which contains Planck's constant h, yet does *not* satisfy the third law of thermodynamics.[4] Its derivation and validity are specifically for a gas in the classical region, i.e., devoid of quantum effects that exist at sufficiently low temperatures.

2. Quantum ideal monatomic gases of either bosons or fermions *do* satisfy the third law of thermodynamics, with the entropy approaching zero in the zero-temperature limit, as illustrated in Fig. 4.5(b).

Assuming that a system's energy is monotonically related to its temperature, the Boltzmann entropy, in the form of Eq. (3.3), implies that

$$S(E) = k \ln \Omega(E) \to 0 \text{ for } E \to E_{\min}. \tag{6.9}$$

where E_{min} is the minimum possible internal energy, namely, the ground state. Many authors have written that if a macroscopic system has a non-degenerate ground state, namely, with $\Omega(E_{\min}) = 1$, then the system satisfies the Planck form of the third law of thermodynamics. Others have observed that no experiments get down to the ground state energy and really, the entropy must depend on the energy-level spectrum above the ground state.

However, in 1981, Michael Aizenman and Elliott Lieb showed that for *some* models of lattice systems, where particles can be only at well-defined lattice sites, a non-degenerate ground state is sufficient for the third law of thermodynamics to hold. It is not necessary to consider the low lying excited states of the energy-level spectrum.[5]

6.4 ENTROPY UNITS, DIMENSIONLESS ENTROPY

Given that the units of entropy, $J K^{-1}$, come solely from the constant k in Eq. (3.3), it is natural to ask why this is the case.[6] Is there something deep hidden here? Clausius discovered thermodynamic entropy, and expressed its change in a slow, reversible heating process as the heating energy divided by temperature, $dS = dQ/T$, as described in Sec. 8.1.2. Given the standard

---

[4]It is the classical limit of a quantum ideal gas of bosons or fermions. The appearance of h enables the Sackur-Tetrode entropy to join seamlessly with the low temperature behaviour of entropy.

[5]M. Aizenman & E. H. Lieb, "The third law of thermodynamics and the degeneracy of the ground state for lattice systems," J. Math. Phys. **24**, 279–279 (1981).

[6]Historically, entropy was measured in the "entropy unit, " $1 \text{ e.u.} = 4.184 \text{ J K}^{-1} \text{ mol}^{-1}$.

units joules and kelvins for energy and temperature, the units of entropy follow.

The Boltzmann entropy, $S(E) = k \ln \Omega(E)$ (Eq. (3.3)), is interesting in that all the physics is contained in the dimensionless value of $\Omega(E)$, the number of accessible microstates with system internal energy E. This is a property of the quantum energy-level spectrum, determined by the intermolecular forces in the system, which differ from system to system. The entropy units arise from the Planck constant, which relates the joule and kelvin. Note that the energy-level spectrum is for the *total* system and not individual molecules.

In the next four subsections, I raise and answer a number of relevant questions related to the units of entropy.

6.4.1 Entropy's weird dimensions

Concavity. The dimensions $J K^{-1}$ obviously relate energy (in joules) to temperature (in kelvins). We saw in Secs. 3.3.1-3.3.2 that the slope of the concave-downward curve of entropy *vs.* internal energy is the inverse temperature, $(\partial S/\partial E)_V = 1/T$. This implies the units joules/kelvin for entropy in agreement with the Clausius entropy.

Boltzmann's constant and entropy. The kelvin temperature can be defined by a dilute gas, where molecular interactions are negligible, and

$$\frac{PV}{N} = kT \equiv \tau. \tag{6.10}$$

The proportionality constant, Boltzmann's constant k is fixed by the requirement that at the temperature of the triple point of water, $T = T_{tr} = 273.16\,\mathrm{K}$.

Carnot cycle. Recalling Fig. 2.11, we examine a reversible Carnot cycle with a monatomic ideal gas working substance. The energy input at the higher reservoir temperature T_+ is $Q_+ = NkT_+ \ln(V_d/V_c)$, and the energy input (which is negative) at the lower reservoir temperature T_- is $Q_- = NkT_- \ln(V_b/V_a) < 0$. For a monatomic ideal gas, along the two adiabatic paths, $T_-^{3/2}/V_b = T_+^{3/2}/V_c$ and $T_-^{3/2}/V_a = T_+^{3/2}/V_d$. Dividing the latter equation by the former, we obtain $V_a/V_b = V_d/V_c$, and Eq. (2.14) follows, namely, $Q_+/T_+ + Q_-/T_- = 0$ or $Q_+/T_+ = |Q_-|/T_-$. This suggests potential significance for the ratio Q/T, which has units $J K^{-1}$. Clausius's work showed that $Q_+/T_+ + Q_-/T_- = 0$ was a statement that $\Delta S = 0$ for the cycle.

6.4.2 Dimensionless entropy

In Eq. (3.3), $S(E) = k \ln \Omega(E)$, the system has internal energy E consistent with its temperature and pressure. The entropy S is proportional to the system's number of particles N so it cannot be tabulated numerically in handbooks or databases. Instead, the entropy per mole, $S_{\mathrm{mol}} = S/n$, where

$n = N/N_A$, and N_A is Avogadro's number, is tabulated. I prefer to work with the number of molecules N rather than the number of moles. Define the *dimensionless entropy*,

$$S_{\text{dim}} = \frac{S}{k} \quad \text{dimensionless, but size dependent.} \tag{6.11}$$

The dimensionless entropy satisfies the property,

$$(\partial S_{\text{dim}}/\partial E)_V = \frac{1}{kT} = \frac{1}{\tau}, \tag{6.12}$$

where, τ is tempergy, defined in Eq. 3.14. Going one step further, define σ, the dimensionless entropy *per particle*,

$$\sigma \equiv \frac{S_{\text{dim}}}{N} = \frac{S_{\text{mol}}}{R}, \quad \text{where } R = N_A k, \tag{6.13}$$

the universal gas constant $R = 8.3145 \, \text{J mol}^{-1}$. The kelvin can be viewed as an energy, i.e., $1 \, \text{K} = 1.3807310223 \, \text{J}$, and tempergy and internal energy have the same units. However they are very different entities physically. Internal energy is extensive and represents a stored energy. Tempergy is intensive, has energy units, and is *not* related to a stored system energy in general, as I discuss below.

Key Point 6.9 *If temperature had been defined historically as an energy, entropy would have been dimensionless by definition and we might never have encountered the Kelvin temperature scale or Boltzmann's constant.*

6.4.3 Numerics

Dimensionless entropy per molecule. What are typical values of σ, the dimensionless entropy per molecule? Consider graphite, with molar entropy $S_{\text{mol}} = 5.7 \, \text{J K}^{-1} \text{mol}^{-1}$ at standard temperature and pressure. Equation (6.13) implies the dimensionless entropy per particle, $\sigma = 50.68$. Similarly, one finds that diamond has $\sigma = 50.29$, while lead has $\sigma = 57.79$. At $T = 1 \, \text{K}$, solid silver has $\sigma = 8.5 \times 10^{-5}$, which is consistent with σ approaching zero for T *(and τ)* $\to 0$. It turns out that the dimensionless entropies per particle of monatomic solids are typically < 10, as illustrated in Fig. 6.6.

The Sackur-Tetrode equation, (4.4), for a monatomic gas implies the dimensionless entropy per particle,

$$\sigma = -1.16 + 1.5 \ln(M) + 2.5 \ln(T), \tag{6.14}$$

where M is the atomic mass in mass units. At the temperature 298.15 K, Eq. (6.14) gives $\sigma = 15.2$ for helium gas, and $\sigma = 21.2$ for the for the much heavier radon. Typically, as shown in Fig. 6.6 and Table 6.1, the dimensionless entropy values of most monatomic solids is in the range $15 - 25$. Figure 6.6(a), also

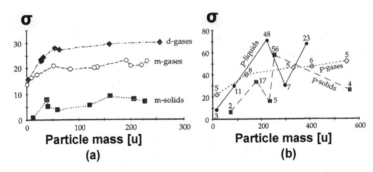

Figure 6.6: σ vs. particle mass, in mass units u. (a) Monatomic solids (solid rectangles), monatomic gases (open circles) and diatomic gases (solid diamonds). (b) Polyatomic solids (solid squares), polyatomic liquids (solid circles), polyatomic gases (open diamonds). Numbers at data points are numbers of atoms per molecule. Lines connecting data points are included to help identify materials of a given type. (Reproduced from H. S. Leff, "What if entropy were dimensionless?," Am. J. Phys. **67**, 1114–1122 (1999), with the permission of the American Association of Physics Teachers.)

shows that for monatomic solids and gases, and diatomic gases, entropy tends to increase with molecular mass. In Fig.6.6(b), water is the polyatomic liquid with 3 atoms and (approximate) mass 18 u. The much more complex and heavier polyatomic liquid $C_{16}H_{32}$ has mass 224 u and 48 atoms per molecule. The polyatomic gases each have approximately the same number of atoms per molecule, and the entropy of these compounds increases with mass.

Key Point 6.10 *Referring to Fig. 6.6 and Table 6.1, at room temperature and standard atmospheric pressure, monatomic solids typically have the lowest entropy values for given molecular mass. Monatomic gases have higher entropy values and the entropy of diatomic gases tends to be higher yet. There is a tendency for substances with larger numbers of atoms per molecule to have higher entropy values. Such molecules offer more ways to store energy, and have more accessible states. When the entropy dance, namely, the system continually jumping from one state to another, involves more states, the entropy is larger. This is temporal spreading.*

Numerical values and the Boltzmann entropy. From the entropy expression $S = k \ln \Omega$ and the definition of the dimensionless entropy per molecule σ, Eq. (6.13), $\sigma = \ln \Omega^{1/N}$, where N is the number of molecules. Inverting the latter, we find

$$\Omega = e^{N\sigma} \tag{6.15}$$

In Table 6.2, $N = 10^{19}$ for values of σ from 10^{-10} to 100.

Key Point 6.11 *Even for very small σ values, the number of states implied by the Boltzmann entropy is surprisingly large. For $10^{-10} \leq \sigma \leq 100$,*

Table 6.1: Typical ranges of dimensionless entropy per molecule σ by substance type. Some *atypical* values (not shown) lie outside these ranges.

Type	Range of σ values
Monatomic solids	$0.3 - 10$
Monatomic gases	$15 - 25$
Diatomic gases	$15 - 30$
Monatomic and polyatomic solids	$10 - 60$
Polyatomic liquids	$10 - 80$
Polyatomic gases	$20 - 60$

Table 6.2: Number of accessible states Ω *vs.* σ when $N = 10^{19}$

σ	Ω
10^{-10}	10^{10^9}
0.1	$10^{10^{18}}$
1.0	$10^{10^{19}}$
10	$10^{10^{20}}$
100	$10^{10^{21}}$

$\Omega = 10^{10^x}$ *with* $9 \leq x \leq 21$. *A nondegenerate ground state is defined as one with* $\Omega = 1$. *The results here suggest that at cryogenic temperatures, the number of accessible states is much larger than one.*[7]

6.4.4 Physical interpretation of tempergy

Ideal gas. First, it is known from kinetic theory that for a very dilute classical monatomic gas, the average energy per atom is $3\tau/2$, so τ is a rough indicator of average molecular energy.

Photon gas. In Sec. 5.3, we saw that the internal energy of a photon gas at temperature T in volume V is $E = 7.58 \times 10^{-16} VT^4$, while the number of photons is $N = 2.03 \times 10^7 VT^3$. This means that the average energy per photon can be written $E/N = 2.7\tau$, so τ is a reasonably good indicator of the average photon energy. However, these examples are special cases, and in general tempergy does *not* indicate an average energy per particle.

---

[7]When $\sigma = 10^{-10}$ and $N = 10^{19}$, $\Omega = 10^{10^9}$, the entropy $S \approx 10^{-14}$ J/K.

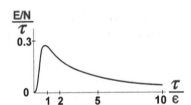

Figure 6.7: Energy per particle divided by temperature for N independent, two-state particles at temperature T, as described in the text.

Bose-Einstein gas. Another example is an ideal Bose-Einstein gas, which I discussed in Sec. 4.1.1. We had seen in Eqs. (4.3) and (4.4) that the Sackur-Tetrode entropy equation for an ideal gas holds when the volume per particle, v, is much greater than the thermal wavelength cubed, i.e, $v \gg \lambda^3$ with $\lambda^3 = (h^2/(2\pi mkT))^{3/2}$. A helpful result from statistical mechanics is that below the critical temperature, the internal energy per *excited* particle of an ideal Bose-Einstein gas is $E/N_{ex} = 0.77\tau$. If we combine this with Eq. (4.6) and (4.7), the energy per particle is[8]

$$\frac{E/N}{\tau} = \frac{1}{\tau} \times \frac{E}{N_{ex}} \times \frac{N_{ex}}{N} = 0.77 \times \left(\frac{\tau}{\tau_c}\right)^{3/2} = \frac{0.77\tau^{3/2}}{\tau_c} \qquad (6.16)$$

Thus $(E/N)/\tau$ is not constant, but rather is proportional to $\tau^{3/2}$. Tempergy itself is not a good measure of the internal energy per particle for an ideal gas of bosons below the critical temperature.

Particles with two states. Next, consider a system with N independent, two-state particles, each with possible energies 0 and $\epsilon > 0$. For high temperatures, i.e., with $\tau/\epsilon \gg 1$, the fraction of particles with energy ϵ approaches $1/2$. That is, the number of excited states with energy ϵ approaches $1/2$. Thus, $N_{ex}/N \to 0.5$, $E/N = N_{ex}\epsilon/N \to 0.5\epsilon$, and the energy per particle $(E/N)/\tau \to 0$ for $\tau \to \infty$. Figure 6.7 shows that $E/N < 0.28\tau$ for all values of τ. Thus, for this simple model of two-state independent particles, τ is not a good indicator of the average energy per particle.

Dimensionless entropy and τ. We know from Eqs. (6.11) and (6.12) that for an isothermal heat process at constant volume, $\Delta S_{dim} = Q/\tau$, so if $\Delta S_{dim} = 1$, then $Q = \tau$.

Key Point 6.12 *The energy τ is required to increase a system's dimensionless entropy by one at constant temperature. A generalisation is:[9] For a macroscopic system with thermal inertia (heat capacity) $C \gg k$, the tempergy τ is the energy needed to increase the dimensionless entropy by unity. This holds for all but the lowest achievable temperatures.*

---

[8]The critical tempergy $\tau_c = kT_c = k\,h^2(2\pi\,m)^{-1}(2.612V/N)^{-2/3}$

[9]H. S. Leff, "What if entropy were dimensionless?," Am. J. Phys. **67**, 1114–1122 (1999).

CHAPTER 7

Language & Philosophy of Thermodynamics

CONTENTS

7.1   THE LANGUAGE OF WORK & HEAT

7.1.1   Thing vs. process

We saw earlier that in the first law of thermodynamics, Eq. (1.9), $\Delta E = Q + W$, where Q and W represent energies transferred by heat and work *processes* respectively. What words should we use to describe these quantities? We speak of the work done by an object on another object, which is unobjectionable.

Key Point 7.1 *W is the energy transferred to a system by a work process. Once the process is over, the transferred energy becomes stored internal, kinetic, and/or potential energy. The language of work is not problematic. However, the language of heat has endured long-standing and continuing language difficulties.*

As observed in Sec. 1.2, in the now defunct caloric theory, heat was envisaged as a fluid that could neither be created nor destroyed. Unfortunately, remnants of the view of heat as a *thing* lives on in the English language. This causes a good deal of confusion. In 1957, American ecologist and philosopher, Garrett Hardin, best known for his warnings about world overpopulation, observed difficulties with the language of heat.[1]

Hardin pointed out that from Greek antiquity to the work of Robert Boyle, confusion reigned because 'heat' was assigned to the wrong category, namely, that of a *substance*. As such, heat should have weight, and British physician, George Fordyce, asserted that heat did have weight. Count Rumford's experiments initially supported Fordyce's claim, but subsequently found and corrected experimental errors. That led him to conclude that heat is *not* a substance.[2] In summary: Heat is related to verbs rather than nouns.

However, still today, heat is typically treated as a substance, with language like "the heat transferred" and "I feel the heat." For this reason, John Jewett referred to heat as "the most misused word in modern language."[3] Robert Romer has argued strongly that *heat is not a noun*.[4] Yet we continue to see vestiges of caloric theory in books and scientific articles, as well as to hear it in common talk. Some familiar examples are, *the ocean holds a lot of heat; heat rises, a well-insulated house will lose less heat in winter, and keep the door closed so the heat does not go out.* Romer stated that caloric theory thinking continues to contaminate the minds of beginning students, and also those of too many professors.

Thermodynamicist Mark Zemansky put it this way,[5] "Heat and work are methods of energy transfer, and when all flow is over, the words heat and work have no longer any usefulness or meaning." He pointed out that the result of such a transfer is a changed internal energy of the system. After the heat and work process, only the system's internal energy has meaning. There is no way to know how much of the internal energy came from heat and how much came from work. Yet, Romer accepts terms like "transfer of heat" and "flow of heat," because the wording makes it clear that heat is a shorthand for energy in transit.

Key Point 7.2 *I suggest using language like "object A heats object B, or vice versa," where possible. Terms such as "heating energy" or "heat energy," and any noun that clearly implies that it is energy that is transferred is preferable to using heat as a noun.*

---

[1] G. Hardin, "The threat of clarity," Am. J. Psych. **114**, 392–396 (1957).

[2] See Fig. 1.6 and the discussion of Rumford's cannon ball boring experiment.

[3] J. W. Jewett, "Energy and the Confused Student III: Language," The Physics Teacher **46**, 149–153 (2008).

[4] R. H. Romer, "Heat is not a noun," Am. J. Phys. **69**, 107–109 (2001).

[5] M. W. Zemansky, "The use and misuse of the word 'heat'," The Physics Teacher **8**, 295–300 (1970).

Key Point 7.3 *For any heat process where Q is the energy transferred, once the process is complete, the transferred energy is part of the stored internal energy. All that remains is internal energy, without any trace of the history leading to it. The same comments hold for work. To reiterate: work and heat represent processes, and the thing that is transferred is energy.*

A notable misuse of the word *heat* is the term *heat capacity*, $C_x = mc_x$. Here, c_x is the system's specific heat capacity, defined in Eq. (6.3) as the heat capacity C_x per unit mass. The *heat capacity* C_x determines a system's temperature change ΔT with the gain or loss of energy Q in a process at constant x, e.g., constant volume $(x = V)$ or constant pressure $(x = P)$, namely,

$$\Delta T = \frac{Q}{C_x} = \frac{Q}{mc_x}. \tag{7.1}$$

For a given amount of added energy Q, a substance with higher C_x has a lower temperature change. In a sense, C_x acts as *thermal inertia*.

Key Point 7.4 *As observed earlier, an appropriate name for C_x is thermal inertia. Unfortunately, the term heat capacity is so commonly used and ingrained within the physics and engineering culture, it is not likely for thermal inertia to be accepted widely by teachers, textbook authors, and researchers.*

The last example in this section is the nebulous term *thermal energy*. Some people use that term to describe Q in the first law of thermodynamics, while others use the same term to describe internal energy.

Key Point 7.5 *The term thermal energy is acceptable as a descriptor for Q. Thermal suggests temperature and heat. Some authors define thermal energy to be that portion of a macroscopic system's internal energy that can change in a heat process, which is acceptable if the definition is made clear.*

7.1.2   Analogy: Bank transactions & W, Q, E

Work and heat, being processes, differ from internal energy, which is stored. A good analogy associates bank checks, deposits, and withdrawals with work, and ATM (automatic teller machine) transfers with heat.[6]

1. A bank check, deposit, or withdrawal entails a process where money is transferred from or to a bank account. The *thing* that is transferred is money.

2. An ATM transfer entails a process that transfers money from or to the latter bank account. Again the *thing* transferred is money.

---

[6]G. D. Peckham & I. J. McNaught, "Heat and work are not 'forms of energy'," *Journal of Chemical Education* **70**, 103–104 (1993).

3. The balance in the bank account is the analog to the internal energy.

The analogy to the first law of thermodynamics for say a one month time period is this:

(Net checks, deposits, withdrawals) + (Net ATM transactions)
$$= \text{(Change in the value of the bank balance)}. \quad (7.2)$$

Each term is measured in money units, the analogy to joules. Just as Q and W disappear after a thermodynamic process, with only the resulting internal energy remaining, the same is true for monthly checks, deposits, withdrawals and ATM transactions. At month's end, only the account balance remains.

Key Point 7.6 *Checks, deposits, withdrawals and ATM transactions are processes and are not forms of money, just as heat and work are processes rather than forms of energy. A bank account's balance represents stored money, just as the internal energy of a system is a form of stored energy.*

7.1.3   More about defining heating energy

The concept of work is familiar from classical mechanics, where it is defined as the force applied to an object multiplied by the resulting displacement along the direction of that force. For a gas with pressure P, undergoing a tiny volume change dV, the work done *on* the gas *by* the piston is $dW = -PdV$. Note that $dW > 0$ for a compression because $dV < 0$, and $dW \leq 0$ for an expansion with $dV > 0$. Emanating from mechanics, work differs from heat, which is a purely thermodynamical entity.

Key Point 7.7 *A heat process can occur only with systems that store internal energy. There is no internal energy in classical mechanics of point particles and rigid bodies, so there can be no heat process.*

Heat is often described, albeit not defined, as "the energy transferred across a system's boundary because of temperature difference." This is not a definition of *heat*, but rather a description, because it does not prescribe an operational way to measure Q quantitatively.

Key Point 7.8 *An acceptable description of heat is this: A heat process can be induced at a system's boundary by a temperature difference (finite or infinitesimal), resulting in the transfer of energy Q, called heating energy.*

Key Point 7.9 *Notably, in thermodynamics, one also studies reversible isothermal volume changes, where there is zero temperature difference between the system and reservoir. A truly universal definition of Q must be consistent with reversible isothermal processes.*

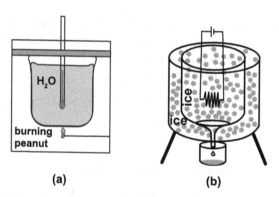

Figure 7.1: (a) Simple calorimeter to measure the heating energy output of a burning peanut. (b) Ice calorimeter to measure the heating energy output of a powered electric resistor. Melted ice drips into a cup.

One can define *heat* in either of two primary ways. One uses the first law of thermodynamics, as explained in Sec. 2.2, where heating energy Q was defined in terms of work energy W. That definition has the advantage that it applies to reversible isothermal processes. The other way defines heat calorimetrically, when energy is transferred from a higher to lower temperature region to heat an object. For example, a peanut might be burned under a beaker containing a mass m of water, as depicted in Fig. 7.1(a). Measuring the water's resulting temperature rise ΔT, enables calculation of the energy Q released during the combustion, as given by (7.1), $Q = mc_p\Delta T$. Both heat and work are involved because as the peanut burns some of the energy released goes into work done expanding the heated constant-pressure air surrounding the peanut. In the first law of thermodynamics, $\Delta E = Q + W$, the total work *on* the peanut is $W = -P_{\text{atm}}\Delta V$. The enthalpy change is $\Delta H \equiv \Delta(E + P_{\text{atm}}V) = Q$.

An alternative is to use an ice calorimeter, which entails a measurement of the mass of ice maintained at $0\,^\circ\text{C}$ that melts when it receives heating energy Q, as depicted in Fig. 7.1(b). An electric resistor with resistance R powered by a battery runs for time t. The "Joule heat" is $Q = IVt$, where $I = $ current [A] and $V = $ voltage [V]. This energy melts a mass Δm of ice, where $Q = L_{\text{melt}}\Delta m$, and L_{melt} is the *latent heat* of ice, measured in J kg^{-1}. The melted ice drips into a cup, whose mass is monitored to measure Δm. Note that a heat process between two systems occurs only at the boundary separating those systems. In the case of the ice calorimeter, this boundary is the surface of the resistor.

It is not always obvious that an energy transfer process is unambiguously heat or work. Suppose laser light impinges on a system and is absorbed, generating an increase in the system's temperature. The laser light is monochromatic and for a cleverly chosen frequency can be used to excite electrons selectively from one energy state to a higher one. This suggests a work process. On the other hand, laser light is radiation, one of the forms of what is commonly referred to as *heat transfer*.[7] Laser light is very different from thermal radiation in a photon gas, which has a mix of photon energies that depends on the temperature.

Key Point 7.10 *The gist of this discussion is that classifying laser light as work or heat is ambiguous.*[8]

7.1.4 Isothermal, reversible volume changes

The common thermodynamics example of reversible isothermal heating of a gas at temperature T requires the gas to maintain contact with a constant temperature reservoir at the same temperature as the gas. As pointed out, this is inconsistent with the often used view that "heat is an energy transfer from higher to lower temperature." Here are two ways to deal with this dichotomy using *purist* and *utilitarian* views. In the *purist* view, the word *reversible* connotes an infinitesimally slow volume change that occurs without friction or dissipation. The forward and reversed processes entail work and heating processes during which work W is done *on* the gas and energy Q is absorbed *by* the gas. For an isothermal compression, the gas transfers energy to the reservoir and $Q < 0$, with work $W > 0$ done *on* the gas. The purist description is idealistic in that it entails mental constructs not achievable in a laboratory. Yet it leads to a useful theoretical framework in which all quantities are well defined. It can provide reasonably good approximations to the behaviour of real physical systems. Using this view, nonzero heating and cooling processes with $Q \neq 0$ can occur while both the gas and reservoir maintain the same constant temperature. The *utilitarian* view is that reversible processes are an idealisation that is unachievable in practice, and a heating process necessarily requires a temperature difference.

Outside the purist and utilitarian views, John Norton[9] has observed that for any process, there must be some deviation from equilibrium, i.e., states he calls *near-equilibrium states* must be involved even for reversible processes. Norton writes that the "...driving forces are so delicately balanced around equilibrium that only a very slight disturbance to them can lead the process to reverse direction." He points out that these envisaged processes proceed ultra-slowly, following a succession of near-equilibrium states. It is not possible for them be be equilibrium states or there could be no process.

Near-equilibrium states are not part of the purist view because they deviate from equilibrium states, and they are not part of the utilitarian view because

---

[7]I avoid this terminology, which suggests that heat is a thing rather than a process.

[8]This is discussed in G. Laufer, "Work and heat in the light of (thermal and laser) light," Am. J. Phys. **51**, 42–43 (1983).

[9]Norton, J. D. "Thermodynamically reversible processes in statistical physics," Am. J. Phys. **85** 135–45.

they are not achieved in a laboratory. One cannot even assign thermodynamic variables like temperature and pressure to them. Nevertheless, the concept of near-equilibrium states provides a useful mental picture that can aid our understanding of a reversible process.

The utilitarian view is that because a reversible process must proceed infinitely slowly by definition but the known age of the Universe is finite, no reversible process can have ever occurred! All real processes occur in finite time and the utilitarian view is that reversible processes can be approximated satisfactorily by sufficiently slow, but finite, processes in laboratory experiments.

In the purist view, there is no *wiggle room* for deviations from pure idealisations, i.e., purists follow pure thermodynamics in full detail. The utilitarian view is not bound by the precise definitions, language and strict logic of pure thermodynamics. Instead, it entails mental models wherein quantities such as temperature are imagined to have meaning even for non-equilibrium states. It is assumed that such an approach will approximate what happens in a laboratory. Utilitarians use thermodynamic equations developed using purist thermodynamics, but with utilitarian approximations and interpretations in mind. Table 7.1 provides a summary of the purist and utilitarian views of reversible isothermal volume changes.

Key Point 7.11 *From a purist's viewpoint, a reversible, isothermal volume change occurs with zero-temperature difference between a reservoir and system. This contradicts the common statement that a heat process occurs through a temperature difference. The definition of heat in terms of work in Sec. 2.2 can be used by the purist. A utilitarian's view of an isothermal volume change is that, despite the word isothermal, such a process must involve a small temperature change between reservoir and system.*

7.1.5 Work and heat for friction processes

In 1984, Bruce Sherwood and W. H. Bernard published an expository article that illuminated the roles of work and heat during sliding friction.[10] They considered two blocks in mechanical contact, with the bottom block fixed in place and the top block sliding over it at constant velocity. In Fig. 7.2(a),the force needed to maintain constant velocity is f, which implies that the friction force on the upper block is -f. The external work done by the rightward force when the block moves through displacement d is $W_{ext} = fd$. We know from experience that both blocks get warmer during the process, so this is a thermodynamics problem, rather than one in pure mechanics. Thus, the first law of thermodynamics applies.

---

[10]B. A. Sherwood & W. H. Bernard, "Work and heat transfer in the presence of sliding friction," Am. J. Phys. **52**, 1001–1007 (1984).

Table 7.1: Purist *vs.* Utilitarian Views for a Reversible Isothermal Process. (Adapted from H. S. Leff & C. E. Mungan, "Isothermal heating: purist and utilitarian views," Eur. J. Phys. **39**, 045103 (2018). © European Physical Society. Reproduced by permission of IOP Publishing. All rights reserved.)

Item	Purist View	Utilitarian View
Equilibrium states	Allowed	Not specified
Near-equilibrium states	Only equilibrium states allowed	Allowed
Processes	Quasistatic; *only* equilibrium states	Not restricted
Process speed	Infinitesimal; a process takes forever	Slow and finite
$\|T_{\text{reservoir}} - T_{\text{system}}\|$	Zero, to satisfy the strict isothermal condition	finite (infinitesimal is unmeasurable)
Strict logic and precise language	Yes	Not necessary
Interpretations	Consistent with logic	Not restricted

For the two block system, the external work is simply $W_{ext} = fd$, the work done by the rightward force on the upper block, because the leftward force of the wall on the lower block acts through zero displacement. If we apply the first law of thermodynamics to the two-block system, we get

$$\Delta(E_+ + E_-) = (Q_+ + Q_-) + fd, \qquad (7.3)$$

where $+$ and $-$ denote *upper* and *lower*, respectively. Assuming negligible energy exchange with the surroundings, if one block heats, the other one cools, so $Q_+ = -Q_-$. Additionally, symmetry dictates that $Q_+ = Q_-$. These two results imply $Q_+ = Q_- = 0$. Thus, Eq. (7.3) reduces to $\Delta(E_- + E_+) = fd$. There is no reason that ΔE_+ should differ from ΔE_-, so

$$\Delta E_+ = \Delta E_- = fd/2. \qquad (7.4)$$

Apply the first law to the upper block, $\Delta E_+ = fd + W_{\text{fric}}^+$, and use Eq. (7.4), to obtain

$$W_{\text{fric}}^+ = -fd/2. \qquad (7.5)$$

We can understand how this comes about using the model in Fig. 7.3. A view of the surfaces with a microscope shows them to be jagged, and when a tooth of one surface gets stuck on a tooth of the other surface, there is a friction force. In the model, both teeth bend until subsequent motion breaks the contact between them. Thus, the force does not act through the full displacement d, but rather, through an average displacement $d_{\text{eff}} < d$. This continues, for

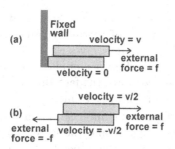

Figure 7.2: (a) An upper block slides with constant velocity v over an identical lower block affixed to a wall. A rightward external force f and a friction force $-f$ act on the upper block. (b) The system viewed in the centre of mass frame. It is assumed that there is a sufficient vertical normal (perpendicular to the surface) force to produce horizontal friction.

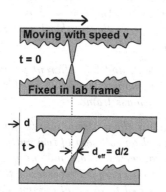

Figure 7.3: Model of friction for a block being dragged over another. Shown is one tooth from each block dragging one another. Many pairs of teeth drag one another simultaneously in macroscopic blocks.

a large number of teeth, giving rise to an *average* friction force f, with the friction work on the upper block being

$$W_{\text{fric}}^+ = -f d_{\text{eff}} = -f d/2. \qquad (7.6)$$

Applying the first law to the fixed lower block gives $\Delta E_- = W_{\text{fric}}^- = f d/2$, which is consistent with

$$W_{\text{fric}}^+ = -W_{\text{fric}}^-, \qquad (7.7)$$

because when the upper block does positive work on the lower block, the lower block does negative work on the upper block. Microscopically, at a point of contact of two teeth, the blocks exert equal and opposite forces on one another through the *same* displacement.

It is instructive to examine the two-block system in the centre of mass frame, which moves rightward with velocity $v/2$ relative to the lab. In this frame the upper block moves rightward a distance $d/2$ while the lower block moves leftward the same distance, so the final relative displacement is d, as in the lab frame. The first law of thermodynamics, applied in the centre of mass frame to the two-block system, gives Eq. (7.3), $\Delta E_+ = \Delta E_-$ once again, and $Q_+ = Q_- = 0$.. Then the first law applied to the upper block is $\Delta E_\pm = W'_{\text{fric}} + fd/2$. The prime $'$ connotes the fact that work is path-dependent and the path differs in the centre of mass frame.

Internal energy change, being *internal*, cannot depend on the frame of reference and thus $\Delta E'_\pm = fd/2$, and the latter equation implies $W'_{\text{fric}} = 0$. This is so because as the tip of the upper tooth moves leftward with $d_{\text{eff}} = -d/2$ in Eq. (7.6), the block itself moves rightward by $d/2$. This dictates a centre of mass displacement of $d'_{\text{eff}} = 0$, which implies $W'_{\text{fric}} = fd'_{\text{eff}} = 0$.

Key Point 7.12 *This example teaches us that*

1. *The warming of each block comes from the work by friction forces, and not from a heat process. Another view is that the work done by the external rightward force on the two-block system is dissipated and stored as internal energy.*

2. *Work is path dependent, and the different paths in the lab and centre of mass frames result in non-zero friction work in the lab frame and zero work in the centre of mass frame.*

3. *Zero work by friction in the centre of mass reference frame is reminiscent of the symmetric head-on collision of two putty pieces in Fig. 2.3. There, kinetic energy was converted to internal energy, while here, external work is converted to internal energy. Processes with friction are irreversible, and as work is dissipated to internal energy, entropy is generated.*

7.2 LINKS BETWEEN THERMODYNAMICS LAWS

7.2.1 Thermal equilibrium & zeroth law

For any thermodynamic system, there are three relevant types of equilibrium. The first is mechanical equilibrium, with the sum of all forces acting on the system adding to zero. For a macroscopic system in the absence of external forces like gravity, mechanical equilibrium implies a uniform pressure throughout the volume. The second type is chemical equilibrium, which is important when different chemical species react with one another. The third, which is of paramount interest here, is thermal equilibrium.

If a system is in thermal equilibrium, all parts of it are at the same temperature. If the system can exchange energy with the environment (assumed to be the surrounding atmosphere), the system's equilibrium temperature and pressure are those of the atmosphere. If thermal equilibrium does not exist, the state of the system will change until a state of thermal equilibrium is reached.

In a 1787 set of lectures, Joseph Black stated his principle of thermal equilibrium, namely "...all bodies communicating freely with each other, and exposed to no inequality of external action, acquire the same temperature, as indicated by a thermometer. All acquire the temperature of the surrounding medium." Nearly 100 years later, in his 1872 book *Theory of Heat*, James Clerk Maxwell recognised the important concept of thermal equilibrium, writing, "Bodies whose temperatures are equal to that of the same body have themselves equal temperatures." These observations led to the zeroth law of thermodynamics, named by Ralph Fowler and Edward Guggenheim who suggested it as a discussed in Sec. 2.5.1, needed *law* in their 1939 book on statistical mechanics.[11]

Key Point 7.13 *The zeroth law addresses the existence of thermodynamic equilibrium states. However, it is the Clausius statement that assures the establishment of such states, which are essential in thermodynamics. Thus, the Clausius statement lies at the very heart of thermodynamics.*

7.2.2 Heating & direction of energy flow

Thermal equilibration between two objects with initially different temperatures requires that the hotter object heat the cooler one until they are both at the same temperature. This happens only if heating energy naturally flows from hotter to colder objects. As the hotter object loses energy, it cools, and as the cooler object absorbs an equal energy, it gets warmer. This proceeds until there is temperature equality. Such behaviour was assumed for the now obsolete caloric, as described in Sec. 1.2, and is the crux of the Clausius statement of the second law of thermodynamics, in Sec. 2.4.

It is interesting to ask what would happen if energy flowed from cold to hot.[12] In the remainder of this section, I'll imagine a cold to hot world called the C2H world. I assume that absorption of energy increases temperature, as it does in our hot-to-cold world. This imagined world behaves strangely.

1. If hot and cold objects are put into thermal contact, the hot one becomes hotter and the cold one gets colder! There is no temperature equilibration.

2. Because two systems do not equilibrate, it is not possible to reach equilibrium states.

3. Being unable to reach an equilibrium state means that it is not possible to measure temperature.

---

[11]A subtle difference between Black's statement and the zeroth law is that the latter is based on objects A, B, and C being in equilibrium with one another pairwise *a priori*. In contrast, Black alluded to the bodies "acquiring" the same temperature, which is more akin to the second law of thermodynamics and the process of equilibration.

[12]H. S. Leff & R. Kaufman, "What if energy flowed from cold to hot? Counterfactual thought experiments," Phys. Teach., in press.

4. By definition, the Clausius statement of the second law is not obeyed.

5. The first law cannot be satisfied because it requires connecting initial and final equilibrium states for which internal energy E exists.

6. The third law of thermodynamics (Sec. 6.1) is not satisfied because it requires the concept of temperature.

Key Point 7.14 *In the imagined C2H world, the science of thermodynamics as we know it would not exist. This shows the importance of the Clausius statement of the second law, which assures that systems will equilibrate with one another and with their environment. The Kelvin-Planck statement of the second law requires a cyclic process with at least an initial (and final) equilibrium state of the system's working substance, which is assured by the Clausius statement. Indeed, the Clausius statement, along with energy conservation, are indispensable elements of thermodynamics.*

7.2.3 Linkage between first and second laws

The first law of thermodynamics contains several assumptions:

1. Energy is conserved, meaning that any internal energy change ΔE must come from the sum of the heating energy Q and work energy W supplied.

2. Equilibrium states with internal energies $E_{initial}$ and E_{final} exist.

3. Q and W are well defined and measurable for the process leading from the *initial* to *final* state.

Key Point 7.15 *We conclude that:*

1. *The first law requires initial and final equilibrium states, the postulate of the state function internal energy, conservation of energy, and the definitions of heat and work. The zeroth law solidifies the concept of equilibrium and the Clausius form of the second law assures that equilibrium can be established.*

2. *From Sec. 2.4, assuming the first law of thermodynamics holds, the Clausius statement implies the Kelvin-Planck statement of the second law which, in turn, implies the existence of the entropy function S (this is shown in Sec. 8.1.2) and the principle of entropy increase.*

3. *While this book was in production, Richard Kaufman and I discovered a logical argument showing that the existence of initial and final equilibrium states in the first law implies the Clausius statement of the second law, i.e., the first law implies the second law of thermodynamics. This is a heretofore overlooked linkage between the first and second laws.*

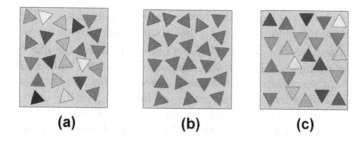

(a)   **(b)**   **(c)**

Figure 7.4: (a) 22 triangles that have differing, seemingly unpredictable angular orientations and shadings. (b) 22 triangles that all have the same shading but with differing angular orientations. (c) 22 triangles that are randomised by shading, each with one vertex pointing either up or down. Among (a)–(c), which is more *ordered*? Which is more *disordered*?

7.3   THE LANGUAGE OF ENTROPY

7.3.1   More about metaphors

Disorder Because the *disorder metaphor* goes back to Helmholtz and Boltzmann, it is given more credence than it deserves. Most commonly, we tend to associate disorder with the irregularity of *spatial* arrangements, which we humans can often recognise. But spatial order can be a tricky thing to discern. For example, see Fig. 7.4, parts (a)–(c). Do you find one of the figures more *ordered* than the other? I some ways, I think part (c) looks most ordered and (a) looks most disordered. However, without a clear definition of disorder, it is impossible to say with authority whether any of these types of randomness gives more or less disorder. Different people could reasonably come to different conclusions, which points to a major problem with the disorder concept: It is *subjective*. Most important, thermodynamics entails energy, and simple images such as Figs. 7.4(a–c) do not address the role of energy at all. Different shades of grey could designate different energy intervals, but a workable definition of *disorder* would still be needed.

An illustrative example is the heating of ice and the resulting liquid water. Consider the melting of one kilogram (kg) of ice to water, which is sometimes viewed as being more disordered than the crystalline ice, for which molecules are arranged regularly. That regularity exists to a lesser degree in the liquid, which takes on the shape of its container or simply spreads out into a thin surface layer. When the liquid water is heated and vaporised, the water vapour is seen as more disordered than the liquid, i.e., the vapour becomes less ordered spatially. Each molecule of vapour spreads throughout a larger volume, unlike the liquid, whose volume does not change significantly as it is heated. Because the numerical entropy per kg of water vapour exceeds that of liquid water, which exceeds that of ice, the disorder metaphor seems to work. This suggests that the disorder metaphor is valid, but some strong caveats are needed.

The situation is far too subtle for a one-word descriptor like disorder to suffice. For example, if 1 kg of water vapour is expanded to twice its volume at a constant temperature, its entropy increases. But has the vapour really become more disordered? Its molecules are farther apart but what definition of disorder would lead to the conclusion that disorder increases with increasing volume? Further, 2 kg of ice has twice the numerical entropy of 1 kg of ice at the same temperature. Given that its molecules are just as regularly arranged as for the 1 kg sample, in what sense is it more disordered? What definition of disorder supports this *and* the example above of the expansion of a gas?

If expansion is believed to generate disorder, how do we explain the fact that as water is heated from 0°C to 4°C, its volume *decreases*, while its numerical entropy *increases*? If the process is done in reverse, cooling the water, there is expansion, with a corresponding entropy decrease! This is counter to the notion that expansion leads to higher entropy. The unfortunate disorder metaphor is typically applied inconsistently, ignoring existing counterexamples. The bottom line is that disorder is a very poor metaphor in general for understanding entropy.

In 1944, Karl Darrow wrote an article in the American Journal of Physics entitled, "The concept of entropy." Among other things, he examined examples for which entropy could be associated with disorder, and observed that disorder is not always useful, as indicated by a free expansion of a gas in a constant-temperature environment. Entropy increases, but one cannot unambiguously associate more disorder with its state. He wrote, "Would anyone spontaneously say that a gas in a two-litre bottle is more disorderly than the same gas at the same temperature in a one-litre bottle?" In a discussion of the free expansion in their book, *Atmospheric Thermodynamics*, Craig Bohren and Bruce Aldrich wrote that increased volume only means increased disorder if one *defines* disorder that way. However, doing so is tantamount to defining disorder via entropy, rather than the other way.

The term disorder has been criticised by a variety of others writers, including Herbert Dingle, who described the view that entropy is disorder as "a most inessential visualisation which has probably done much more harm than good." P. G. Wright examined various examples of real phenomena and concluded, "...there is no clear correlation between propositions about entropy and anything intuitively obvious about disorder." My good friend, the late organic chemist Frank Lambert critically assessed usage of the disorder metaphor in chemistry textbooks, and successfully convinced authors of dozens of general and physical chemistry textbooks to remove statements relating entropy with disorder from new editions of their books. Lambert, who promoted the view that entropy is a measure of "energy dispersal" rather than disorder, died in December 2018 at age 100. Hopefully, others will carry on his crusade. Recently, Daniel Styer observed that entropy does not always mean *disorder, uniformity,* or *smoothness.* Using specific examples, he showed why, and gave some history of the lore that entropy is *disorder.*[13]

Key Point 7.16 *Although the term disorder is appropriate to interpret the entropy of a paramagnetic spin system, it can hinder understanding for gases, liquids and solids. No definition of disorder that works consistently as a general entropy metaphor is known, and perhaps none is possible. The term disorder bypasses and hides the intimate connection between energy and entropy.*

Upon the death of J. Willard Gibbs, a list of subjects intended for supplementary chapters to Gibbs' *Equilibrium of Heterogeneous Substances* was discovered. One of the topics was "On entropy as mixed-up-ness." Mixed-up-ness sounds a lot like disorder and Gibbs, who had considerable mathematical skills, might have been able to solidify its meaning. But he did not live to bring this to fruition. It is best to avoid the term *disorder.*

Key Point 7.17 *Recognising the language difficulties that pervade thermodynamics, Mark Zemansky wrote the following poem for physics teachers:*[14]

> *Teaching thermal physics is as easy as a song*
> *You think you make it simpler, when you make it slightly wrong*

Some people have told me that disorder and uncertainty mean the same thing to them. They see no difference in the two concepts. A common definition of disorder is *a lack of order or regular arrangement,* which suggests a physical property of the system. In contrast, uncertainty, defined as: *doubt,* is a property of the information known about a system. Thus, the terms are very different.[15]

Energy spreading The metaphor of spreading, discussed earlier in Sec. 2.4, is based explicitly upon space, time, and energy. Space is intimately involved because energy tends to spread spatially to the extent possible. The view here is that when hot and cold objects are put in thermal contact, energy spreads *equitably* between them. When an object is in thermal equilibrium at constant temperature, its quantum state varies temporally, i.e., the system's occupied state "dances" (spreads) over accessible quantum energy states.

Key Point 7.18 *The concept of spatial spreading of energy can be helpful as an interpretive tool for thermodynamic processes. Similarly, temporal spreading over microstates can be useful for interpreting entropy in a specified thermodynamic state. In both cases, there is a direct linkage between energy and entropy. As energy is redistributed spatially, total entropy increases. When*

---

[13]D. Styer, "Entropy as Disorder: History of a Misconception," The Physics Teacher **57**, 454–458 (2019).

[14]M. W. Zemansky, "The use and misuse of the word 'heat'," Phys. Teach. **8**, 295–300 (1970).

[15]A personal view of disorder and uncertainty is that in the morning, my hair, the little that still remains, is very disordered, and my level of *uncertainty* about that is zero.

there is temporal spreading over more microstates, entropy is higher than when that temporal spreading is over fewer accessible states.

In 1862, Rudolf Clausius proposed a function, which he called *disgregation*, which I discussed in Sec. 2.6.4. He envisaged molecules in constant motion, and viewed that motion as tending to "loosen the connection between the molecules, and so to increase their mean distances from one another." That mental picture was similar to what here is called spatial energy spreading. Clausius promoted the concept of disgregation, prior to his 1865 article introducing entropy.

Notably, Kenneth Denbigh used the spreading idea to describe irreversibility. He cited three behaviours that display divergence toward the future: a branching towards a greater number of distinct kinds of entities; a divergence of particle trajectories from one another; and a spreading over an increased number of states of the same entities. These statements entail space and time and although they do not refer specifically to energy, they can be interpreted in terms of it. His point was that every macroscopic natural process is essentially a process of mixing, namely, a tendency of constituent particles to intermingle, e.g., in a diffusion process. Denbigh observed that the spatial mixing of particles is in fact a mixing or sharing of their total energy; i.e., what I call *spatial energy spreading*.

Key Point 7.19 *When particles move and mix, they carry with them their translational, rotational, and vibrational stored energies. They spread those energies and exchange energy with other particles.*

Multiplicity Multiplicity, which was promoted by Ralph Baierlein,[16] is related to missing information, which was discussed in Sec. 2.6.5. It is a descriptor for the total number of states Ω that are available to a system with a given system energy (or a range of energies, $\Delta E \ll E$). The Boltzmann entropy is the logarithm of the *multiplicity*.

Optiony. This metaphor, due to R. M. Swanson,[17] has the same meaning as *multiplicity*, i.e., the number of options among the possible microstates equals the multiplicity.

Freedom. This is yet another term that focuses attention on the number of microstates of a specific macrostate. When the number of microstates is larger, a system has more freedom of choice.[18]

Along with the Boltzmann entropy expression, the latter metaphors can help us understand why for example 2 kg of copper has twice the entropy of 1 kg of copper. Suppose at a given temperature (which implies an average

---

[16]R. Baierlein, "Entropy and the second law: A pedagogical alternative," Am. J. Phys. **62**, 15–25 (1994).

[17]R. M. Swanson, "Entropy measures amount of choice," J. Chem. Ed. **67**, 206–208 (1990).

[18]This term seems to be due to Garey L. Bertrand, who evidently never published his work, which is cited in J. B. Brissaud, "The meanings of entropy," Entropy **7**, 68–96 (2005).

energy), 1 kg of copper has Ω_1 states available to it. We can treat 2 kg of copper as two 1-kg samples, the combination of the two 1-kg blocks make up a 2-kg sample. Because either 1-kg block can be in any of Ω_1 microstates while the other also has the same number of accessible microstates, the combination 2-kg system has $\Omega_2 = \Omega_1 \times \Omega_1$ microstates. Because $\ln \Omega_2 = \ln \Omega_1 + \ln \Omega_1 = 2 \ln \Omega_1$, it follows that $S_2 = 2S_1$. This clarifies why a 2 kg block has larger entropy than a 1 kg one.

Key Point 7.20 *The metaphors multiplicity, missing information, optiony, and freedom can help understand the meaning of entropy in an equilibrium state. Only the energy spreading metaphor conveys the notion that thermodynamic processes entail a spatial redistribution of energy, and that thermodynamic equilibrium corresponds to a temporally dynamic equilibrium at the microscopic level. The energy spreading metaphor for a process describes a change, which is consistent with Clausius's meaning of entropy, namely, "in transformation."*

CHAPTER 8

Working, Heating, Cooling

CONTENTS

8.1 THE VALUE OF CYCLES

8.1.1 What is a cycle?

The term *cycle* refers to a thermodynamic process that takes a *working substance*, typically a gas, from an initial equilibrium state through a succession of other states (equilibrium and/or non-equilibrium), ending back in the *initial* equilibrium state. Because the working substance begins and ends in the same equilibrium state, (a) by definition, it has undergone a cyclic process, and (b) because the working substance has undergone zero net change, all changes caused by the process are confined to the environment. The environment is assumed to be a constant temperature and pressure atmosphere and to have one or more work sources, e.g., weights on pulleys that can be lowered or raised.

Only if the cycle is quasistatic can it be drawn on a pressure-volume (PV) diagram. Recall that in Sec. 2.4.2 I showed that a quasistatic path need not be reversible. In Fig. 8.1, it is denoted by a solid curve. That figure shows a rightward upper path ab, with the working substance doing positive work, and a leftward lower path ba, with the working substance doing negative work, i.e., having positive work done on it by an external work source. The figure shows that more positive work than negative work is done *by* the working substance. The net work, $W_{\text{net}} = \oint P dV > 0$, done in this clockwise cycle is positive, a characteristic of a *heat engine*. The grey shaded area equals W_{net}.

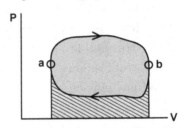

Figure 8.1: A cyclic process from initial state a to intermediate state b along the upper path and back to the initial state along the lower path. The work done by the working substance (shaded area= $\oint P dV$) equals the area under the upper path ab less the cross-hatched area.

8.1.2 How Clausius used cycles

Rudolf Clausius used an arbitrary cycle to prove his so-called *Clausius inequality*. Figure 8.2 is inspired by a derivation of that inequality presented by Enrico Fermi in his little book on thermodynamics.[1] Fermi's work follows the basic ideas of Clausius in a more accessible, modern way.

1. A reservoir at temperature T_0 exchanges energy with an arbitrary number of reversible Carnot heat engines (ten are shown in the figure, with the limit of infinitely many in mind.).

---

[1]E. Fermi, *Thermodynamics*, (Dover Publications, Inc., 1936), Ch. IV.

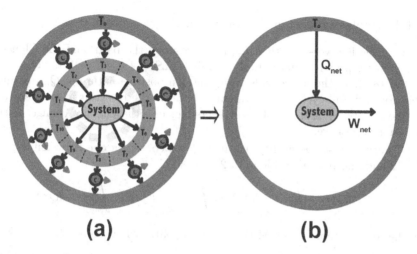

(a)   **(b)**

Figure 8.2: The net result of the combination of reversible Carnot cycles and reservoirs in (a) is shown in (b). This is because each Carnot cycle ends in its initial state and each intermediate reservoir gains and loses the same amount of energy.

2. The ith such heat engine exchanges heating energy with the T_0 reservoir, and with an *intermediate* reservoir at temperature T_i. It also includes a work source, namely a pulley with a weight hanging from it that can be raised or lowered.

3. The ith intermediate reservoir exchanges energy with the system undergoing the cyclic process, such that the net energy exchange with reservoir i is zero.

4. The heat engines are designed to generate the heating and cooling needed to take the system through its path. The limit of an infinite number of heat engines is envisaged (see Eq. (8.1) below).

5. In one cycle of the system, the only nonzero heating process is with the T_0 reservoir. Suppose Q_{net} denotes the heating energy *from* that reservoir, and W_{net} is the net amount of work done *on* the various pulley-weight systems, as in Fig. 8.2b.

6. The heating energy Q_{net} added to the system from the reservoir at T_0, cannot be positive or the net result would be the transfer of energy from a single-temperature reservoir at T_0 and the production of an equal amount of work, in violation of the Kelvin-Planck form of the second law of thermodynamics, as shown in Fig. 8.2. Thus, $Q_{net} \leq 0$.

Key Point 8.1 *Energy* $Q_{net} = T_0 \sum_{i=1}^{n} Q_i/T_i$ *is received from the T_0 reservoir. The condition $Q_{net} \leq 0$ implies that as the number of Carnot cycles increases without bound,*

$$\sum_{i=1}^{n} \frac{(Q_i)}{T_i} \leq 0 \rightarrow \oint \frac{dQ}{T} \leq 0 \quad \text{Clausius inequality.} \qquad (8.1)$$

8.1.3 Implications of cycles for entropy

If the cycle is reversible, then $dQ/T = dS$ along the path, and the Clausius inequality Eq. (8.1) reduces to an equality, $\Delta S = \oint dQ/T = 0$, i.e, $Q_{net} = 0$ in Eq. (8.1). In the irreversible case, the non-positive nature of Q_{net} assures that for each full cycle of the system in Fig. 8.1 the reservoir at temperature T_0 gains entropy if the cycle is irreversible; i.e., the entropy change of the universe is positive.

Clausius chose the cyclic path in (8.1) to consist of one irreversible path $1 \rightarrow 2$ and another reversible path $2 \rightarrow 1$, so

$$\int_{irrev,1\rightarrow2} dQ/T + \int_{rev,2\rightarrow1} dQ/T = \int_{irrev,1\rightarrow2} dQ/T - \int_{rev,1\rightarrow2} dQ/T \leq 0.$$

In the second step, I reversed the reversible path, which equals $S_2 - S_1$, so

$$S_2 - S_1 \geq \int_{irrev,1\rightarrow2} dQ/T. \tag{8.2}$$

If the process is an adiabatic one, the right side is zero and $S_2 \geq S_1$. This is the principle of entropy increase.

Key Point 8.2 *The Clausius inequality (8.1) is linked with the Kelvin-Planck statement of the second law of thermodynamics. That inequality leads to the grand principle of entropy increase for the universe, for which the right side of (8.2) is zero. The system undergoing the cycle in the Clausius inequality is arbitrary. Because of the cyclic process, and all changes are confined to the environment. The cycle is simply a tool to learn how the world behaves.*

8.1.4 PV & TS diagrams

Useful models of heat engine and refrigeration cycles use combinations of paths that are at least approximately achievable in real devices. Although pressure *vs.* volume (PV) plots are common, temperature *vs.* entropy (TS) plots are used normally only in advanced books. However, given the importance of entropy generally and its central role in this book, and most important, the usefulness of entropy as a tool for understanding cycles, I include TS diagrams here. Figure 8.3 shows several several typical quasistatic thermodynamic paths on both PV and TS plots. These are:

1. Path 1-2, constant pressure (isobaric) expansion. This can be accomplished using a container with a freely floating piston ceiling. The weight of the piston divided by its area is the external pressure. Slow heating then results in the isobaric expansion 1-2. As the system expands at constant pressure, both its entropy and temperature increase as shown in Fig. 8.3(b). The entropy increase can be attributed to increased spatial

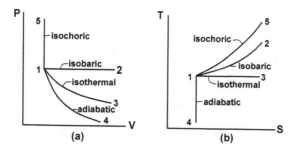

Figure 8.3: Quasistatic isobaric, isothermal, adiabatic, and isochoric curves on PV and TS plots.

spreading of energy, and increased energy, i.e., more energy that can spread, as the temperature rises.

2. Path 1-3, constant temperature (isothermal) expansion, has the system in thermal contact with a constant-temperature reservoir. A very slow expansion will then be approximately isothermal, with the reservoir supplying heating energy as that system does positive work on an external load. In order for the expansion to approach the quasistatic ideal, the load must decrease as the expansion occurs. The system's entropy increases with volume, as energy spreads spatially.

3. Path 1-4, reversible adiabatic (isentropic), approximated by a slow expansion of an insulated gas. As the gas expands, it does work on a (decreasing) external load as its pressure drops. Entropy change for this process is $dS = đQ/T$ with $đQ = 0$, so the entropy is constant. The temperature and pressure lie below those of the isothermal and isobaric curves because no energy is added from a reservoir as the gas does positive work.

4. Path 1-5, constant volume (isochoric) heating. Because the volume is fixed it is the heating that causes the entropy increase. With the volume fixed, a *higher* temperature is needed for the system to achieve the same volume and entropy as for the isobaric expansion.

8.2   EXAMPLES OF CYCLES

8.2.1   Reversible Carnot cycles

Reversible Carnot Clockwise Cycle. The most well known cycle is the clockwise *reversible* Carnot cycle. This serves as a model for real heat engines that operate between high (T_+) and low (T_-) temperatures. In all Carnot cycles, the energy sources and sinks at these temperatures are assumed to be constant-temperature reservoirs. For a dilute gas the Carnot cycle appears as

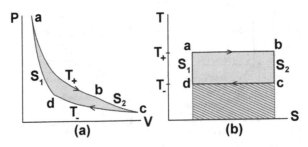

Figure 8.4: Clockwise reversible Carnot cycle: (a) PV plot. (b) TS plot. The crosshatched area is Q_-; the shaded + crosshatched areas equal Q_+.

in Fig. 8.4, accomplished in principle by the steps shown in Fig. 2.14. It is apparent that the TS plot is much simpler that the PV plot, involving only horizontal and vertical paths that make a rectangle for *any* working substance.

Along the Carnot cycle's isothermal paths ab and cd, the heating energy gained by the working substance is $dQ = TdS$, and applying this to path ab, $Q_{ab} = T_+ \int_a^b dS > 0$, which is the area under the horizontal line ab in Fig. 8.4(b). This is the sum of the areas of the shaded and crosshatched rectangles. I'll adopt the notation $Q_+ \equiv Q_{ab} > 0$. Similarly, $Q_{cd} = T_- \int_c^d dS < 0$, and I'll define $Q_- \equiv |Q_{cd}| > 0$, the crosshatched area under path cd. Furthermore, for the entire cycle $\Delta E = 0$, so $\Delta E = Q_+ - Q_- - W_{by} = 0$, and the work by the working substance per cycle is $W_{by} = Q_+ - Q_-$ (shaded area of $abcd$).

Key Point 8.3 *On the TS diagram, the heating input is the area under path ab and the heating output is the crosshatched area under path cd. The shaded areas on the PV and TS diagrams represent the net work done per cycle.*

Reversible Carnot Counterclockwise Cycles. The reversible, clockwise Carnot cycle can be run in the reverse direction, which reverses the path arrows, as in Fig. 8.5(a) and (b). Heating energy is gained *by the system* at the lower temperature T_- and *ejected from the system* at the higher temperature T_+; i.e., the lower temperature reservoir loses energy while the upper temperature reservoir gains energy.

The reversible counterclockwise (CCW) cycle has the characteristics of a refrigerator, whose working substance (refrigerant) takes in energy from the refrigerator compartment at low temperature and rejects it to the kitchen at a higher temperature. It is similarly characteristic of an air conditioner, which removes energy from the space being cooled and typically ejects energy to the warmer outdoors. In addition, such a cycle models a heat pump that removes energy from cool *outdoor* air and delivers energy to the higher-temperature *indoor* space being heated. These are explored in more detail in Sec. 8.5.

For the CCW Carnot cycle, it might appear that *heat* energy flows *uphill*, from lower to higher temperature. That is not true. The temperature rise of

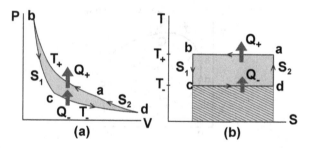

Figure 8.5: Counterclockwise Reversible Carnot cycle: (a) PV plot. (b) TS plot. The crosshatched area $= Q_-$; the shaded + crosshatched areas $= Q_+$.

the working substance (refrigerant), comes from the adiabatic compression *work* process da in Fig. 8.4. That is a *work* process, NOT a heat process! In the CCW Carnot cycle, the heating energy added *to* the working substance from the T_- reservoir keeps the temperature constant as the gas expands.

8.2.2 Efficiency measures

Thermal Efficiency of Heat Engines. Heat engines generate work to perform a desired task, using energy from a high temperature reservoir. A common measure of efficiency for a heat engine is the ratio of the work per cycle, W_{by}, by the working substance to the energy input *from* the high temperature reservoir:

$$\text{Thermal efficiency of heat engine} \equiv \eta = \frac{W_{by}}{Q_+} \qquad (8.3)$$

The reversible Carnot heat engine is important for several reasons. First, it uses two reservoirs, which simulate high and low temperature regions in steam engines and electricity generating plants. Second, there are two adiabatic segments, so all entropy changes occur along the isothermal segments. Third, as for all cycles, there is zero entropy change of the working substance during each cycle, so all net changes occur in the environment.

For the reversible Carnot heat engine, the entropy changes of the high and low temperature reservoirs are $\Delta S_+ = -Q_+/T_+$ and $\Delta S_- = Q_-/T_-$ respectively. The entropy change of the universe per cycle is

$$\Delta S_{univ} = \frac{-Q_+}{T_+} + \frac{Q_-}{T_-} = 0 \text{ for the reversible cycle.} \qquad (8.4)$$

Thus, $Q_-/Q_+ = T_-/T_+$ and

$$\eta_{car} = \frac{W_{by}}{Q_+} = 1 - \frac{Q_-}{Q_+} = 1 - \frac{T_-}{T_+} \qquad \text{Reversible Carnot cycle.} \qquad (8.5)$$

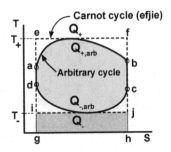

Figure 8.6: An arbitrary *reversible* clockwise cycle *abcda*. An infinite number of reservoirs heat the working substance along the upper path *ab*. Along the lower path *cd*, energy is sent to an infinity of lower temperature reservoirs. The latter paths are separated by reversible adiabatic paths *bc* and *da*. (Adapted from H. S. Leff, "Reversible & irreversible heat engine & refrigerator cycles," Am. J. Phys. **86**, 344–353 (2018) with permission of the American Association of Physics Teachers.)

If the Carnot cycle is *irreversible*, then Eq. (8.4) is replaced by $\Delta S_{\text{univ}} = Q_-/T_- - Q_+/T_+ > 0$, and Eq. (8.5) is replaced by

$$\eta_{\text{irr}} < 1 - \frac{T_-}{T_+} = \eta_{\text{car}} \quad \text{Irreversible Carnot cycle.} \tag{8.6}$$

The efficiency inequality (8.6) holds also for the large class of *reversible* cycles in Fig. 8.6. The highest and lowest temperatures are T_+ and T_-. In the temperature-entropy diagram, Fig. 8.6: (i) Q_-, the area of rectangle *ijhgi*, is less than $Q_{-,\text{arb}}$, which is the area of the shape *dchgd*; (ii) Q_+, the area of the rectangle *efhg* is greater than $Q_{+,\text{arb}}$, the area of shape *abhga*. Thus, $Q_- < Q_{-,\text{arb}}$ and $Q_+ > Q_{+,\text{arb}}$. It follows that $Q_{-,\text{arb}}/Q_{+,\text{arb}} > Q_-/Q_+ = T_-/T_+$ and thus,

$$\eta_{arb} \leq \eta_{\text{car}} = 1 - \frac{T_-}{T_+} \quad \text{(Carnot theorem) .} \tag{8.7}$$

Key Point 8.4 *Carnot's major achievement was finding that the reversible Carnot cycle has the highest possible efficiency, $\eta_{\text{car}} = 1 - T_-/T_+$ for any heat engine cycle operating between reservoirs with temperatures (T_-, T_+).*

Coefficient of Performance for Heating and Cooling. Referring to Fig. 8.5, several points should be clarified:

1. Counterclockwise cycles can be used for heating, cooling or both. If used for heating the specific desired task is to deliver energy Q_+ to the high temperature region. In contrast, for cooling, the desired task is to remove energy Q_- from the lower temperature region.

2. To accomplish either heating or cooling, the counterclockwise cycle requires positive external work on the working substance each cycle.

This is reflected in Fig. 8.6 by the fact that the work along the counterclockwise path dcba is negative (the working substance does negative work). Another view is that the magnitude of the negative work along path *dab* in Fig. 8.5(a) exceeds the positive work along *bcd*; also the magnitude of the negative Q_+ exceeds the positive Q_- in Fig. 8.5(b).

3. Recognising that external work is needed to perform these tasks, we define the positive external work

$$W_{on} \equiv -W_{by}. \tag{8.8}$$

Given the possibility of performing heating or cooling with counterclockwise cycles that require such external work, it is common to define a *heating coefficient of performance* for a "heat pump" by

$$COP_h \equiv \frac{Q_+}{W_{on}}. \tag{8.9}$$

The heating COP is discussed further in Sec. 8.5. The cooling *coefficient of performance* for a refrigerator or air conditioner is defined by

$$COP_c \equiv \frac{Q_-}{W_{on}}. \tag{8.10}$$

8.2.3 Reversible & irreversible Otto cycles

Clockwise Otto cycle. Historically, from at least Carnot's work, heat engines were an important motivator for the development of thermodynamics. In 1876, the German engineer Nikolaus Otto built a 4-stroke internal combustion engine, the familiar engine type used in pre-electric automobiles. The Otto cycle takes in fuel (petrol), mixes it with air in a carburetor, and burns the mixture in cylinders fitted with movable pistons. The combustion produces hot, high pressure gas that pushes the pistons, doing work. Coupling via a transmission turns the wheels of the vehicle. The combustion products are then exhausted and the process repeats itself with a new mixture of petrol and air. Note the potentially useful mnemonic connection between *Otto* and *auto*.

The model used for this cycle is illustrated in Fig.8.7. Path *ab* connotes the constant volume heating of the working substance. This is a convenient simplification of reality, where a petrol-air mixture is ignited and a chemical reaction releases the energy that heats the gas. In contrast, for the reversible model, an infinite number of reservoirs is needed to effect path *ab*. For the irreversible quasistatic case with reservoirs at the maximum and minimum temperatures $T_b \equiv T_+$ and $T_d \equiv T_-$, path *ab* entails transfer of energy from the high temperature reservoir to the working substance, bringing it slowly, and irreversibly, from state *a* to state *b* The input and output transfers are defined to be nonnegative: $Q_{in} \geq 0$ and $Q_{out} \geq 0$.

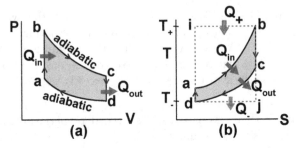

Figure 8.7: Reversible Otto cycle with constant volume and constant-entropy paths: (a) pressure *vs.* volume diagram; (b) temperature *vs.* entropy diagram, and circumscribing Carnot cycle (dashed lines).

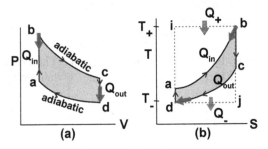

Figure 8.8: Irreversible Otto cycle: (a) *p vs. V* diagram; (b) *T vs. S* diagram, circumscribing Carnot cycle (dashed lines). Thick arrows connote energy exchanges from and to the T_+ and T_- reservoirs.

Path *cd* cools the gas, either reversibly or irreversibly and quasistatically. In an actual internal combustion engine, after expansion *bc*, the combustion products are exhausted to the surrounding air, and new outside air is compressed along path *da*. In the closed-cycle model in Fig. 8.7, the working substance is compressed along path *da* after being cooled along *cd*.[2]

Key Point 8.5 *The reversible Otto cycle does not typify an actual internal combustion engine because it requires an infinite number of reservoirs along paths ab and cd. The irreversible quasistatic cycle uses only two reservoirs.*

The irreversible quasistatic Otto cycle with two reservoirs is illustrated in Fig. 8.8. The irreversible cycle has energy $Q_{in} > 0$ transferred along path *ab* each cycle from the high temperature reservoir at T_+, with energy $Q_{out} > 0$ delivered to the low temperature reservoir at T_-. Thick arrows depict these irreversible energy flows. The work done per cycle is $W = Q_{in} - Q_{out}$. It is evident from Fig. 8.7(b) that $Q_{in} < Q_+$ and $Q_{out} > Q_-$. The entropy change

---

[2]Actual internal combustion engines are not true cycles because the fuel-air mixture never returns to its original state. Rather there is combustion, and exhaust of the products.

of the universe per cycle is solely that of the reservoirs

$$\Delta S_{\text{univ}} = \Delta S_{\text{res}} = Q_{\text{out}}/T_- - Q_{\text{in}}/T_+ > Q_-/T_- - Q_+/T_- = 0. \quad (8.11)$$

The terms $Q_-/T_- - Q_+/T_- = 0$ present the entropy change of a reversible Carnot cycle operating between T_+ and T_-. The thermal efficiency of *both* the reversible and irreversible Otto cycles are less than a Carnot cycle operating between T_+ and T_-:

$$\eta_{\text{otto}} = 1 - Q_{\text{out}}/Q_{\text{in}} < 1 - Q_-/Q_+; \text{ i.e., } \eta_{\text{otto}} < \eta_{\text{car}}. \quad (8.12)$$

Key Point 8.6 *Two observations are: (1) The reversible and irreversible Otto cycles have lower thermal efficiencies than the reversible Carnot cycle that operates between the highest and lowest Otto cycle temperatures, because paths ab and cd have variable temperatures. (2) The reversible and irreversible and quasistatic cycles have the same thermal efficiency, η_{otto}, which is determined by the ratio, $Q_{\text{out}}/Q_{\text{in}}$. The efficiency η_{otto} is independent of the entropy change in Eq. (8.11), which entails a knowledge of both Q_{in} and Q_{out}, not simply the ratio $(Q_{\text{in}}/Q_{\text{out}})$. Thus, the same efficiency holds for the reversible and irreversible, quasistatic cycles. The second law efficiency introduced in Sec. 8.3 accounts for irreversibility.*

Counterclockwise Otto Cycle The Otto cycle model arose to provide a way to understand the internal combustion engine and there is no reason to believe that the corresponding counterclockwise cycle is useful. Still we can envisage a variant of the reversible cycle in Fig. 8.7 with all arrows reversed. In that case, Q_{in} becomes Q_{out} and *vice versa*.

Key Point 8.7 *For the reversible counterclockwise Otto cycle, Q_{in} and Q_{out} require an infinite number of reservoirs to achieve paths dc and ba. Although one can robotically calculate a coefficient of performance for a refrigerator or heat pump, this model is not useful because it does not move energy from a low to high temperature region. The reversible counterclockwise Otto cycle is not an acceptable model for a refrigerator or a heat pump.*

If the counterclockwise Otto cycle is *irreversible and quasistatic*, as shown in Fig. 8.9(b), reservoirs are needed at the temperatures of states a and c. The highest and lowest temperatures at states b and d are reached adiabatically. Along path dc, the input energy Q_{in} comes solely from the higher-temperature reservoir, and the output energy Q_{out} along ba goes to the lower temperature reservoir. Surprisingly, the counterclockwise cycle moves energy from higher to lower temperature, rather than the reverse! The external work W_{on} is dumped to the lower temperature reservoir, consistent with $W_{\text{on}} = Q_{\text{out}} - Q_{\text{in}} > 0$. This results from the assumption that state c has a higher temperature than state a. The reverse situation is possible, as described here:

Key Point 8.8 *The irreversible, quasistatic counterclockwise Otto cycle in Fig. 8.9 is neither a heat pump nor a refrigerator because it takes energy from a higher temperature and ejects it to a lower temperature. The counterclockwise*

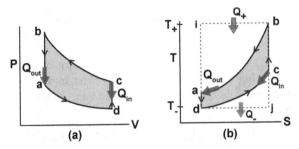

Figure 8.9: Counterclockwise quasistatic irreversible Otto cycle, with reservoirs at temperatures T_a and T_c.

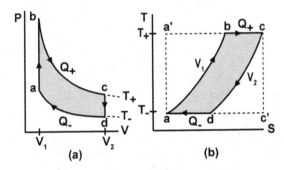

Figure 8.10: Clockwise Stirling cycle, with alternating isothermal and constant volume paths on (a) a pressure-volume diagram and (b) a temperature-entropy diagram. The dashed rectangle $aa'cc'a$ is a comparison reversible Carnot cycle.

variant of a heat engine need not be an acceptable refrigerator or heat pump.[3] *However, if the Otto cycle had $T_a > T_c$, the CCW cycle would behave like a proper heat pump or refrigerator, moving energy from lower to higher temperature, and is an acceptable model.*

8.2.4 Reversible & irreversible Stirling cycles

Clockwise Stirling cycle. The first Stirling heat engine was built in 1816 by Robert Stirling. Its thermodynamic cycle consists of alternating isothermal and constant volume (isochoric) paths, with heating processes along all four paths, as shown in Fig. 8.10. The working substance of a Stirling cycle is typically air, which is readily available and inexpensive. The working substance moves back and forth continually between colder and warmer chambers. Unlike the Otto cycle, which models an *internal* combustion engine, the air is heated by an *external* source at temperature T_+. This source

[3]R. H. Dickerson and J. Mottman, "Not all counterclockwise thermodynamic cycles are refrigerators," Am. J. Phys. **64**, 413 (2016).

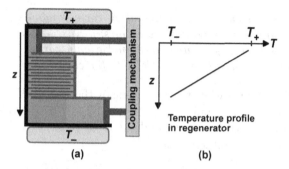

Figure 8.11: (a) Regenerator for a Stirling cycle. (b) Temperature profile of the regenerator. (Adapted from Schroeder, Daniel V., An Introduction to Thermal Physics, p. 133. ©2000. Reprinted by permission of Pearson Education, Inc., New York, NY.)

can be solar energy, and cooling can entail energy transfer to an external sink at the lower ambient temperature T_-.

The efficiency of the engine is enhanced by what Stirling called an *economiser*. Nowadays that is commonly called a *regenerator*. This regenerator stores energy from a previous cycle, feeds it to the working substance in one part of the cycle, and recovers it in another. After the isothermal expansion bc, Fig. 8.10 shows that the energy absorbed by the regenerator during cd is recycled to the gas along ab. With an assumed *perfect* regenerator (a fictitious idealisation) that recycles all the stored energy, the engine operates solely by exchanging energy with higher and lower temperature external reservoirs.

Figure 8.11(a) shows symbolically how a regenerator works. After some initial number of cycles, a temperature profile is established in the regenerator as the gas moves first upward, then downward, through the regenerator. Along path cd, the upper piston moves left while the lower piston moves right, maintaining the fixed volume V_2 in Fig. 8.10. As the hot gas winds its way through the generator, it loses energy to the regenerator and its temperature decreases. This process causes the regenerator to develop a temperature profile, as illustrated in 8.11(b). Then the gas is compressed along path da while in thermal contact with the low temperature reservoir.

The gas then returns through the regenerator, flowing from cold to hot, with the lower piston moving left and the upper piston moving right maintaining volume V_1. This is path ab, along which, the cold gas heats slowly and quasistatically. Next, the gas is expanded along path bc while in thermal contact with the hot reservoir. This completes one cycle.

Key Point 8.9 *We take the Stirling engine to be the cycle per se plus the regenerator. The energy exchanges with the regenerator are solely internal. If the cycle is reversible and 100% energy recovery were possible using an ideal* *regenerator, the efficiency would be that of the corresponding Carnot cycle with the same high and low temperature reservoirs:* $\eta_{perfect\ stirling} = \eta_{car}$. *Practical Stirling engines have regenerators that enable only partial energy recycling, and have efficiencies well below the Carnot efficiency.*

Without perfect regeneration, the work done by the cycle is $W = Q_{abc} - |Q_{cda}|$, with energy input Q_{abc}, and $\eta_{stirling} = 1 - |Q_{cda}|/Q_{abc}$, while $\eta_{car} = 1 - |Q_{c'a}|/Q_{a'c}$. Figure 8.10(b) shows that $|Q_{cda}| > |Q_{c'a}|$ and $Q_{a'c} > Q_{abc}$, so $\eta_{stirling} < \eta_{car}$. Thus, $\eta_{stirling} < \eta_{car}$.

Key Point 8.10 *Figure 8.10(b) shows that without regeneration, heating and cooling occur for all temperatures between T_- and T_+, requiring an infinite number of reservoirs. Thus, the reversible Stirling heat engine cycle is not a good model of a real heat engine.*

Key Point 8.11 *The irreversible, quasistatic Stirling heat engine operates with only two reservoirs, and has the same PV and TS diagrams as in Fig. 8.10. Heating along path abc comes solely from the higher temperature reservoir, while cooling along cda sends energy solely to the low temperature reservoir. With only two reservoirs this is a more realistic model than the reversible Stirling heat engine. As found for the Otto cycle, the thermal efficiency is the same for the reversible and irreversible, quasistatic cases (see Key Point 8.6).*

Figure 8.10(b) shows that the rejected energy along path cda, $Q_{cda} > Q_-^{car}$ and the heating of the gas along path abc, $Q_{abc} < Q_+^{car}$. Note that the latter Carnot energy is the area under the horizontal line jbc in Fig. 8.10(b). Using Eq. (8.5) and the inequality, $Q_{cda}/Q_{abc} < Q_-^{car}/Q_+^{car}$, it follows that $\eta_{stirling} < \eta_{car}$, consistent with Carnot's theorem, Eq. 8.7. Here $\eta_{stirling}$ is the efficiency of a clockwise Stirling heat engine cycle operating between T_+ and T_-.

Counterclockwise Stirling cycle. Although the Stirling cycle was invented as a heat engine, its counterclockwise counterpart has since achieved considerable practical success as a refrigerator. This device takes in heating energy Q_- at temperature T_- and rejects energy Q_+ at T_+.

Key Point 8.12 *The reversible counterclockwise cycle requires a continuum of reservoirs between the highest and lowest temperatures, and is not an acceptable model of a refrigerator or heat pump.*

The quasistatic *irreversible* counterclockwise Stirling cycle takes in heating energy Q_{adc} along path adc. However, part of that energy comes from the low temperature reservoir along ad and from the high temperature reservoir along dc. Similarly it rejects heating energy Q_{cba} along cba, with Q_{bc} going to the high temperature reservoir and Q_{ba} to the low temperature reservoir.

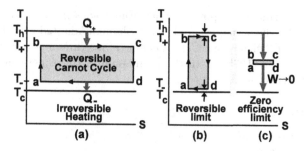

Figure 8.12: (a) NCA cycle (b) Zero power limit $T_+ \to T_h, T_- \to T_c$ (c) Zero power limit $T_+ \to T_-$ and $W \to 0$. Temperatures T_h and T_c are fixed.

Key Point 8.13 *The irreversible counterclockwise Stirling cycle exchanges energy in both directions with each reservoir, and it is not a good model of a traditional refrigerator or heat pump.*

8.2.5 Irreversible Carnot engine

In 1975, Frank Curzon and Boye Ahlborn introduced physics teachers to an irreversible heat engine cycle sometimes referred to as the Novikov-Curzon-Ahlborn, or NCA, cycle.[4] That cycle had been introduced to the engineering community earlier by Novikov. It consists of a reversible Carnot cycle that operates with isotherms at temperatures T_+ and $T_- < T_+$, and exchanges energy irreversibly with reservoirs at temperatures T_h and T_c, with $T_c < T_- < T_+ < T_h$. Figure 8.12(a) shows the cycle and energy transfers Q_+ and Q_- for each cycle.

Note that the quasistatic Carnot cycle requires infinite time per cycle, but finite-time energy exchanges with the reservoirs. To justify the model, we can envisage the cycle to have slow, but finite, rates of energy transfer along the isotherms bc and da, and very slow adiabatic volume changes along the adiabatic paths ab and cd that approximate a quasistatic Carnot cycle, but are consistent with nonzero energy transfer rates. This is a utilitarian view, as discussed in Sec. 7.1.4.

Two limiting zero-power cases are shown in Fig. 8.12(b) and (c). Figure 8.12(b), with $T_+ \to T_h$ and $T_- \to T_c$, brings the maximum work, maximum efficiency Carnot limit that takes an infinite time, and thus generates zero power. Figure 8.12(c) is the zero-work, zero-efficiency limit $T_- \to T_+$, where energy simply flows from $T_h \to T_c$, and the power generated is zero. For intermediate values of T_- and T_+, there is a power maximum. Curzon and Ahlborn provide the algebraically complex steps needed to find the conditions for that maximum. Their work inspired much interest, which led to an

---

[4]F. L. Curzon & B. Ahlborn, "Efficiency of a Carnot engine at maximum power output," Am. J. Phys. **43**, 22–24 (1975).

Table 8.1: Performance of power plants. Efficiencies $\eta_{car}, \eta^\star$, and η_{pp} are for Carnot and NCA heat engines, and power plants. (Adapted from F. L. Curzon & B. Ahlborn, "Efficiency of a Carnot engine at maximum power output," Am. J. Phys. **43**, 22–22 (1975), with the permission of the American Association of Physics Teachers.)

Power Plant	$T_-[K]$	$T_+[K]$	η_{car}	$\eta^\star$	η_{pp}
West Thurrock (U.K.) Coal-Fired Steam Plant	298	838	54.1%	40%	36%
CANDU (Canada) PHW Nuclear Reactor	298	573	48.0%	28%	30%
Larderello (Italy) Geothermal Steam Plant	353	523	32.3%	17.5%	16%

entire field of study called finite time thermodynamics.[5] Curzon and Ahlborn assumed that the energy transfers at T_+ and T_- are proportional to the temperature differences between the working substance and the reservoirs, i.e., $Q_h = \alpha(T_h - T_+)$ and $Q_c = \beta(T_- - T_c)$. The parameters α and β are assumed to be proportional to the thermal conductivity and transfer times. The details are beyond the scope here, and I will simply cite the results. The temperatures that maximise the power satisfy,

$$T_-^\star/T_+^\star = (T_c/T_h)^{1/2}, \tag{8.13}$$

and the optimal energies per cycle from the high temperature reservoir and to the low temperature reservoir are,

$$Q_h^\star = \frac{\alpha\beta}{(\alpha + \beta)} T_h^{1/2}[T_h^{1/2} - T_c^{1/2}] \text{ and } Q_c^\star = \frac{\alpha\beta}{(\alpha + \beta)} T_c^{1/2}[T_h^{1/2} - T_c^{1/2}]. \tag{8.14}$$

These values imply the efficiency at maximum power,

$$\eta^\star = 1 - Q_c^\star/Q_h^\star = 1 - (T_c/T_h)^{1/2} = \eta_{nca} \quad \text{(NCA efficiency).} \tag{8.15}$$

Key Point 8.14 *The Novikov-Curzon-Ahlborn irreversible heat engine has a number of notable features:*

1. *Although perhaps fortuitous, the efficiency $\eta^\star$ is a reasonably good predictor of the thermal efficiency of existing fossil fuel and nuclear electrical generating power plants, as indicated in Table 8.1.*

2. *The NCA heat engine has heating through finite temperatures at both the high and low temperature ends, and its isothermal expansion and compression segments occur in finite time.*

---

[5]Andresen, B., Salamon, P., & Berry, R. S., "Thermodynamics in finite time," Phys. Today **37**, 62–70 (September 1984).

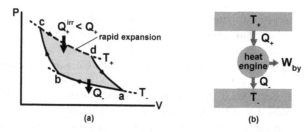

Figure 8.13: (a) Irreversible Carnot cycle with rapid expansion along *cd*. (b) Two-reservoir heat engine.

3. *Each cycle of the NCA engine generates nonzero entropy $\Delta S > 0$.*

4. *The thermal efficiency $\eta_{nca} = 1 - (T_c/T_h)^{1/2}$ depends only on the two reservoir temperatures and satisfies $\eta_{nca} < \eta_{car}$.*

5. *During one cycle, the entropy change of the universe comes solely from changes in the reservoirs, and is positive:*

$$\Delta S_{univ} = \frac{\alpha\beta}{(\alpha + \beta)} \frac{[T_h^{1/2} - T_c^{1/2}]^2}{(T_h T_c)^{1/2}} > 0. \tag{8.16}$$

The Novikov-Curzon-Ahlborn efficiency η_{nca} arises in other contexts, including the reversible Otto and Joule cycles and other *reversible* cycles operating under maximum-work conditions.[6]

8.3 IRREVERSIBILITY & 2ND LAW EFFICIENCY

The NCA cycle is one way to have an irreversible heat engine. Another way is to replace the slow expansion *cd* with a rapid one, as in Fig. 8.13(a). The initial and final states are the same, so the internal energy change ΔE is unaltered. However, because the piston moves rapidly away from the expanding gas, the gas does less work $W_{by,irr}$ than for the reversible case, but with the same internal energy change, $\Delta E = Q_+^{irr} - W_{by}^{irr}$. Because $W_{by}^{irr} < W_{by}$, the heating energy must be lower than for the reversible case, i.e., $Q_+^{irr} < Q_+$. This results in a lower thermal efficiency than for the reversible case, $\eta_{irr} = 1 - Q_-/Q_+^{irr} < \eta_c$ and nonzero entropy production. Similar results are found if other parts of the Carnot cycle are done quickly rather than quasistatically.

Key Point 8.15 *For the irreversible cycle in Fig. 8.13(a), less work is done on the gas along the irreversible cycle's upper path cd, and less heating energy input from the upper reservoir than for the Carnot cycle. This causes lower*

---

[6]H. S. Leff, "Thermal efficiency at maximum work output: new results for old heat engines," Am. J. Phys. **55**, 602–609 (1987).

entropy decrease of the upper reservoir with the same entropy increase of the lower reservoir. The result is that the universe gains entropy $\Delta S_{univ}^{irr} = -Q_+^{irr}/T_+ + Q_-/T_- > 0$, so $Q_-/Q_+^{irr} > T_-/T_+$, and the thermal efficiency $\eta_{irr} = 1 - Q_-/Q_+^{irr} < 1 - T_-/T_+$, consistent with Carnot's theorem.

We saw in Sec. 8.2.3 that the reversible Otto cycle, with variable-temperature paths, has a thermal efficiency less than a Carnot cycle operating between the Otto cycle's maximum and minimum temperatures. We also saw that irreversible but quasistatic Otto and Stirling cycle efficiencies are the same as for their reversible counterparts. The reason for this non-intuitive behaviour was explained in Key Point 8.6.

In order to better characterise the efficiency of cycles, physicists and engineers have sought a new measure of efficiency that accounts for limitations imposed by the second law of thermodynamics. In 1975, the American Institute of Physics sponsored a study that introduced the second law efficiency to the physics community.[7] Consider a heat engine cycle that operates between reservoir temperatures T_+ and T_-, absorbs Q_+ each cycle and ejects Q_-, as illustrated in Fig. 8.13(b). The work per cycle by the working substance is $W = Q_+ - Q_-$, and the entropy change of the universe ΔS_{univ} per cycle is $\Delta S_{univ} = Q_-/T_- - Q_+/T_+ \geq 0$, consistent with the second law of thermodynamics. The entropy change is solely from the reservoirs because the working substance has zero entropy change per cycle. Multiplying ΔS_{univ} by T_-/Q_+ and doing some algebra, we obtain

$$\eta_{car} - \eta = \frac{T_- \Delta S_{univ}}{Q_+} > 0 \quad \text{Carnot's theorem.} \quad (8.17)$$

Key Point 8.16 *Equation (8.17) holds for any irreversible two reservoir heat engine. For those cycles, Carnot's theorem holds because $\Delta S_{univ} > 0$. In contrast, the reversible Otto and Stirling cycles obey Carnot's theorem because they have variable temperature paths. Equation (8.17) does not hold for a reversible cycle that has variable temperature paths, or for the irreversible quasistatic Otto and Stirling cycles for which $\eta = 1 - Q_{out}/Q_{in} \neq 1 - Q_-/Q_+$. In those cases, η does not account for irreversibilities in the two-reservoir heat engines.*

An efficiency measure ϵ that accounts for irreversibility can be defined in terms of *exergy* (or *available energy*), which was introduced in Sec. 3.5.2. For a work-generating process that requires actual work W_{by}, the input exergy can be written, $B_{in} = W_{by} + B_{lost} = W_{by} + T_0 \Delta S_{univ}$. In this case, the minimum

---

[7]K. W. Ford, *et al.*, ed., *Efficient Use of Energy: A Physics Perspective*, AIP conference proceedings 25 (AIP, New York, 1975). See also T. V. Marcella, "Entropy production and the second law of thermodynamics: An introduction to second law analysis," Am. J. Phys. **60**, 888–895 (1992).

possible exergy needed to produce the work W_{by} equals the work W_{by} itself. The *second law efficiency* ϵ is defined as,

$$\epsilon \equiv \frac{B_{min}}{B_{in}} = \frac{W_{by}}{W_{by} + T_0\Delta S_{univ}} = \frac{1}{1 + (T_0\Delta S_{univ}/W_{by})} \leq 1, \qquad (8.18)$$

where T_0 is the temperature of the surroundings.

Key Point 8.17 *Several points are worth emphasising.*

1. *Unlike the thermal efficiency, which is defined in terms of work and heat, the second law efficiency ϵ accounts, via inclusion of ΔS_{univ}, for limitations imposed by irreversibilities.*

2. *For given Q_{in}, the second law (ϵ) and the thermal (η) efficiencies, are both decreasing functions of the entropy change of the universe, ΔS_{univ}.*[8]

3. *The definition (8.18) can be applied to the irreversible quasistatic Otto and Stirling cycles. Because, $\Delta S_{univ} > 0$, $\epsilon < 1$. For the reversible Otto, Stirling, or any reversible cycle, $B_{min} = W_{by} = W_{rev}$, and $\epsilon = 1$.*

8.4 COMBINED-CYCLE HEAT ENGINES

Combined-cycles consist of two heat engines. One, a gas turbine, does work W_{gt}, with thermal efficiency $\eta_{gt} = W_{gt}/Q_+$ and output heating energy $(1 - \eta_{gt})Q_+$. The output heating energy is then used to power a steam turbine, producing more work W_{st}. The main principle of combined-cycle operation is that the gas turbine's heating output temperature is high enough to power the steam turbine. Gas turbines require less metal than steam turbines, and (expensive) heat-resistant metals can be used, enabling higher inlet temperatures, $1200 - 1700\,\mathrm{K}$. Typical steam turbine inlet temperature are as high as $\sim 900\,\mathrm{K}$, and they can operate usefully at the gas turbines exhaust temperatures of $700 - 900\,\mathrm{K}$.

Combined-cycles have become increasingly popular for electricity generation because, with larger high to low temperature ratios, they offer higher efficiencies than are possible otherwise. Gas turbine engines can burn a variety of fuels and can be brought online in $5 - 30$ minutes. They also can be used for electricity *peaking*[9] and emergency power generation. At the end of 2018, natural gas-fired combined-cycle power plants provided almost 90 percent of total natural gas-fired generation in the United States.

---

[8]This monotone dependence of efficiency with entropy change is limited to cycles employing two reservoirs. See H. S. Leff & G. L. Jones, "Irreversibility, entropy production, and thermal efficiency,"'Am. J. Phys. **43**, 973–980 (1975).

[9]Increased power generation to meet high power demand at certain times of the day or year.

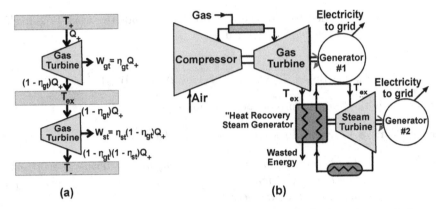

(a)   **(b)**

Figure 8.14: (a) Energy flows during a specified time interval for a combined-cycle heat engine. (b) Power plant configuration. (Reproduced from H. S. Leff, "Thermodynamics of combined-cycle electric power plants," Am. J. Phys. **80**, 515–518 (2012), with the permission of the American Association of Physics Teachers.)

A combined-cycle plant's energy flows are shown in Fig. 8.14(a), and a specific power plant configuration is in Fig. 8.14(b). Referring to the labelling in Fig. 8.14(a), the efficiencies of the gas turbine and steam turbine cycles are,

$$\eta_{\text{gt}} = \frac{W_{\text{gt}}}{Q_+} \quad \text{and} \quad \eta_{\text{st}} = \frac{W_{\text{st}}}{(1 - \eta_{\text{gt}})Q_+}. \tag{8.19}$$

These equations can be used to obtain the overall efficiency of a combined-cycle engine,

$$\eta_{\text{cc}} = \frac{W_{\text{gt}} + W_{\text{st}}}{Q_+} = \eta_{\text{gt}} + \eta_{\text{st}} - \eta_{\text{gt}}\eta_{\text{st}}. \tag{8.20}$$

Key Point 8.18 *The configuration in Fig. 8.14(b) shows the gas and steam turbine engines powering two separate electric generators. It shows wasted energy at the linkage between the gas turbine's exhaust and the steam turbine's inlet temperatures, which are not included in Eqs. (8.19) and (8.20). Equation (8.20) can be used to graph η_{st} vs η_{gt} for various values of the combined-cycle efficiency η_{cc}, as shown in Fig. 8.15(a). The shaded rectangle shows that combined-cycle efficiencies above 0.50 are possible when η_{st} is as low as 0.20 if the corresponding η_{gt} is at least 0.38.*

Using the η_{gt} and η_{cc} curves in Fig. 8.15(b) with Eq. (8.20), one can calculate the implied steam turbine efficiency η_{st}. The average of the values implied for temperatures 1473 K and 1673 K is shown as a dashed curve, with the dotted curve being the Novikov-Curzon-Ahlborn efficiency η_{nca}, representing the steam turbine with $T_- = 395$ K in Eq. (8.15), chosen to give good agreement with the η_{st} curve.

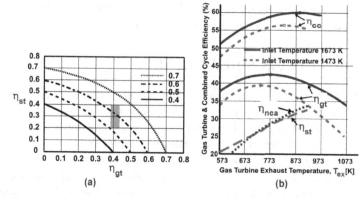

Figure 8.15: (a) Steam turbine efficiency η_{st} vs. gas turbine efficiency η_{gt} for combined-cycle efficiencies $\eta_{cc} = 0.4, 0.5, 0.6, 0.7$. The shaded rectangle identifies the approximate region of current-day combined-cycle efficiencies. (b) Thermal efficiency for the combined-cycle (η_{cc}), gas turbine (η_{gt}), steam turbine (η_{st}), and Novikov-Curzon-Ahlborn (η_{nca}) efficiencies vs. gas turbine exhaust temperature. (Same reference and permissions as Fig. 8.14.)

Key Point 8.19 *In March 2018, the Chubu Electric Nishi-Nagoya power plant Block-1, powered by GE's 7HA gas turbine, was recognised by GUINNESS WORLD RECORDS™ as the world's Most efficient combined-cycle power plant, based on achieving* $\eta_{cc} = 63.08$ *percent gross efficiency. Our graphs show that for* $T_{ex} \approx 900\,K$, *the maximum combined-cycle efficiency* $\eta_{cc} = 0.60$ *comes about with a gas turbine efficiency* $\eta_{gt} = 0.41$ *and steam turbine efficiency* $\eta_{st} = 0.32$. *Note that the combined-cycle's maximum efficiency occurs for higher exhaust temperatures, and in order to maximise* η_{cc}, *the gas turbine must operate below its own efficiency maximum.*

If a Carnot cycle operates between an ambient temperature of 298 K and 1073 K, $\eta_{car} = 0.82$ and $\eta_{nca} = 0.58$. Current combined-cycle efficiencies are between the latter two efficiencies.

I close this section with some observations. (1) A gas turbine's blade tips have tangential speeds exceeding 1700 km/h (472 m/s), greater than the speed of sound in air, and are subject to high temperatures. This requires extremely strong and heat resistant materials. (2) Combined-cycle power plants show how relatively straightforward thermodynamics, together with advances in the thermal properties of materials and clever design, can dramatically increase the efficiency of electricity generation. (3) Modern combined-cycle electric generating plants have changed the landscape of electric power generation and will likely continue to do so for many years. (4) For teachers, the combined-cycle power plant offers a practical, socially relevant example of thermodynamics, suitable for inclusion, at least in concept, in general physics and upper division thermodynamics, and courses on energy and environment.

These cycles are excellent examples of large-scale energy machines, efficiency, and the capture and recycling of wasted energy.

8.5 LORD KELVIN'S HEAT PUMP

In 1852 William Thomson proposed a novel heating device to pump energy from a lower-temperature region, say the winter outdoors, to a higher-temperature region. He claimed that one could burn a fuel (coal in those days), extract energy Q_- from the cold winter outdoors, and deliver a greater energy $Q > Q_-$ to the interior space being heated.

That idea was startling at the time because it appears at first glance that this is ruled out by the Clausius form of the second law of thermodynamics: *It is impossible for energy to flow spontaneously, that is, with no other effect, from a lower to higher temperature region.* However, the energy Q_{fuel} from a fuel can generate work W_{on} on a working substance to extract energy Q_- from the cold outdoor air at temperature T_- and deliver energy $W_{\text{on}} + Q_-$ to the interior space. Thomson envisaged a new quantity of working substance, namely air, being used for each cycle. Following the *Kelvin cycle* in Fig. 8.16, there is no violation of the second law because the energy flow in the proposed warming machine is *not* spontaneous. It requires the combustion of a fuel and operation of a heat engine to do work, as explained here.

1. A mass of outdoor air in state a at atmospheric pressure P_a and outdoor temperature T_- is taken into a metal cylinder, and is expanded isothermally from state a to state b with lower pressure P_b, while energy Q_- is transferred from the outdoor air.

2. The air is then transported at constant pressure and temperature to a well-insulated cylinder that is indoors, where it is compressed adiabatically to state c. In that state, its pressure is once again P_a, and the temperature is $T_+ > T_-$.

3. The heated air is delivered to the desired space as cooler air is pushed out, keeping the room's air pressure constant. Processes 1–4 are then repeated.

Assuming a monatomic, diatomic ideal gas to simulate air, we can estimate the amount of energy that is pumped from lower to higher temperature. The energy input from the outdoors along path ab is $Q_- = NkT_a \ln(V_b/V_a)$ and the energy delivered during path ca is $Q_P = C_P(T_c - T_a)$. On isotherm ab, $T_a = T_b$ and on isobar ca, $P_a = P = b$ and $V_c/V_a = T_c/T_a$. On the isentropic bc, $T_b V + b^{\gamma-1} = T_c V_c^{\gamma-1}$.[10]

---

[10]γ is the ratio of the heat capacity at constant pressure to that at constant volume, which is $(5/2)/(3/2) = 5/3$ for a diatomic gas. For classical ideal gases, the constant pressure heat capacity is $C_P = \gamma(\gamma - 1)$.

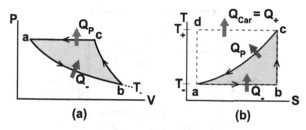

Figure 8.16: (a) A PV plot of a counterclockwise Kelvin cycle. (b) A TS plot of the same cycle. The dashed rectangle is the circumscribing Carnot cycle. The shaded areas represent the external work done on the working substance.

Some algebra shows that the heating coefficient of performance COP_h that was introduced in Sec. 8.2.2, is

$$COP_h = \frac{Q_P}{W_{on}} = \frac{1}{1 - Q_-/Q_P} = \frac{1}{1 - \ln(T_c/T_a)/(T_c/T_a - 1)}. \quad (8.21)$$

Thomson used the temperatures $T_a = 283K$ and $T_c = 300K$, which implies $COP_h = 35$; i.e., the energy delivered at the higher temperature is 35 times larger than the work required to operate the heat pump. In Thomson's day, the most likely way to generate the work W was to run a steam engine, for which $W_{on} \approx 0.10 \times |\Delta H_{coal}|$ (heat of combustion of coal) because steam engines were about 10% efficient. Therefore, a more telling ratio is

$$COP_{h,net} = \frac{Q_P}{|\Delta H_{fuel}|} \approx 3.5. \quad (8.22)$$

Key Point 8.20 *Although direct combustion heating with a furnace was possible, William Thomson knew it was better to use the fuel to power a steam engine to run his warming machine. This extracted energy Q_- from a colder region and combined it with the work W_{on} by the steam engine—providing more heating energy than from the fuel directly.*

In Fig. 8.16(b), the circumscribing reversible Carnot cycle is shown as a dashed rectangle $abcda$. The first law of thermodynamics applied to one cycle of Kelvin heat pump is $\Delta E = 0 = Q_- - Q_P + W$, so $W = Q_P - Q_-$. Thus, the heating COP for the Kelvin heat pump is $COP_h^{kel} = Q_P/(Q_P - Q_-) = 1/(1 - Q_-/Q_P)$. For the Carnot cycle, $COP_h^{car} = 1/(1 - Q_-/Q_{car})$. Both the Kelvin and Carnot heat pumps have the same energy input Q_- at T_-, while the output of the Kelvin and Carnot heat pumps are Q_P and Q_{car}, respectively, with $Q_{car} > Q_P$. From this, it follows that

$$COP_h^{kel} < COP_h^{car}. \quad (8.23)$$

Thomson's warming machine was evidently never built. However modern vapour-compression systems enable heating with energy transferred from the environment together with energy provided by an electric motor.

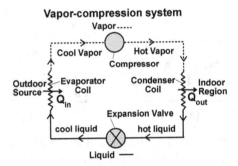

Figure 8.17: Schematic diagram of a liquid refrigerant's flow through an outdoor evaporator coil, then compression of the vapour, then through a condenser coil, heating the cooler indoors, and flow of the liquid through an expansion valve, completing the cycle. The sequence is repeated *ad infinitum*.

Key Point 8.21 *Modern-day heat pumps are very different from the reversible Kelvin cycle and follow a cycle similar to that in Fig. 8.17. They operate between hot and cold temperature regions using a working fluid (refrigerant) that is compressed by a motor, liquefied in a condenser coil, expanded through a narrow nozzle, and vaporised in an evaporation coil. This is shown in Fig. 8.17. Heating of the refrigerant occurs twice: (i) by the warmer outside air when the refrigerant is cold, and (ii) by a compressor motor via a work process, making the refrigerant hotter than the space to be heated. Despite the inappropriate name "heat pump," all heat processes are downhill.*

Heating energy transfers occur through coiled tubes through which the working fluid flows. As described already, to accomplish indoor heating, one coil is placed outdoors and the other indoors. In contrast, kitchen refrigerators have their cold coils *inside*, next to the food compartment, while the hot coil is outside the enclosure, heating the kitchen. That is why kitchens tend to be relatively warm.[11]

Key Point 8.22 *In areas where the water table is within feasible drilling depth, groundwater, which typically maintains a nearly constant temperature that reflects the average outdoor temperature, can be used as the cold source.[12]*

Figure 8.18 labels the relevant energy flows per cycle. For the heat pump, the $COP_h = (Q_+/W)$. However, a more meaningful measure of the overall

---

[11]For years I told students of U.S.A President Harry Truman's dictum, "If you can't stand the heat, stay out of the kitchen." I then said, "He really understood thermodynamics!"

[12]In the USA, groundwater temperatures range from $2 - 25\,°C$. In the UK, where ground temperatures (not groundwater) are typically $8 - 13\,°C$, the cold coil for a heat pump can be embedded $\sim 1\,m$ below the surface.

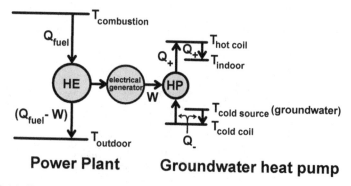

Power Plant   **Groundwater heat pump**

Figure 8.18: Diagram of the path of a vapour-compression system's refrigerant.

efficiency, including that of the *electrical generating plant* (HE in the figure), is $COP_{net} = (Q_+/Q_{fuel})$. Writing the thermal efficiency of the power plant as $\eta_{elec} = (W/Q_{fuel})$, it follows that

$$COP_{net} = \eta_{elec} \times COP_h. \qquad (8.24)$$

Nowadays, the highest efficiency heat pumps have heating COPs of ~ 3. If electricity generation efficiency is $\eta_{elect} \approx 0.4$, then $COP_{net} \approx 1.2$.

8.6 COOLING & CRYOGENICS

8.6.1 Cooling techniques

Since prehistoric times, humans have used fire to heat, cook, and provide lighting. Heating with fire enabled us to produce steel and forge tools. As heat-resistant materials became available, pottery and china were possible. Unlike heating, the ability to cool was limited to ice produced naturally during winter months in regions with sub-freezing winter weather.

Inventor and engineer Jacob Perkins, received the first patent in 1835 for a vapour-compression refrigerator. Similar patents were issued soon thereafter in other countries. Ammonia, ether, or alcohol were typical refrigerants. These were toxic and could leak, and their use was limited initially to large-scale plants to produce *harvested ice* for meat-packing plants, breweries and others.

Subsequently, more compact and effective refrigeration techniques were developed. Each time better refrigerators were produced, new physics could be learned, e.g., if a gas gets cold enough, it will condense into a liquid. Researchers had learned that by reducing the pressure of water, it could be made to boil at a temperature less than 100 °C.[13] Cooling of gases and

---

[13]The boiling point of water on Mount Everest at altitude 8,848 m (29,029 ft) is 71 °C.

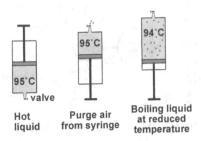

Figure 8.19: A liquid boiling at room temperature by pressure reduction.

liquids can occur in several ways. First, If there is a lower-temperature system available, put the gas or liquid in contact with it. Second, a gas can be cooled *mechanically* by having it expand slowly and adiabatically, as in the adiabatic expansion segment of a Carnot cycle. The degree of cooling is limited by the physical size of the container housing the gas. A variant is to have the gas flow through a narrow nozzle, experiencing a sharp pressure drop. Similarly, a liquid can also be cooled mechanically by lowering its pressure adiabatically, bringing it to a boil. With the advent of vacuum pumps, this could be accomplished by pumping the vapour above the liquid. Boiling reduces a liquid's internal energy, and lowers its temperature further.

An excellent physics demonstration illustrates the effect of lower pressure on the boiling point. It uses a plastic syringe (without a needle). A valve is attached to the needle adaptor, as illustrated in Fig. 8.19(a), to prevent flows of water and/or air. The four-step process is: (1) Open the valve and pull the plunger out to draw some warm water, say at 95 °C at normal atmospheric pressure, into the syringe, then close the valve (Fig. 8.19(a)). (2) Point the needle adapter upward and open the valve. Then depress the plunger to force existing air (and a bit of water) out, and close the valve. (3) Pull the plunger out to reduce the pressure on the water. If a sufficiently low pressure is achieved, the boiling point will shift to a value below 95 °C, and the liquid will begin to boil. This is shown in Fig. 8.19(c). (4) As the liquid boils, the water cools and when the temperature gets below the boiling point at that reduced pressure, the boiling will cease. Figure 8.19(c) shows a temperature reduction of 1 °C,

The phenomenon of cooling under a pressure reduction was discovered accidentally by Louis Cailletet in Paris in 1877. He had been trying to liquefy ethylene by subjecting it to high pressure and then cooling it. However, during one such trial, the high pressure ruptured the vessel housing the gas, and the rapid expansion briefly formed a faint mist in the glass tube. Cailletet

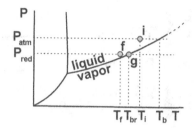

Figure 8.20: Labeled segment of the pressure-volume phase diagram, Fig. 4.7, described in the text.

concluded correctly that the sudden pressure drop had somehow cooled the ethylene, condensing some of it, whereupon it quickly vaporised into the air.

Such cooling can be visualised using the modification, Fig. 8.20, of the pressure-volume phase diagram, Fig. 4.7. Suppose the target system is initially at atmospheric pressure, P_{atm}, with boiling point T_b, and the initial system state is i at a temperature $T_i < T_b$. A glass of water at the dinner table satisfies these conditions, as does a cup of hot, but not boiling, water. If the pressure is lowered to the reduced value P_{red} in the vicinity of state g, the liquid-vapour line is reached and the water begins to boil at the reduced temperature T_{br}. As boiling progresses, the temperature falls further to $T_f < T_{br}$. The path followed is uncertain and not shown in Fig. 8.20.

Cailletet and Geneva physicist Raoul Pictet independently and with differing techniques, had tried to liquefy gases by a sequence of steps that included (as mentioned), (i) compressing the gas to many atmospheres of pressure; (ii) cooling the resulting high density gas using the coldest energy sink available, e.g., boiling ethylene at 170 K; and (iii) cooling the high pressure gas by forcing it through a narrow *expansion valve* (explained further below). Neither Cailletet nor Pictet were able to produce a sufficient liquid volume to aid other experiments or to cool other systems.

8.6.2   Joule-Thomson process

The Joule-Thomson (J-T) process forces a gas under pressure through a porous plug, i.e., a material with small pours, like a wad of steel wool, that impedes flow of a gas or liquid. A J-T process is shown in Fig. 8.21. Initially, the gas occupies volume V_1 to the left of the plug, with gas pressure and temperature P_1 and T_1, respectively. An external force $F_1 = P_1 A$ acts on a frictionless piston of cross-sectional area A, forcing the gas through the plug.

The objective is to have the gas moved to the right chamber with reduced pressure $P_2 < P_1$ and temperature $T_2 < T_1$. We assume the thickness of the porous plug can be made negligible so all the gas originally in the left chamber ends in the right one. Cooling is not guaranteed and indeed, heating is a possible outcome, as I will explain. If we apply the first law of thermodynamics

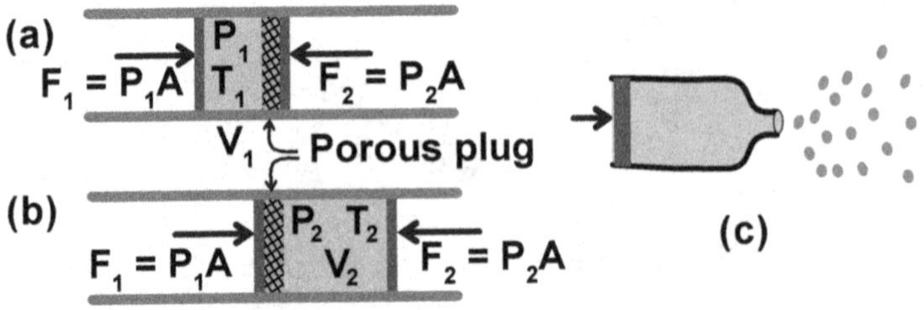

Figure 8.21: Depiction of a porous plug expansion. (a) The gas, initially in chamber 1, is forced through a porous plug. (b) The gas is finally in chamber 2, with $P_2 < P_1$, $V_2 > V_1$, and $T_2 < T_1$. (c) Gas is forced through a narrow nozzle, experiencing pressure drop, which forms liquid droplets.

to the gas, for the process $\Delta E = E_2 - E_1 = Q + W = 0 + P_1 V_1 - P_2 V_2$. Here, $Q = 0$, assuming the walls of the system are well insulated.

$$E_1 + P_1 V_1 = E_2 + P_2 V_2 \implies H_1 = H_2, \text{ (constant-enthalpy process)}. \quad (8.25)$$

For this constant-enthalpy process, it is useful to define a *Joule-Thomson coefficient* μ_{JT}, defined as

$$\mu_{JT} \equiv \frac{T_2 - T_1}{P_2 - P_1} \to \left(\frac{\partial T}{\partial P}\right)_H \text{ for given values of } P \text{ and } T. \quad (8.26)$$

For an infinitesimal reversible volume change, the first law of thermodynamics is $dE = dQ - PdV = TdS - PdV$. Using $d(PV) = PdV + VdP$, the first law can be written, $dE = TdS - d(PV) - VdP$. Thus, for an infinitesimal, adiabatic J-T process, $dH = dE + d(PV) = TdS - VdP = 0$, which implies

$$dS = -(V/T)dP > 0 \text{ because } dP < 0. \quad (8.27)$$

Key Point 8.23 *The J-T process is irreversible, and is consistent with the second law of thermodynamics.*

It is shown in thermodynamics books[14] that the J-T coefficient, Eq. (8.26), can be written as

$$\mu_{JT} = \frac{T(\partial V/\partial T)_p - V}{C_p}. \quad (8.28)$$

The J-T coefficient can be positive, negative, or zero. For an ideal gas, $\mu_{JT} = 0$, showing that any heating or cooling effect is attributable to intermolecular forces. For nitrogen gas, a J-T expansion cools the gas for $T < 621\,\text{K}$, while helium cools only for $T <\sim 45\,\text{K}$. Figure 8.22 shows an apparatus that does repeated J-T expansions. During a J-T expansion: (i) The gas experiences

---

[14]See, e.g., Callen, H. B. *Thermodynamics and an Introduction to Thermostatistics* (John Wiley & Sons, New York, 1985), p. 165.

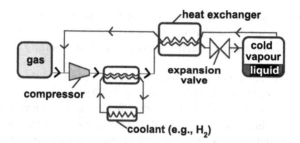

Figure 8.22: Repeated Joule-Thomson cycles, each throttling gas through an expansion valve. Some gas is liquefied, with unliquefied, cold vapour returned for another pass through the expansion valve, which produces more liquid.

a rapid drop in pressure and the boiling point becomes $T_b' < T_b$. (ii) If the reduced boiling point T_b' is lower than the gas temperature, the gas will boil. (iii) Boiling requires energy input to break bonds going from liquid to vapour, and the gas cools as it boils.

In 1885, Zygmunt Wróblewski, who with Karol Olszewski had liquefied oxygen in 1883, measured a value of 33 K for the critical point of hydrogen, and had indications that the boiling temperature was approximately 23 K. However, he was unable to induce sustained boiling.[15] In 1898, James Dewar, inventor of the Dewar flask (see Fig. 1.22), succeeded in liquefying hydrogen, and reported a boiling point of 20 K. Helium was the last element to be liquefied in a landmark experiment by Heike Kamerlingh Onnes in 1908. For example helium boils at 4.22 K at atmospheric pressure, but if its pressure is reduced by a factor of about 15, say, to 0.066 atm, and maintained at that level, its temperature will drop to about 2.2 K. Recall that the temperature is determined by the vapour pressure and *vice versa*. Cooling by rapid boiling requires continuous pumping down of the liquid with a vacuum pump. As this happens the vaporisation energy (*latent heat*) is supplied by the liquid itself, and it is this energy transfer that leads to the cooling effect. Kamerlingh Onnes reduced the helium's temperature from 4.2 K to approximately 1 K by continuous pumping.

Key Point 8.24 *The basic techniques described here have been used to liquefy all the stable elements. A system can always be cooled to the coldest temperature energy sink available. To cool further, one must use an adiabatic process. For a gas, this can be an adiabatic expansion in a cylinder, or expansion through a porous plug (Joule-Thomson process), or through a nozzle, as in Fig. 8.21(a)-(c). These techniques can be fine tuned in a number of ways: (i) The J-T process can be repeated multiple times, as in Fig. 8.22. (ii) The liquid of one material can be used to pre-cool a substance prior to liquefaction.*

---

[15] Unfortunately, while studying other properties of hydrogen in 1888, Wróblewski was severely burned in a lamp oil accident, and died soon thereafter.

(iii) The liquid coolant temperature can be lowered by pumping off the vapour to promote boiling and temperature decrease. (iv) A cascade of processes (i–iii) can be used with substances having successively lower boiling points.

In their book on cryogenic engineering,[16] Timmerhaus and Reed divide progress in cryogenics (temperatures $\lesssim$ 130 K) into three eras. The first, *thermodynamics fundamentals*, was from the primary development of thermodynamics in about 1850 until approximately 1900, ending with the liquefaction of hydrogen and some refinements of experimental techniques. The work of Kamerlingh Onnes, described in the next section, began what Timmerhaus and Reed called the second era of cryogenics, *improved liquefaction technology*. This lasted until about 1950. The third era, *application development*, which includes adiabatic demagnetisation (Sec. 8.6.5), is still in progress. Notably, Kamerlingh Onnes applied his skills to further cryogenic investigations, one of which was his discovery of superconductivity in mercury.

8.6.3   Liquid helium-4

After James Dewar liquefied hydrogen, there was one element that had not been liquefied, helium, the least reactive element. The isotope Helium-4 (^{4}He) which comprises 99.9998% of naturally occurring helium, has a complete outer shell of electrons. It resists accepting another electron or forming covalent bonds with other atoms. The forces between helium atoms are extremely weak, and only slowly moving ^{4}He interact significantly with nearby atoms, forming liquid clusters. As a result, helium does not liquefy until it reaches the ultra-cold temperature of 4.22 K. Temperatures at or below 4 K are called *ultra-cold.*

Reaching that low temperature required enormous experimental skills, excellent equipment, and an ability to use substantial amounts of liquid hydrogen for pre-cooling. Also needed was the availability of helium, which occurs as alpha decay (i.e., helium nuclei) within earth's crust, and must be mined. Kamerlingh Onnes extracted it from monazite sand, which Heike's brother Onno Kamerlingh Onnes, director of the Commercial Information Office in Amsterdam, had helped obtain from America. The sand was heated, making the individual grains explode, releasing the gas. Four chemists worked for several months removing impurities, and on July 10, 1908 Heike Kamerlingh Onnes successfully liquefied helium. He was awarded the Nobel Prize in 1919 for his investigations on the properties of matter at low temperatures which led to the production of liquid helium.

The Leiden laboratory (see Fig. 8.23) was unique in the ability to do experiments within a few degrees of absolute zero. After liquefying helium, Onnes tried to reach the assumed vapour-liquid-solid triple point for ^{4}He,

---

[16]Timmerhaus, K. D. & Reed, R. P. E. *Cryogenic Engineering* (Springer, New York, 2007).

Figure 8.23: Many visitors came to the Leiden Cryogenic Laboratory after the liquefaction of helium-4. In this 1919 photo are: Paul Ehrenfest, Hendrik Antoon Lorentz, Niels Bohr and Heike Kamerlingh Onnes, with the helium liquefier in the background. (Reprinted with permission from National Museum Boerhaave Leiden, Leiden.)

but did not find one, and we now know that helium does not have a vapour-liquid-solid triple point! It is the only element lacking such a triple point.

Despite experimental skills, Kamerlingh Onnes did not discover the fact that at temperatures between 1.77 K and 2.19 K, ^{4}He is transformed from a normal liquid into a "superfluid" phase. He did see indications of an abnormal density maximum in the neighbourhood of 2.19 K, but his conservative ways kept him from making this public without ironclad evidence. He did not expect that a new phase of matter was lurking in his lab, and it was not until 1927 that Willem Keesom and Mieczyslaw Wolfke discovered the "lambda transition" in liquid helium, a name suggested by the specific heat curve at the transition temperature, Fig. 8.24(a), which is reminiscent of a mirror image of the Greek letter λ. Note that ^{4}He's specific heat satisfies the third law of thermodynamics: $c \to 0$ for $T \to 0\,\mathrm{K}$.

The superfluid transition is an extreme quantum phenomenon that occurs dramatically as the liquid's temperature is lowered by pumping on normal liquid helium. At the transition, the normal rapid bubbling that accompanies boiling suddenly stops, and the liquid becomes mysteriously still. Even after this work, which suggested the existence of a phase transition, it took 11 more years for the discovery that liquid helium becomes a superfluid (He II), made independently by Pyotr Kapitsa in Moscow, and independently by John F. Allen and Donald Misener at the University of Toronto.

The remarkable work of Kamerlingh Onnes began the field of low temperature physics. A related major finding was the discovery by Onnes of superconductivity in mercury and other metals at sufficiently low temperatures, i.e., the electrical resistance falls to zero, giving rise to persistent

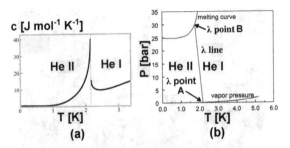

Figure 8.24: (a) Specific heat of ^{4}He in the neighborhood of the lambda point. (b) Pressure-temperature phase diagram of ^{4}He. (Pane (a) is made available under the Creative Commons CC0 1.0 Universal Public Domain Dedication. Pane (b) is printed with permission under the terms of the GNU Free Documentation License.)

currents without power input. Another was that helium did not solidify under its own vapour pressure. The strange behaviour of helium is attributable to: (i) the very weak forces between helium atoms, causing its low boiling point, 4.2 K, and (ii) helium's low mass, which leads to substantial zero-point vibrations, that deter freezing. Theorists have predicted that at sufficiently high pressures and low temperatures, solid helium-4 is achievable. At currently achievable pressures, ^{4}He remains a liquid arbitrarily near $T = 0$ K. The phases of ^{4}He are visualised in Fig. 8.24(b). In liquid helium at low temperature, the product mT of mass × temperature becomes small enough that the thermal wavelength, Eq. (4.3), $\lambda = h^2/(2\pi/mkT)^{1/2} \geq$ mean interparticle distance. Wave functions of adjacent atoms overlap, and the atoms are strongly correlated and lose their individual identities.

Example. The superfluid has a host of remarkable properties, one of which is illustrated in Fig. 8.25(a). In (a), two chambers with equal masses of He II at the same temperature $T_L = T_R < 2.19$ K are connected by a fine capillary tube, so narrow that superfluid, but not normal helium can flow through. Both chambers have floating-piston ceilings. In (b), the right chamber is put in contact with a hotter reservoir at temperature $T_R > T_L$, which causes superfluid to flow from left to right to equalise the pressure. The left floating piston drops and the right one rises. In both chambers there is a mixture of normal and superfluid ^{4}He, with a greater fraction of normal fluid on the warmer right.

The superfluid is in its ground state and carries *zero entropy*. albeit with nonzero internal energy, so the entropy of the helium mixture equals the entropy of its normal component. To maintain the equilibrium fraction f_L in Fig. 8.25(b), energy must flow from the helium on the left to the reservoir as superfluid flows from the left chamber. Similarly, as superfluid becomes normal fluid on the right, energy flows from the reservoir at temperature T_R, to maintain the equilibrium fraction f_R. The energy flows are illustrated in Fig. 8.25(b). Superfluid passes through the capillary without measurable

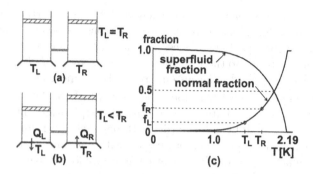

Figure 8.25: Two identical chambers contain the same mass of superfluid ^{4}He. (a) The chambers have equal temperatures. (b) The right chamber has a higher temperature, and fluid flows from left to right. (c) The equilibrium fractions of normal and superfluid ^{4}He in a mixture of the two for $T < 2.19\,\mathrm{K}$.

resistance. As energy flows from the higher temperature reservoir, superfluid becomes normal fluid. The increased pressure lifts the floating piston. Net energy $(Q_R - Q_L)$ flows from the hotter to the colder reservoir. This creates a height difference between the pistons, converts superfluid to normal ^{4}He, and heats it from $T_L \to T_R$. This process is irreversible because energy is exchanged through finite temperature differences. The second law of thermodynamics guarantees that the net entropy change of the helium plus that of the reservoirs is non-negative. **End of example.**

8.6.4   Helium-3

One of the best examples of the difference between Bose-Einstein and Fermi-Dirac systems becomes clear comparing the low temperature behaviours of ^{4}He and ^{3}He. A good review of the differences and of the road to discovering extraordinary properties of ^{3}He has been written by Nobelist David. M. Lee.[17]

For temperatures below the lambda temperature $T_\lambda = 2.19\,\mathrm{K}$, ^{4}He behaves as two fluids, normal and superfluid, the latter exhibiting superflow through pores and cracks. The superfluid has zero entropy, and does not interact with the walls of the container dissipatively. In contrast, the normal component transports (heating) energy efficiently and has nonzero viscosity, and can exchange energy with container walls. The superfluid liquid ^{4}He behaviour is a Bose-Einstein condensation phenomenon, as described in Sec. 4.1.1.

In contrast, ^{3}He atoms are fermions, because each atom contains an odd number of spin $1/2$ elementary particles. Its three nucleons are accompanied by two electrons with opposite spins, and has net spin $= 1/2$. At millikelvin temperatures, there is insufficient energy to excite electron spins, but the

---

[17]D. M. Lee, "The extraordinary phases of liquid ^{3}He," Rev. Mod. Phys. **69**, 645–665 (1997).

proton and neutron spins *can* be excited. This noteworthy behaviour is possible because the magnetic moments of electrons and nucleons are inversely proportional to their particle masses. The nucleon mass is ~ 1836 times more massive than electrons, so their magnetic moments are $\sim 1/1800$ times smaller than each electron's, and adjacent energy levels are more closely spaced.

We can compare ^{3}He with a superconductor, which displays the remarkable property of zero conductivity below a critical temperature. The theory that explains this, put forth by John Bardeen, Robert Schrieffer, and Leon Cooper in 1957, is based upon electrons pairing up with one another to form *Cooper pairs* that behave as bosons. One can compare a Cooper pair with two rock and roll dancers with similar, *correlated* dancing moves even when they are separated by other dancers. The strong correlation of Cooper pairs has them marching in lock step. Can similar pairing occur in ^{3}He, a fermion liquid?

In a superconductor at low enough temperatures, electrons form pairs when a net attractive force, even a weak one, exists. As an electron moves through a lattice of positive ions, regions of higher density ions form, which attract other electrons. The lattice is central in establishing an attractive force between electrons, countering the normal strong Coulomb repulsion. This comes from the dynamic response of the positive ions forming a crystal, and is consistent with fact that the superconducting transition temperature depends on the atomic mass of the lattice's isotope (the *isotope effect*). At a superconductor's critical temperature T_c, it becomes energetically favourable to form pairs.

Helium-3 liquefies at temperature 3.2 K. Because of ^{3}He's relatively strong interatomic forces, it seemed likely that if Cooper pairs formed in liquid ^{3}He, they would differ from the Cooper pairs in superconductors. The strong short-range repulsion of the ^{3}He prevents such pairing, and eventually it was found that superfluid ^{3}He pairing with *odd* angular momentum ℓ properly describes the properties of superfluid ^{3}He. The physics is complex, but the gist is that at low enough temperatures, it becomes energetically favourable to form Cooper pairs with nonzero spin and odd orbital angular momentum. Superfluid ^{3}He forms Cooper pairs even without a lattice.

In his Nobel lecture, Lee describes his path toward his prize-winning work, which led to the discovery that ^{3}He has a density maximum at $T \approx 0.32$ K. He invoked the Clausius–Clapeyron[18] equation (4.16), to argue that the entropy per particle (or per mole) of the liquid will be less than that for the solid at low temperatures, but larger than the solid entropy at high temperatures, and at the temperature where the $s_{liq} = s_{sol}$, the melting curve has a minimum. Figure 8.26(a) shows this behaviour, and additionally, the molar entropy of the solid approaches $\sim 0.69R \approx R \ln 2$ for high temperature, suggesting that the entropy is primarily from randomised spin-1/2 nuclei.

Figure 8.26(a) shows the pressure temperature phase boundary (melting line). Note that the solid and liquid ^{3}He entropy replace the vapour

---

[18]Benoit Paul Émile Clapeyron was one of the founders of thermodynamics.

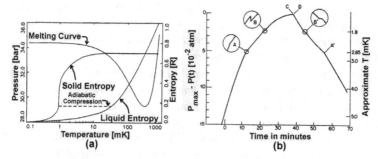

Figure 8.26: (a) Melting pressure of ^{3}He *vs.* temperature, and the entropy of liquid and solid ^{3}He *vs.* temperature on a semilog plot. (b) Compression followed by decompression of ^{3}He, first cooling and then heating it. (Pane (a) is reprinted with permission from D. M. Lee, "The extraordinary phases of liquid 3He," Rev. Mod. Phys. **69**, 645–665 (1997). Copyright 1997 by the American Physical Society. Pane (b) is reprinted with permission from D. D. Osheroff, R. C. Richardson & D. M. Lee, "Evidence for a New Phase of Solid He3," Phys. Rev. Lett. **28**, 885–888 (1972). Copyright 1972 by the American Physical Society.)

and liquid entropies for the common evaporation and condensation in the Clausius–Clapeyron, Eq. (4.16), i.e.,

$$\left(\frac{dP}{dT}\right)_{\text{melting line}} = \frac{(s_{\text{liq}} - s_{\text{sol}})}{(v_{\text{liq}} - v_{\text{sol}})}. \tag{8.29}$$

For $T \lesssim 0.315\,\mathrm{K}$, Fig. 8.26(a) shows that $(s_{\text{liq}} - s_{\text{sol}}) < 0$, and because $(v_{\text{liq}} - v_{\text{sol}}) > 0$, $dP/dT < 0$. Also, when $T \gtrsim 0.315\,\mathrm{K}$, $(s_{\text{liq}} - s_{\text{sol}}) > 0$, while $(v_{\text{liq}} - v_{\text{sol}}) > 0$, and $dP/dT > 0$.

Key Point 8.25 *The behaviour of the phase boundary (melting) curve is unique in that at temperatures where its slope is negative, the addition of heating energy is required to freeze the ^{3}He liquid! The strange behaviour of liquid ^{3}He for temperatures less than 0.315 K includes the property of it cooling under an adiabatic pressure increase—which makes a technique called Pomeranchuk cooling possible. This requires cooling liquid ^{3}He to at least 0.315 K, followed by compressing the cold liquid adiabatically. Cooling under pressure is consistent with the negative slope of the melting curve in Fig. 8.26(a) for $T < 0.315\,K$.*

Prior to 1971, no cryogenic laboratory had achieved temperatures low enough to see superfluidity in liquid ^{3}He. The advent of the ^{3}He-^{4}He dilution refrigerator was a major advance, using the property that in ^{3}He-^{4}He mixtures, cooling occurs as the two components separate. The dilution refrigerator made it possible to achieve and maintain temperatures of $\sim 10\,\mathrm{mK}$. From that temperature, Pomeranchuk cooling could decrease the ^{3}He's temperature further. Figure 8.26(b) shows the process of Pomeranchuk compression for about 37 s, lowering the temperature of the ^{3}He, followed by decompression for about 31 s, raising its temperature.

In Fig. 8.26(b), the circled blowup A shows a discontinuity in the slope at temperature 2.65 mK during cooling, and a sudden pressure drop at B followed by the previous slope. Similar discontinuities appear during decompression at A$'$ and B$'$. This behaviour signified discontinuities in $(dP/dT)_{\text{phase boundary}}$ on the pressurisation curve of liquid ^{3}He in equilibrium with solid ^{3}He at nearly 3.44 MPa in a compressional cooling cell. Subsequent investigation using nuclear magnetic resonance showed that the A and B features were associated with dynamic magnetic effects in the liquid.

Together with experiments by others, this supported the formation of superfluid phases at A and B. For example, the motion of a vibrating wire was observed to change dramatically at the temperatures and pressures of A and B, indicating major changes in flow properties. The evidence amassed led to the conclusion that two new superfluid phases of ^{3}He had formed. Douglas Osheroff, Robert Richardson and David Lee were awarded the 1996 Nobel Prize in Physics for their work. Formation of the superfluid phases requires temperatures about one thousand times smaller than for superfluidity of ^{4}He.

8.6.5 Adiabatic demagnetisation

Cryogenic laboratories use salts containing paramagnetic ions to achieve ultra-low temperatures. An example is ferric ammonium alum, with a net spin of 5/2. Such paramagnets have spin entropy vs. temperature curves similar to those in Fig. 8.27(a).

Key Point 8.26 *If we begin with magnetic field intensity H_2 and temperature $T(H_2)$, and slowly and adiabatically decrease H from H_2 to $H_1 < H_2$, and if the path followed has constant spin entropy, the path can be denoted by a leftward horizontal arrow that terminates on the H_1 curve at temperature $T(H_1) < T(H_2)$. Bringing the external field intensity to zero reversibly would take the system to the $H = 0$ curve, with the same entropy value it had throughout, and with final temperature $T_0 < T(H_1) < T(H_2)$. This cooling by adiabatic demagnetisation is used to achieve ultra-low temperatures.*

Our discussion so far has focused on the entropy of the spin system, namely, that of the magnetic moments only, without regard for other contributions to the total entropy. If the paramagnetic salt is a solid, there are also entropy contributions from lattice vibrations and possibly from electrons. Additionally, the goal is usually to cool a body, X, which also contributes to the total entropy. Reversible adiabatic demagnetisation assures constancy of the *total* entropy of the system, and not of the spin system alone. Therefore the graph

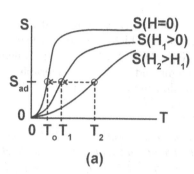
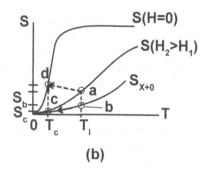

Figure 8.27: (a) Entropy S vs. temperature curves for a paramagnetic spin system system for various magnetic field values. (b) Entropy S of the spin system and entropy S_{X+0} of all other constituents.

in Fig. 8.27(a) does not describe real adiabatic demagnetisation correctly, and it is well to rethink the experimental setup.

Key Point 8.27 *Magnetic cooling is usually used only after mechanical refrigeration techniques have been exhausted, i.e, helium has been liquefied and has been pumped down, so its temperature is lowered from 4.22 K to $T_i \approx 1 K$.*

Let T_i be the temperature of the paramagnetic salt and all other bodies including X (the body being cooled) in a uniform magnetic field $H > 0$. The entire system is thermally insulated, and H is slowly brought to zero. This slow, pure-work process is nearly reversible, and the total entropy is approximately constant. It is helpful to view the system as the paramagnetic spin system plus the paramagnetic salt's lattice, plus the body X and any other entropy-bearing parts. I label everything other than the spin system by $X+0$. As the full system is cooled, the entropy S_{X+0-} must decrease to offset the entropy increase of the spin system, to maintain constant total entropy.

This is depicted in Fig. 8.27(b), which shows three entropy vs. temperature curves. Two of these, labeled $S(H > 0)$ and $S(H = 0)$, are for the pure spin system considered above, before and after the demagnetisation, respectively. The third curve is labeled S_{X+0}, the net entropy of the lattice, electrons and system X to be cooled, which is not affected appreciably by the value of H. The initial state of the spin system is a and that of everything else is b. As H is reduced to zero, the spin system attains state d, and everything else attains state c. The final temperature of the entire system is T_c. Entropy constancy means that $S_a + S_b = S_c - S_d$.

Key Point 8.28 *With initial temperature $T_i \approx 1K$, adiabatic demagnetisation involving electron-spin paramagnetism can reduce the temperature to the millikelvin (mK) region. This diminution of temperature by a factor of $1/1000$ is indeed impressive!*

Even more extreme cooling, to the microkelvin (μK) temperatures, can occur using nuclear adiabatic demagnetisation, which involves nuclear spin paramagnetism. This is explained in Key Point 8.29 The experimental challenges encountered in adiabatic demagnetisation are considerable, and interested readers are directed to the excellent exposition in G. K. White's book.[19]

Key Point 8.29 *Nuclear spin adiabatic demagnetisation is effective when the initial temperature T_i is in the mK region, and the magnetic field is high enough to reduce the spin system entropy significantly. This is possible because nuclear magnetic moments are about 10^3 times smaller than those of electrons. Thus there are many accessible magnetic states at low temperatures and energies, and substantial entropy. This nuclear spin system entropy is first lowered by the field, and then brought to zero (approximately) reversibly,. The nuclear spin entropy increases while the entropy and internal energy of body X are reduced. Temperatures in the microkelvin (μK) region can be reached.*

---

[19]White, G. K., *Experimental techniques in low temperature physics* (Oxford University Press, Oxford, 1979).

CHAPTER 9

Sanctity of the 2nd Law of Thermodynamics

CONTENTS

9.1 MAXWELL'S DEMON

9.1.1 Statistical nature of the 2nd law

Birth and evolution of the demon. The great physicist James Clerk Maxwell made fundamental advances in multiple fields, perhaps the most notable being electricity and magnetism, where the fundamental set of equations is commonly referred to as Maxwell's equations. Near the end of his 1871 book, *Theory of Heat*, Maxwell wrote in a section called "Limitation of the second law of thermodynamics":

> *One of the best established facts in thermodynamics is that it is impossible in a system enclosed in an envelope which permits neither change of volume nor passage of heat, and in which both the temperature and the pressure are everywhere the same, to produce any inequality of temperature or of pressure without the expenditure of work.*

Maxwell's comment can be interpreted to mean that nature is *fair*, apportioning energy *equitably* for macroscopic systems in equilibrium.[1] A human or automaton could generate regions of higher and lower temperature and/or pressure by doing external work. Maxwell clarified his intent:

1. As long as we operate macroscopically, we cannot influence individual molecules of which matter is made. Maxwell imagined a "being" who could actually follow molecules one by one and also had human-like intelligence.

2. Such a being could observe and control the molecules of a gas without influencing their energies. Having obtained his speed distribution in 1860, Maxwell had a clear mental image of gas molecules. He knew that they do not have uniform speeds, and he divided the gas into two halves, A and B, with a central partition containing a small hole.

3. The intelligent *being* observes the molecules individually, as depicted graphically in Fig. 9.1. It selectively opens and closes the hole, allowing faster molecules to pass from A to B, and slower ones to move the other way, keeping the number of molecules fixed in each chamber

4. Maxwell's being does not exchange energy with molecules, but simply opens and closes a small door in the separating partition to generate a temperature difference between chambers.

5. Maxwell concluded that the *being* would violate the second law of thermodynamics. Most important, Maxwell observed that without an intelligent being, by simply leaving the little door in the partition open

---

[1]The concept of equity is used in this book in Secs. 2.4 and 3.3.3.

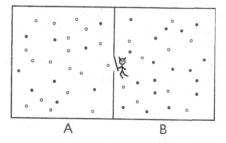

Figure 9.1: Depiction of Maxwell's "intelligent being," sorting molecules according to their speeds, resulting in more high speed (solid circles) molecules in chamber B and more lower speed (open circles) molecules in chamber A.

for a while, there was a nonzero probability that fluctuations would lead to the same temperature difference that the being achieved. Maxwell's main point was: *the second law of thermodynamics is a statistical law, not an absolute one.*

The imaginary intelligent being works in the microscopic world, which differs considerably from our macroscopic world. Ironically, with modern technology, it is now possible to control systems at the nanometer size scale. This is not *despite* fluctuations, but rather, *because* of them! It turns out that at the nanoscale, a probabilistic approach is still used, albeit somewhat differently.(See the discussion of Brownian ratchets in Sec. 9.4.)

Naming the demon. In 1874, William Thomson introduced the term "Maxwell's intelligent demon." He observed that Maxwell envisaged this fictitious character as a "being endowed with free-will and fine enough tactile and perceptive organisation to give him the faculty of observing and influencing individual molecules of matter." About five years later, Thomson wrote that he intended the term *demon* to be a creature of imagination who possessed specific mechanical skills, differing from actual animals only in its smallness and agility. To Thomson, the purpose of the demon was to aid us in understanding the *dissipation of energy* in nature. He gave as examples, selectively operating on atoms to cause one-half a closed jar of air or a bar of iron to become hot and the other cold, and sorting molecules in a mixture to reverse the natural process of diffusion.

Key Point 9.1 *Maxwell's thought experiment dramatised the fact that the second law is a statistical principle that holds nearly all the time for a system composed of many molecules, i.e., a macroscopic system. Although there is a nonzero probability that molecular transfers similar to those achieved by the demon will occur if a hole in a separating partition is simply left open for a while, that probability is exceedingly small for a macroscopic system.*

Maxwell's dissatisfaction with the term "demon." Maxwell evidently was not satisfied with Thomson's view of his intelligent being, including the term "demon." He clarified that what Thomson called a demon was no more than a small being that could open and close valves so as to sort molecules without exchanging energy with them. Maxwell emphasised that his being was nothing more that a *valve*, which is very different from being a demon. He also reiterated that his intended purpose was to show that the second law of thermodynamics had only a "statistical certainty." Maxwell also pointed out that less intelligent demons could operate a one-way valve, to allow molecules to pass only from A to B, thereby increasing the pressure in B and decreasing it in A, which would enable the performance of work, say, lifting a weight with a pulley. In summary, Maxwell wrote, "Call him no more a demon but a valve like that of the hydraulic ram, suppose."

Although Thomson's term *demon* was an incorrect characterisation of Maxwell's intent, it pervades dictionaries, textbooks, and research articles. For example, the Second Edition (1989) of The Oxford English Dictionary, under its entry for James Clerk Maxwell, defines Maxwell's demon as a being imagined by Maxwell to be able to sort molecules so as to achieve a temperature difference, "in contradiction to the second law of thermodynamics." Fortunately, the *Webster's Third New International Dictionary* definition properly cites Maxwell's intention to "illustrate limitations of the second law of thermodynamics."

Key Point 9.2 *It is emphasised that Maxwell did not intend to suggest a violation of the second law, but rather to reveal its statistical nature. The common emphasis on the possible violation of the second law of thermodynamics has been adopted by many subsequent researchers. In this sense, Maxwell's demon became a puzzle to be solved. Researchers asked, "If such a demon cannot defeat the second law, then why not? If it can defeat the second law, how does that affect the status of the second law?"*

Key Point 9.3 *Maxwell did not relate his mental construction to entropy, and initially misunderstood Clausius's definition of entropy, i.e., he adopted a non-standard definition in early editions of his Theory of Heat, namely, the entropy of an isolated system cannot increase, but whose changes must be less than or equal to zero. Despite that early confusion, Maxwell's demon has helped reveal an important linkage between entropy and information. Unfortunately, Maxwell's death in 1879 (at age 48) prevented him from knowing the substantial impact of his thought experiment.*

9.1.2 The Szilard engine

Leo Szilard introduced his famous model more than 60 years after Maxwell introduced his *intelligent being.* Szilard simplified the problem in two ways: He reduced the macroscopic gas of Maxwell to a single molecule, and the demon's sorting was based on its *location* rather than its *speed.* Szilard's engine is

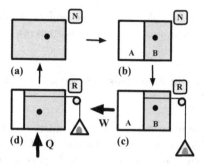

Figure 9.2: The four-step Szilard engine described in the text. The small rectangles show the memory state at each step. N is the standard state and (R,L) denote that the molecule is in the (right, left) sides of the partition.

depicted in Fig. 9.2, and it acts as a *pressure demon*. It creates a pressure differential that can be used to do work on an external load, e.g., by lifting a weight as shown in parts (c) and (d) of Fig. 9.2.

Step 0 of the Szilard engine cycle has the entire volume V of a cylinder is available to the molecule, as shown in Fig. 9.2(a). The small, upper-right rectangle labeled N represents the demon's memory state N, which stands for *no data*. The four steps of the engine are:

1. $(a) \rightarrow (b)$. A partition is placed into the cylinder, dividing it into two equal chambers. The memory state remains N.

2. $(b) \rightarrow (c)$. The demon determines the side in which the molecule resides. The molecule is shown in the right in Fig. 9.2(b) and (c). The demon's measurement is recorded and the memory state is now R.

3. $(c) \rightarrow (d)$. The partition is replaced by a piston, and the recorded result is used to couple the piston to a load, namely a weight (shown here pulling the piston rightward) that is lifted by work W. In this process, the gas pushes the piston to the left end of the cylinder. The energy lost by the gas is replenished by a heat process moving energy $Q = W$ from an energy reservoir at constant temperature to the gas during the volume expansion. The memory state remains R.

4. $(d) \rightarrow (a)$. The partition is removed and the demon's memory state is changed to N. In addition, the pulley configuration constitutes a *second* memory of the side the molecule was in after piston placement in (b) and the pulley must be removed. This erases that second memory.

Key Point 9.4 *Repeating the process ad infinitum, using the Szilard engine, a Maxwell's demon could convert an arbitrary amount of energy via a heat* *process, to work lifting a weight. This would violate the Kelvin-Planck form of the second law of thermodynamics, converting energy from a single reservoir completely to work.*

There are a number of loose ends and nuances of the Szilard engine that should be mentioned.

1. For the Szilard engine cycle, the load must be varied continually, e.g., using grains of sand, matching the average force on the piston by the molecule to keep the process slow and quasistatic. Szilard and most subsequent researchers did not mention the need for a variable load.

2. Just how the demon determines which chamber the molecule is in after partition placement is not specified.

3. A one-molecule gas is not an everyday occurrence and is not *macroscopic*. Thus, applying normal thermodynamics is questionable.

4. A demon is a physical entity that presumably contains sufficiently many molecules to possess a temperature. If it continually receives energy input via light signals (or other means), its temperature will rise unless it transfers energy to its surroundings. This would confound the basic puzzle, and such energy exchanges are usually neglected.

5. If a lamp is used to generate light signals, photons that miss the demon will heat up the gas and or container walls. Such phenomena threaten the assumption of constant-temperature operation, and most treatments of Maxwell's temperature-demon ignore these details.

Despite such concerns, the Szilard model has been taken seriously and has been the subject of many research articles for over 90 years.

Szilard observed: "One may reasonably assume that a measurement procedure is fundamentally associated with a certain definite average entropy production, and that this restores concordance with the second law. The amount of entropy generated by the measurement may, of course, always be greater than this fundamental amount, but not smaller." He further identified the "fundamental amount" to be $k \ln 2$. His observation was the beginning of information theory. Interestingly, as discussed in the next section, it is now believed it is not *measurement* that must generate sufficient entropy to prevent violation of the second law of thermodynamics, but rather *erasure* of the information recorded by that measurement.

The ingenuity of Szilard's engine is striking. It allows thermodynamic analysis and interpretation, and also entails a binary decision process. Thus, long before the existence of modern information ideas and the computer age, Szilard had the foresight to focus attention on the *information* associated with a binary process. In doing so he discovered the binary digit, or 'bit' of information. Szilard's observation that an inanimate device could effect the required tasks – obviating the need to analyse the thermodynamics of complex human observers, was a precursor to cybernetics.

Key Point 9.5 *Although Szilard did not fully solve the Maxwell's demon puzzle through his one-molecule gas, he identified the three central issues of information gathering as we understand them today—measurement, information, and memory—and he established the underpinnings of information theory and its connections with physics.*

9.1.3 Measurement, memory, erasure

The role of memory in measurement. There is more to this story. A relevant question is: What are the energy and entropy implications of Step 2, the acquisition of information, and Step 4, erasure of information? Before examining them it is well to appreciate that by introducing the concept of information into physics, Szilard's model was at the cutting edge of computer science and established a deep connection between that science with energy and entropy. With Step 1, the demon's measurement of the molecule's left-right location acquires 1 bit of *information*. Here the binary choice is between left and right. A typical dictionary definition of *measurement* is: *a figure, extent, or amount obtained by measuring.* For an example, suppose you measure the length of a pencil and find the result 12.7 *cm*. After you make that measurement, you record it in your memory or on paper. You might forget the result and lose the paper.

Key Point 9.6 *The term measurement implies that the result is recorded on paper, digitally, or another way, and is retrievable. Rolf Landauer published an article entitled, "Information is physical" to emphasise this point.*[2] *Without recording the data, a measurement is incomplete because existing knowledge is no different from that before the measurement.*[3] *The measurement process includes (i) acquisition of data and (ii) a physical memory of the data.*

The entirety of scientific literature, consisting of books, journal articles, databases, conference proceedings and the like constitutes a memory of information established by scientists over the years.

How much energy is needed to measure one bit? Is there a minimum, i.e., *threshold* energy needed to measure a single bit of information? The answer is no. If we record the result 12.7 cm in pencil on a sheet of paper, multiple times, pressing less hard each time, we get a sequence of lighter and lighter impressions, as in Fig. 9.3. There is no known general minimum

---

[2] R. Landauer, "Information is Physical," Physics Today **44**, 23–29 (1991).

[3] If a human observer remembers the measurement and can retrieve it, that constitutes a physical memory.

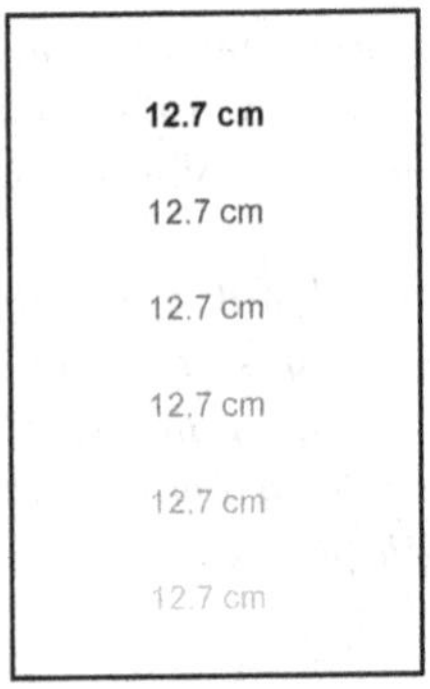

Figure 9.3: A length measurement, recorded in pencil on paper six times, with successively less effort by the writer. There is no known fundamental limit to how much energy is needed to record a measurement.

pressure on the pencil to have a valid record of the information. If we reach what seems to be a minimum, another pencil might enable us to press down even less, still giving a readable result. And if we cannot read it, perhaps using light of a different colour, e.g., ultra violet, would help.

Early researchers took Maxwell's specification of "seeing" molecules seriously and assumed the use of light signals. This requires some kind of light source, which itself requires energy input, and the light beam would exert pressure on each measured molecule, thereby altering its energy. Clearly, the use of light for detection opens the colloquial *can of worms*. Some have considered detecting molecules via their magnetic moments, or by detection of Doppler-shifted radiation from molecules. A prudent view is that one must not prejudice the demon's operation by assuming the use of light signals. That is too restrictive. Is measurement *necessarily* irreversible and does it have a minimum energy cost? It is believed that there is no fundamental energy threshold to acquire and record information.

The task of erasure. Once a measurement has been made, erasure of that data is a nontrivial process. If I measure 12.7 cm, record it in my brain and then tell it to you, it is recorded in *both* our brains. For total erasure, you and I would both have to forget it. That is not necessarily easy or assured. If I suffer total memory loss, you will still have the data in your brain, and erasure is not complete.

More commonly, data might be stored magnetically, with magnetic domains having orientations specific to that data. Erasure of those domains must be done without knowledge of their details, e.g., using a rapidly oscillating and randomly directed magnetic field, or by heating the magnetic media to a temperature sufficiently high to randomly orient the domains. More commonly when dealing with binary digits, e.g., *up* and *down*, erasure could

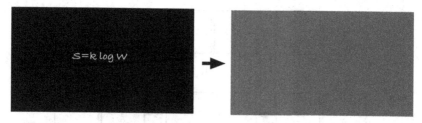

Figure 9.4: Erasure without knowledge of the chalkboard's contents requires coverage of the entire board. After erasure, chalk dust covers the board uniformly, making it grey rather than black.

take all domains to a pre-selected *standard* state such as zero. In this case erasure of the data 100001 would lead to 000000.

A more familiar example is that of erasing a chalkboard that contains an equation, which is a *physical* manifestation of information. The erasure process must cover the entire board because the animate or inanimate erasing entity has zero knowledge of the board's contents. There *could* be writing anywhere. If the erasing entity, e.g., a teacher, did have knowledge of the board's contents, then there would be two memories of the written equation, and erasing the board would leave the other one intact. The entire board must be erased even if the Boltzmann entropy expression $S = k \log W$ is the only thing written on it, as shown in Fig. 9.4. Similarly erasure of a message written in pencil on paper, without knowledge of the contents, requires rubbing the entire paper area with an effective eraser.

Another example of nontrivial memory erasure is destroying a pile of old newspapers. Erasure requires an operation like paper shredding, though with enough time and effort, the resulting set of paper shreds could always be reconstituted. Another way is erasure by incineration. Erasure of a computer's high capacity hard disk requires a lengthy process that covers the entire disk. If a computer's hard drive is damaged and crashes, some or all the information can still reside on the disk and can be read. *That* is an example where erasure, if any, can be partial. Full erasure of a hard drive requires a special effort.

Borrowing results from the theory of computation. The important role of memory and in particular memory erasure came from the work of IBM Watson Research Center researchers Rolf Landauer and Charles Bennett. Landauer had focused on the minimum cost of erasure of one bit of information, and established his famous *Landauer's principle* in 1961. In 1982 Bennett realised that both Landauer's work on erasure and his own work on *reversible computation* were relevant to the Maxwell's demon conundrum.

Key Point 9.7 *Charles Bennett showed, in the context of digital computing, that there is no threshold energy or entropy cost for writing information to a memory or reading it from that memory.*

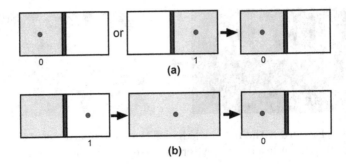

Figure 9.5: (a) Erasure of either 0 or 1 (top line) followed by resetting to the chosen standard state of 0. (b) A way to accomplish erasure via partition removal and resetting via work on the molecule, by pushing the right wall to the centre of the cylinder and adding a new right wall.

To appreciate the nature of erasure, consider a memory system consisting of a cylinder divided into two halves by a central partition, reminiscent of the Szilard engine apparatus. A single molecule in the left chamber represents a 0, and a molecule in the right chamber represents a 1, as shown in Fig. 9.5a. It is assumed here that the cylinder walls with temperature T are in thermal contact with a constant-temperature reservoir at that temperature. How can the erasure and resetting be done? If the initial memory state is 1, a way for the molecule to get to the left half is to remove the partition, then push the right wall to the centre of the cylinder and place a new right wall, as in Fig. 9.5b.

Assuming that in this memory device, the one-molecule gas obeys the ideal gas equations with $N = 1$, the partition removal in Fig. 9.5b increases the entropy of the gas by $k \ln 2$ while the subsequent isothermal compression reduces that entropy back to its initial value, but transfers energy to the reservoir, increasing its entropy by $k \ln 2$. This is an example of:

Key Point 9.8 *Landauer principle*. *Erasure of 1 bit of information requires an entropy generation of at least $k \ln 2$ in the environment. This provides an essential connection between information, energy, and entropy.*

Notice that when the central partition that separates the chambers is removed, according to thermodynamics, the entropy of the one-molecule gas increases. A relevant question is this: If the partition is reinserted, has the entropy been decreased spontaneously? The answer is *no* if we do not know which chamber the molecule is in. Without such knowledge we could not use the reduced volume to do work lifting a weight. Accordingly, it is impossible to state meaningfully that partition placement reduces the entropy. The partition placement brings an uncertainty that muddies the waters and suggests a connection between entropy and information. *This reveals another subtle connection between entropy, information, and energy.*

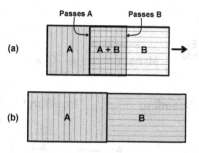

Figure 9.6: Two cylinders begin fully telescoped (not shown) with a mixture of A and B molecules. (a) The right cylinder slides rightward and its left wall allows passage of species A, (vertical lines). The fixed outer cylinder's right wall passes species B (horizontal lines). (b) Species A and B are separated.

A 2-species equilibration model. A very different model with many molecules also demonstrates measurement, memory, memory erasure and resetting.[4] This model is reminiscent of that discussed in Sec. 4.3.3 in connection with the Gibbs paradox and entropy of mixing. It consists of two molecular species A and B. At a specified temperature T, after sufficient time for equilibration, a cylinder of N molecules of species A, will evolve with some of the A molecules becoming type B and some type B molecules transforming to type A. When equilibrium is reached at temperature T, the fractions of each type are $a \geq 0$ and $b \geq 0$ with $a + b = 1$. An example of such species mixing is molecular hydrogen, H_2, which comes in two varieties that depend on their nuclear spin states.

In brief, according to quantum theory, the two atoms making up the diatomic hydrogen, H_2, can either have their nuclear spins parallel (ortho hydrogen) or antiparallel (para hydrogen). These behave differently thermodynamically because for a given rotational energy, ortho hydrogen, with nuclear spin one, has three independent quantum states with the same energy. Para hydrogen, with nuclear spin zero, has only one state. At sufficiently high temperature, in thermodynamic equilibrium, it is three times more likely to find ortho than para hydrogen, and the ratio of ortho to para hydrogen approaches 3 : 1. Letting ortho hydrogen be species A and para hydrogen be species B, $a = 3/4$ and $b = 1/4$.

Suppose we begin with an equilibrium mixture of ortho and para hydrogen with a 3 : 1 ratio. We can in principle separate the two species with a semipermeable membrane that passes only type B molecules, and a second semipermeable membrane that passes only A molecules. Figure 9.6 shows how this can be accomplished by telescoping a cylinder within another cylinder.[5]

---

[4]H. S. Leff & A. Rex, "Entropy of measurement and erasure: Szilard's membrane model revisited," Am. J. Phys. 62, 994-1000 (1994).

[5]A similar procedure was used in Fig. 4.13, where the Gibbs theorem was proved.

(1)   **(2)**

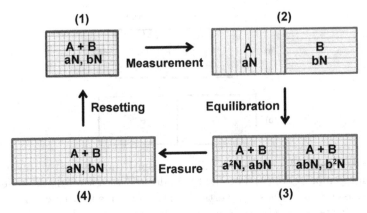

Figure 9.7: (1) Equilibrium mixture of ortho and para hydrogen in volume V, with aN ortho and bN para molecules. (2) Separation by semipermeable membranes doubles the volume. (3) Equilibration brings an equilibrium ortho-para mixture to each chamber of volume V. There are (aN, bN) molecules in the left and right chambers respectively, which constitutes a memory of the original species ratio. Partition removal erases that memory, leading to (4) Resetting by isothermal compression, leads to the initial state in volume V.

Let the initial numbers of type (A, B) molecules be aN, bN respectively, where N is the total number of molecules.

1. **Stage 1.** The initial configuration (1) in Fig. 9.7 is an equilibrium mixture of ortho and para hydrogen with numbers (aN, bN) of type (A, B) molecules with $(a = 3/4, b = 1/4)$.

2. **Stage 2 and process $1 \to 2$.** This shows the system *just* after separation of the species via process $1 \to 2$, which slides an inner cylinder rightward using semipermeable membranes to separate the species, as in Fig. 9.6. The species retain their particle numbers (aN, bN), volume V and temperature T. The separation process requires zero work and there is no heat process, so $Q = W = 0$, $\Delta E_{1 \to 2} = 0$, and the system's entropy change for $1 \to 2$ is $\Delta S_{1 \to 2} = 0$.

3. **Stage 3 and process $2 \to 3$.** Equilibration requires the passage of enough time to bring each gas to the equilibrium ratio of $3 : 1$ relative to the (aN, bN) molecules initially in the (left, right) chambers, i.e., $(a^2 N, abN) = $ (ortho, para) molecules on the left, and $(abN, b^2 N)$ (ortho, para) molecules on the right. The total number of ortho molecules in the left and right chambers is aN and the number of para molecules is bN. No work or heating processes are needed for equilibration, so $\Delta E_{2 \to 3} = 0$. We write the equilibration entropy increase in the left chamber as $S(a^2 N) + S(abN) - S(aN)$. For the right chamber, the entropy change

is $S(abN) + S(b^2N) - S(bN)$. Assuming the ortho and para species entropies are described by the Sackur-Tetrode equation (4.4), doing some algebra shows that the net entropy increase is $\Delta S_{2\to3} = -Nk[a\ln a + b\ln b] > 0$.

4. **Stage 4 and process 3 → 4.** Two kinds of mixing occur upon partition removal. First, ortho molecules from the left mix with para molecules from the right, and para molecules from the left mix with ortho molecules from the right. In each case, both species double their volumes. Using Eq. (4.32), the present notation, the entropy increase from aN ortho and bN para molecules is $\Delta S_{\text{ortho-para}} = Nk\ln 2$. Using Eq. (4.35), ortho-ortho mixing of aN total ortho molecules has $\Delta S_{\text{ortho-ortho}} = aNk[a\ln a + b\ln b + a\ln 2]$. Similarly, para-para mixing of bN molecules has $\Delta S_{\text{para-para}} = bNk[a\ln a + b\ln b + b\ln 2]$. Adding the three entropy changes gives the result $\Delta S_{3\to4} = Nk\ln 2 + Nk[a\ln a + b\ln b]$.

5. **Stage 1 (again) and process 4 → 1.** Process $4 \to 1$ entails resetting, namely, reversible isothermal compression of the equilibrium mixture in volume $2V$ to the initial volume V. External work W is done on the system, sending energy $|Q| = W$ to the constant temperature reservoir. The system's energy and entropy changes are, $\Delta E_{4\to1} = Q + W = 0$ and $\Delta S_{4\to1} = -Nk\ln 2$.

Summary for the cycle 1 → 2 → 3 → 4 → 1. For the full cycle, the system has energy and entropy changes, $\Delta E_{cycle} = 0$, and $\Delta S_{cycle} = -Nk[a\ln a + b\ln b] + Nk\ln 2 + Nk[a\ln a + b\ln b] - Nk\ln 2 = 0$. The concomitant reservoir changes are $\Delta E_{\text{r,cycle}} = W = |Q| = NkT\ln 2$ and $\Delta S_{\text{r,cycle}} = Nk\ln 2$.

Key Point 9.9 *In the cycle of Fig. 9.7, process $1 \to 2$ separates the ortho and para species into separate chambers, and can be interpreted as a **measurement** of the numbers aN, bN, and a **memory** of that information. Process $2 \to 3$, over a long time period, brings an equilibrium ortho-para mixture in each chamber. The numbers of ortho and para hydrogen molecules aN and bN are preserved in the full volume $2V$ of stage 3, which constitutes a lingering **memory**. The **erasure** of that memory is effected with process $3 \to 4$. Resetting, $4 \to 1$, leads back to the original configuration. The binary determination of the number of ortho and para molecules in N hydrogen molecules, requires N bits of information, and the entropy increase of the reservoir implies an entropy generation of $k\ln 2$ per bit, consistent with the Landauer principle for erasure, as in Key Point 9.8.*

Charles Bennett argued using various examples, including a purely mechanical mental model, which he described in a 1987 Scientific American article,[6] that measurement can be done reversibly, and a Maxwell's demon

---

[6]Bennett, C. H., "Demons, engines and the second law," Scientific American, November, 108-116 (1987).

that can operate without entropy generation during measurement is possible. Supporting this is the fact that no work and/or entropy thresholds are known to exist for the needed measurements. Bennett's views are widely accepted.

For the cycle in Fig. 9.7, measurement is reversible, while erasure and resetting results in an increased entropy of a constant-temperature reservoir. This agrees with Bennett's claim that there is no *threshold* energy or entropy cost for writing information to a memory, but erasure is necessarily dissipative. There exists a substantial literature on Maxwell's demon.[7] Critical reviews of aspects of Maxwell's demon have been given by Earman and Norton and Hemmo and Shenker.[8]

Key Point 9.10 *Table 9.1 provides a capsule summary of the life of Maxwell's demon. The study of the demon has led us to fundamental connections between energy and entropy, the focal points of this book. The demon links energy and entropy with information, a concept that had not been appreciated prior to the birth of Maxwell's demon. Erasure of information is special in that it has a threshold energy cost per bit of at least $kT\ln 2$ and generates an entropy increase of at least $k\ln 2$. Measurement, including recording a result, has no known fundamental minimum cost, and can be done with arbitrarily little energy. Landauer's principle has inspired much research since 1961 dealing with the deep connections between physics and information. A reciprocal relation became clear: Maxwell's demon led to the bit of information and to key concepts of information theory, and the science of computation has led to a new understanding of the Maxwell's demon puzzle.*

9.1.4 Maxwell's demon, efficiency, power

Much of the work on Maxwell's demon has been concerned with its exorcism. In a different vein, we can ask how effective a demon can be, regardless of whether it can violate the second law. For example, what rate of energy transfer is attainable by a Maxwell's demon who sorts gas molecules serially, and how much time does it take it to achieve a designated temperature difference, ΔT, across a partition?

It is possible to estimate the demon's effectiveness, and to make numerical estimates based on the energy-time form of Heisenberg's uncertainty principle, as well as by using classical kinetic theory.[9] For a dilute gas at $300\,\mathrm{K}$, the

---

[7]Leff, H.S. & Rex, A. F. *Maxwell's Demon 2: Entropy, Classical and Quantum Information, Computing* (CRC Press, 2002).

[8]J. Earman & J. D. Norton, "Exorcist XIV: the wrath of Maxwell's demon. Part I. From Maxwell to Szilard," Stud. Hist. Phil. Mod. Phys. **29**, 435–471 (1998); "EXORCIST XIV: The Wrath of Maxwell's Demon. Part II. From Szilard to Landauer and Beyond," Stud. Hist. Phil. Mod. Phys. **30**, 1–40 (1999); Hemmo, M. & Shenker, O., "Entropy and Computation: The Landauer-Bennett Thesis Reexamined," Entropy **15**, 3297–3311 (2013). See also the book, Hemmo, M.; Shenker, O., *The Road to Maxwell's Demon* (Cambridge University Press, Cambridge (2012).

[9]Leff, H.S., "Maxwell's demon, power, and time," Am. J. Phys. **58**, 135-142 (1990).

Table 9.1: Phases of the life of Maxwell's demon

Year	Description
1871	The demon was introduced in Maxwell's *Theory of Heat*, to illustrate that the second law of thermodynamics is a statistical law
1911	William Thomson named the *intelligent being* "Maxwell's demon."
1912	Marian Smoluchowski observed that thermal fluctuations would be deadly to a *automatic* device, but left open the possibility that an "intelligent" being might defeat the second law.
1929	Leo Szilard introduced his one-molecule heat engine, utilising a device with a memory, an essential element of intelligence.
1951	Leon Brillouin linked thermodynamic and information entropies.
1961	Rolf Landauer showed that erasure of $1\,bit$ of information in a computer generated at least $kT \ln 2$ of heating to the environment.
1970	Oliver Penrose (independently) discovered Landauer's principle in the context of Maxwell's demon.
1982	Charles Bennett showed that reading and writing digitally could be done with zero dissipation, in contrast with Landauer's principle, and related his and Landauer's findings to Maxwell's demon.

uncertainty principle implies that Power $< 1.5 \times 10^{-6}\,W$. For a gas volume the size of a large room, specifying $\Delta T = 2\,\mathrm{K}$, the demon's processing time turns out to be $> 10^3$ years! With similar assumptions, classical kinetic theory implies a considerably tighter bound, power $< 10^{-9}$ watt, which implies processing times $> 4 \times 10^6$ years. The latter power level is comparable with the average dissipation per neuron in a human brain. These estimates illustrate the practical ineffectiveness of a lone Maxwell's demon using serial processing. Assuming that a temperature difference has been created between two equal portions of a gas, the available energy and efficiency for conversion of this energy to work was examined in Sec. 3.5.3.[10]

9.2 THERMODYNAMICS & COMPUTATION

Computer chip sizes are being continually reduced and processing speeds are being increased. What are the fundamental *thermodynamic* limits to

---

[10]H. S. Leff, "Available work from a finite source and sink: How effective is a Maxwell's demon?," Am. J. Phys. **55**, 701–701 (1987).

computation? Real computers are considerably more dissipative than the Landauer limit $kT \ln 2$ energy dissipated per bit. Neil Gershenfeld has argued that an optimal computer should never need to erase its internal states, and the state transitions should occur no faster than needed in order to avoid unnecessary irreversibility. Further, computation should be no more reliable than needed.

A theorem by Norman Margolus and Lev Levitin implies: (i) the number of elementary logical operations per second that a physical system can perform is limited by that system's energy, and (ii) the amount of information that the system can register is limited by its maximum entropy. This is a good example of the linkage of energy with entropy. Using the Margulos-Levitin theorem, it has been shown that the processing rate cannot be higher than 6×10^{33} operations per second per joule of energy, and that a quantum system with energy E needs time of $\Delta t \geq h/(4E)$ to proceed from an initial state to an independent final state.

Seth Lloyd estimated *ultimate* physical limits on computation in terms of the speed of light c, Planck's constant, the universal gravitational constant G, and Boltzmann's constant k. Lloyd estimated that it will take about 250 years to make up the 40 orders of magnitude in performance between computers that perform (as of year 2000) 10^{10} operations per second on 10^{10} bits, and a 1 kg laptop performing 10^{51} operations per second on 10^{31} bits.

9.3 MORE ABOUT FLUCTUATIONS

9.3.1 Smoluchowski's trapdoor

Maxwell's demon took advantage of the fact that molecules *move*, have varying speeds, exert pressure, and the like. We saw in Sec. 1.1.2 that molecules move randomly in a liquid as evidenced by the visible jiggling of larger suspended particles such as grains of pollen. Because of its random nature, namely, not following a definite, predictable pattern, such Brownian motion brings *fluctuations* relative to the time-averaged values we observe. Thus, in actual thermodynamic systems, there exist (typically small) temperature, pressure, and internal energy fluctuations.

Marian Smoluchowski (1872–1917) was one of the first to introduce the concept of fluctuations in physics. It was he and Albert Einstein who demonstrated that Brownian motion results from molecular movement. Physicist Arnold Sommerfeld wrote, "For Smoluchowski, statistics were the breath of life; the second law represented only an approximate law..." As mentioned, located within a gas, a Maxwell's demon is continually bombarded by gas molecules and by photons from the blackbody radiation field within the container. It can be jostled around by this bombardment, impeding the accuracy of its measuring activities. Smoluchowski pointed out in 1912

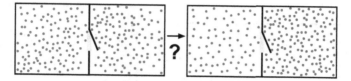

Figure 9.8: Smoluchowski's *unintelligent* sorting demon, intended to allow molecules to move from the left chamber to the right, but not *vice versa*.

that thermal fluctuations would prevent an *automatic* device from operating successfully as a Maxwell's demon.

Such an *unintelligent* sorting demon is depicted in Fig. 9.8. A container with two equal-volume chambers has a trapdoor that opens *toward the right*, as indicated. Initially, equal numbers of molecules, say of argon gas, are in each chamber at the same temperature. Molecules with sufficient speed approaching the trapdoor from the left can push the door open, gaining entry to the right chamber. The door must be spring–loaded to keep it in the closed position most of the time or molecules in the right chamber could move from the left to right chamber. The basic hypothesis was that the configuration in the left panel of Fig. 9.8 would evolve to that in the right panel. If this were to happen, we could use such a device to generate a pressure difference with zero energy and entropy cost, and use that difference to do work, in opposition to the second law of thermodynamics. Notably, this system necessitates neither measurement, information, nor erasure, each of which complicates Maxwell's *intelligent* demon. At first glance, this might seem promising to anyone hoping to overthrow the second law of thermodynamics.

However, as the saying goes, "The devil is in the details." In order for the door to be opened by molecules impinging from the left, the door must have a mass not significantly larger than the molecules themselves. To overcome the force from the spring intended to keep the door closed, the spring must be sufficiently weak. These two requirements mean that the weakly held door itself will be subject to Brownian motion. The door will jiggle, flapping from open to closed and *vice versa* continually, giving molecules from the right chamber as much opportunity to move to the left as those in the left chamber have for moving rightward.

Key Point 9.11 *Brownian motion thwarts the unintelligent demon of Smoluchowski's trapdoor system. The continually flapping door prevents a pressure differential between the chambers in Fig. 9.8. A system in thermodynamic equilibrium with its environment cannot spontaneously generate a pressure difference that would lower the entropy of the universe!*

Modern-day computers enable an interesting test of the findings of Smoluchowski. A mathematically modelled gas can be followed over time, molecule by molecule, simulating the physical situation in Fig. 9.8. Such

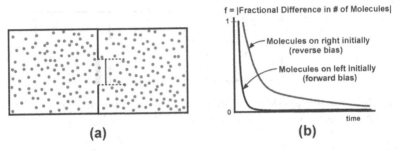

Figure 9.9: Two–chamber gas with sliding trapdoor amenable to computer simulation with 500 molecules in two dimensions, as described in the text. (Adapted from Skordos, P. A. & Zurek, W. H., "Maxwell's demon, rectifiers, and the second law: Computer simulation of Smoluchowski's trapdoor," Am. J. Phys. **60**, 876–882 (1992), with the permission of the American Association of Physics Teachers.)

computer simulations of a trapdoor mechanism, inspired by Smoluchowski's work, were done by Skordos and Zurek in 1992. They led to results that confirm Smoluchowski's conclusion that such a door will not work in thermodynamic equilibrium. The two–dimensional simulation of Skordos and Zurek had a maximum total of 500 molecules. The trapdoor was assumed to slide on frictionless rails, as in Fig. 9.9(a). Door stops restrict the trapdoor to opening rightward but *not* leftward. The simulation assumes that the trapdoor moves horizontally only, at constant speed, and reverses its direction when it touches the door stops. Collisions between the trapdoor and the molecules are assumed to conserve kinetic energy and momentum in the horizontal direction.

The computer simulations were done with various masses for the trapdoor. The results in Fig. 9.9(b) are for door masses of 3-4 times the mass of one molecule. A trapdoor with a mass too low moves so quickly that numerical roundoff error becomes a problem, and a door with too much mass slows the approach to equilibrium and requires longer time averaging. The simulations by Skordos and Zurek typically took several days to keep relative errors below 10^{-10}. Smoluchowski's ideas were tested by starting the simulations with all the molecules either in the left chamber or the right chamber. In these cases, all molecules were begun along the outermost wall of their chamber, the trapdoor had zero initial speed and in the closed position. Then the molecules were released with random initial velocities, and the time evolution of the numbers N_{left} and N_{right}, of molecules in the left and right chambers were monitored until the populations in the two chambers became equal. It was found that the motionless trapdoor acquired kinetic energy and started moving after the first collision with a molecule.

Figure 9.9(b) shows the fractional difference $f \equiv |N_{\text{right}} - N_{\text{left}}|/N$, where $N = (N_{\text{right}} + N_{\text{left}})$, *vs.* time for two cases. When the molecules all begin in the left chamber, there is a forward bias because left→right motion tends to keep the door open (left curve), resulting in a rapid drop $f \to 0$. In contrast,

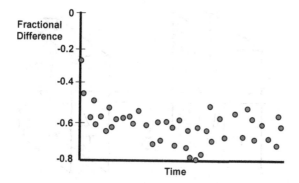

Figure 9.10: Fractional difference $(N_{\text{left}} - N_{\text{right}})/(N_{\text{left}} + N_{\text{right}})$ vs. time, which is negative if the density in the right chamber is higher. Note that here, the direction of forward bias was right to left rather than the reverse. This computer simulation used 170 molecules and the fluctuating fractional difference appears to be settling in at a value of about −0.6. (Adapted from Leff, H.S. & Rex, A. F. *Maxwell's Demon 2: Entropy, Classical and Quantum Information, Computing* (CRC Press, 2002), by permission of the Taylor & Francis Group.)

right→left motion has a reverse bias because molecules moving leftward tend to close the door (right curve), and f approaches 0 less rapidly.

Given that the trap door does not work, as argued over 100 years ago by Smoluchowski, Skordos and Zurek sought an intervention technique that would achieve rectification. They settled on "refrigeration," which removed energy systematically from the trapdoor. Specifically, the trapdoor's speed was reduced by half at regular small time intervals. The energy removed was then redistributed to all gas molecules equally, conserving the system's total internal energy. Rectification was enhanced most when energy removal from the trapdoor was done when the trapdoor was nearly closed, keeping the trapdoor almost closed longer, reducing right→left transitions.

Using a similar refrigeration procedure, Andrew Rex and Ross Larsen independently investigated a computational simulation. With right to left forward bias, they found the *algebraic* fractional difference $(N_{\text{right}} - N_{\text{left}})/N \to -0.6$, as shown in Fig. 9.10. This implies that 80% of the molecules reside in the right chamber (i.e., the fractional difference is $0.2 - 0.8 = -0.6$). The conclusion is that, as expected, refrigeration of the trap door results in rectification.

9.3.2 Feynman ratchet and pawl

A modern discussion of Smoluchowski's ideas was given by Richard Feynman in his well known *Lectures on Physics* Vol. 1. Well known for his creativity, he offered a new look at the problem, envisaging a small paddle wheel immersed

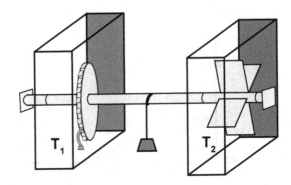

Figure 9.11: Ratchet-pawl (left) linked to a paddle wheel in air (right). The left and right are in contact with reservoirs at temperatures T_1 and T_2. (Adapted from R. P. Feynman, R. B. Leighton, & M. Sands, *The Feynman Lectures on Physics* (Basic Books; New Millennium ed. (2011)), Vol. 1, Fig. 46.1. Reprinted by permission of Basic Books, an imprint of Hachette Book Group, Inc.)

in a fluid (taken here to be air) and a ratchet connected by an axle, with a pawl that allows rotation in only one direction. The molecules in the fluid exhibit random Brownian motion and those collisions can cause the paddle wheel to turn, lifting a weight. The ratchet-pawl and paddle wheel are in thermal contact either with (a) a single reservoir at temperature $T = T_1 = T_2$, or (b) at different temperatures $T_1 \neq T_2$. The pawl prevents rotation in the direction that would lower the weight. Because the paddle wheel turns in only one direction, it appears that it can be used to lift a weight as in Fig. 9.11 using fluctuating forces on the paddle wheel.

Consider the single-temperature case (a) first. Just as the trapdoor in Smoluchowski's system had to have a weak spring and relatively low mass to enable it to be moved by molecular collisions, so must the pawl be sufficiently flexible and of low enough mass to allow the shaft to turn at all. The pawl can jiggle, sometimes enough to allow the ratchet to turn in the wrong direction.

Key Point 9.12 *For $T_1 = T_2$, the likelihood of a work-generating fluctuation by the paddle wheel is the same as that for a backward fluctuation that lowers the weight. There will be no violation of the Kelvin-Planck statement of the second law (Fig. 2.10).*

For case (b), $T_1 \neq T_2$, Feynman analysed the system mathematically. With $T_2 > T_1$. Because of the ratchet and pawl, it is more likely that fluctuations will lift the hanging weight than the reverse. The temperature difference induces hot-to-cold energy flows that result in net positive external work. Feynman went so far as to calculate the efficiency of the device in this mode. However, more than 30 years later, Juan Parrondo and Pep Español pointed out a problem with the ratchet and pawl not considered by Feynman, namely,[11]

[11] J. M. R. Parrondo & P. Español, "Criticism of Feynman's analysis of the ratchet as an engine," Am. J. Phys. 64, 1125–1130 (1996).

Key Point 9.13 *The ratchet-pawl system is coupled to two reservoirs at different temperatures, and cannot be in thermodynamic equilibrium. Rather, a net energy flows from the higher-temperature reservoir through the ratchet-pawl system to the lower-temperature reservoir. This energy flow consists of incoherent, fluctuating motion of the mechanical link between the ratchet-pawl and reservoir on the left and the paddle wheel and reservoir on the right. Effectively, this is a "heat leak" through the device.*

Christian Van den Broeck has observed[12] that modern studies of Maxwell's demons differ from prior ones because advances in technology and experimental techniques make it is possible to manipulate matter at the nanoscale. We can observe nanoscale biological structures out of equilibrium, and are trying to better understand the connections between thermodynamic and probabilistic concepts away from equilibrium. We can also do computer simulations. All of this goes far beyond what Maxwell could envisage.

Maxwell's demon can be compared with a ratchet and pawl and an electrical rectifier, neither of which can systematically transform internal energy from a single reservoir to work. As the demon does mischief it heats up and because it is necessarily small, it is subject to Brownian motion. Its molecules jiggle more and more as it heats up. If this happens, the demon cannot effectively do its job, independent of the need for memory erasure.

9.4   BROWNIAN RATCHETS

9.4.1   Fluctuation phenomena

The philosophical and mathematical formalism that is now called *classical thermodynamics*, which occupies most of this book, was developed primarily in the 18th and 19th centuries, with the invention of the thermometer, the steam engine, and the groundbreaking work by Carnot, Clausius, Mayer, Joule, Colding, Thomson and a host of others. Virtually all of this work is built on the assumption that fluctuations around equilibrium states are negligible.

In contrast, Brownian ratchets operate *because* Brownian motion exists as described in Sec. 1.1.2. In Brownian ratchets, which operate at the nanoscale in biological systems, including the human body, and in tiny man-made devices, there are *directed* diffusion processes, i.e., with preferred directions. These tiny machines operate far from equilibrium with the aid of asymmetric geometry and time-dependent forces that require energy input. Some people use the term *Brownian motor* to connote a ratchet operating against a load on which it does work. Others use that term for a Brownian ratchet, even in the absence of a load. In any case,

Key Point 9.14 *Brownian ratchets do not threaten any laws, and in particular, they operate within the bounds of the second law of thermodynamics.*

---

[12]C. Van den Broeck, R. Kawai & P. Meurs, "Microscopic models of Brownian ratchets," Proceedings of SPIE **5114**, 1–7 (2003).

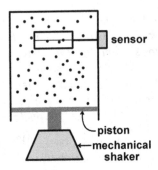

Figure 9.12: A lab experiment where 2000 glass beads are energised by a mechanical shaker that jiggles a piston, keeping the beads in motion. A sensor records the rotational motion of a paddlewheel.

Although fluctuations occur in systems that are in equilibrium, if motion with a preferred direction could occur, the second law of thermodynamics could be violated. In such a case, a Maxwell-demon type of perpetual motion machine such as Smoluchowski's trap door would be possible. A main point is that directed motion of a nanoparticle, called a *Brownian ratchet*, requires that it *NOT* be in equilibrium.

In 2010, to illustrate how a ratchet works, physicists at the University of Twente in The Netherlands constructed a machine, depicted in Fig. 9.12, based on the Brownian ratchet concept.[13] They used 2000 glass beads with diameter 44 mm to simulate a gas. Using a tuneable shaker of variable frequency and amplitude, and a paddle wheel whose paddle surfaces were identically coated, there was no preferred rotation direction. They then introduced an *asymmetry*, placing rubber tape on one side of each paddle of a rotating paddle wheel. Collisions with this softer side converted more of the balls' kinetic energy to internal energy of the paddle, so rotation with the rubberized side leading occurred. The paddle wheel was energised by the gas, which was continually reenergised by the mechanical shaker.

Key Point 9.15 *The main result was that nonzero work was done on the paddle wheel, which rotated in the direction of the soft, rubberised sides. The Twente macroscopic experiment showed how an asymmetry can lead to a directed rotation, and suggests what could happen on the nanometer scale, i.e., $1 - 100\,nm$.*

For the operation of nanoscale machines, *fluctuations, nonequilibrium*, and *asymmetry* are key ingredients that are essential to a Brownian ratchet's operation. An example is the operation of eukaryotic cells in biological systems, generating forces along specific directions, rather than random ones.

---

[13]P. Eshuis, K. van der Weele, D. Lohse & D. van der Meer, "Experimental realisation of a rotational ratchet in a granular gas," Phys. Rev. Lett. **104**, 248001 (2010).

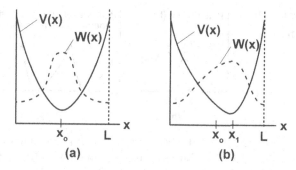

Figure 9.13: Symmetric (a) and asymmetric (b) potential wells $V(x)$ and corresponding probability distributions $W(x)$ for a particle in these wells.

Achievement of non-random directions is essential for the distribution, import and export of materials to internal and external locations. Such cells are also able to apply forces on external objects, i.e., they can do external work. Non-equilibrium states and thermal fluctuations are essential, and equilibrium thermodynamics, used through most of this monograph, is inadequate.

For many years, researchers have dealt with small deviations from equilibrium using so-called linear irreversible thermodynamics. This enabled progress working with transport phenomena such as heat conduction, and better understanding of how thermodynamic forces and fluxes generate entropy. The requirement of deviations from equilibrium being *linear* limited the utility of this framework and, in any case, was inadequate to deal with nanoscale phenomena, for which thermal fluctuations are an integral part.

Key Point 9.16 *Brownian ratchets consist of nanoscale particles that are far from equilibrium. Their motion can be directed by a combination of thermal noise, spatial asymmetry, and judicious use of time-dependent potentials. The idea of using thermal noise to effect transport goes back at least to 1987, with contributions by M. Büttiker, Rolf Landauer (who developed the Landauer principle for memory erasure) and Nicolaas Godfried van Kampen.*

9.4.2 Asymmetry & flashing Brownian ratchet

Imagine a nanoparticle in a symmetric potential well, shown by a solid line in Fig. 9.13(a). The potential minimum occurs at displacement $x(0)$. A low mass nanoscale particle, bombarded from all directions by an enormous number of molecules (typically air or water), will move either left or right with equal likelihood. The particle is expected to spend most time near x_0, and the probability of finding it in $x, x + dx$ is $W(x)dx$ with $W(x)$ looking something like the symmetric dashed curve in Fig. 9.14(a). However, if the potential well is *asymmetric*, as in Fig. 9.13(b), with its minimum at $x_1 > x_0$, it is more likely to find the particle near x_1.

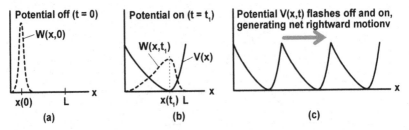

Figure 9.14: (a) Flashing ratchet with potential OFF and a nanoparticle known to have the probability distribution (dashed line) $W(x,0)$ at time $t = 0$. (b) At time $t_1 > 0$, after the potential $V(x)$ comes ON, it is most likely that $x(t_1) > x(0)$. (c) Repeated asymmetric wells that flash ON and OFF periodically, generating net rightward motion.

By definition, the flashing ratchet has a potential that consists of a sequence of identical, repeating asymmetric potential wells of length L. Those potential wells are turned on and off either randomly or at regular intervals. Suppose a nanoparticle, in a sea of molecules that jostle it continually, is known to be in the neighbourhood of $x(0)$ in Fig. 9.14(a). If an asymmetric potential is then turned ON as in Fig. 9.14(b), the nanoparticle is likely to be near $x(t_1) > x(0)$, where the potential is a minimum. For this *dichotomous* situation, when the potential is ON, the particle tends to be right of centre, and when the potential is OFF, the particle moves either leftward or rightward with equal probability. In the repeating asymmetric potential in Fig. 9.14(c), when OFF-ON-OFF cycles are done periodically, it is more likely than not that the nanoparticle drifts rightward. This is a *flashing ratchet*.

Key Point 9.17 *The flashing ratchet requires equipment and a power source to turn the potential ON and OFF. The equipment and power source needed to do this constitutes an external agent. These provide external energy inputs over finite times, flashing ratchets operate irreversibly, dumping heat energy into the environment, and the second law of thermodynamics is not violated.*

The mathematics needed to analyse Brownian ratchets goes beyond the level in this monograph, and I only outline the process, highlighting key parts.

A deterministic example. Consider a single nanoparticle, say, a grain of pollen, of mass m. For simplicity, assume it moves in one dimension and has velocity v, either to the right or left. If a force F_{ext} acts on it, then Newton's second law implies, $F_{ext} = m\dot{v}$, where $\dot{v} \equiv dv/dt$. If the *only* force on the particle is a viscous drag force $-c_d v$, then Newton's second law in the form, mass × acceleration − force = 0 reduces to $m\dot{v} + c_d v = 0$, or

$$\dot{v} + \gamma v = 0 \text{ where } \gamma \equiv c_d/m. \qquad (9.1)$$

The solution to the differential equation (9.1) is,

$$v(t) = v(0)e^{-\gamma t} \text{ with } v(0) \text{ being the initial velocity.} \qquad (9.2)$$

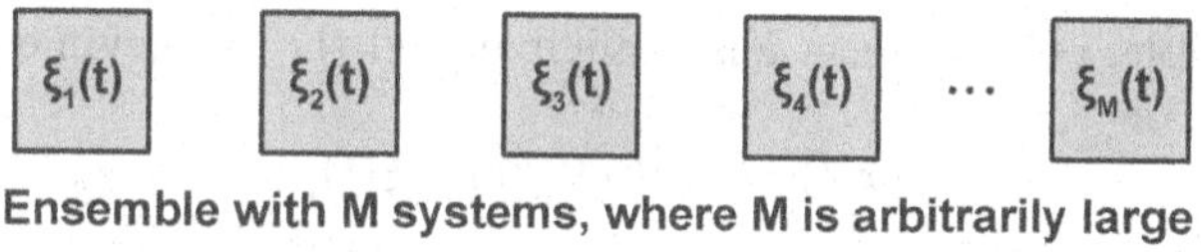

$$\boxed{\xi_1(t)} \quad \boxed{\xi_2(t)} \quad \boxed{\xi_3(t)} \quad \boxed{\xi_4(t)} \quad \cdots \quad \boxed{\xi_M(t)}$$

Ensemble with M systems, where M is arbitrarily large

Figure 9.15: An ensemble of identical macroscopic systems, each following an independent microscopic path, $\xi_j(t)$, as described in the text.

Equation (9.2) shows that motion will be damped, and v will approach zero with a characteristic time $1/\gamma$. The motion is purely *deterministic*, and does *not* apply to a nanoscale situation where m is small enough to be affected by random thermal fluctuations, i.e., when it is a Brownian particle.

Key Point 9.18 *In the absence of thermal agitation from surrounding molecules, a Brownian nanoparticle will have its motion exponentially damped. This is not the case for Brownian ratchets, which feed off thermal fluctuations.*

Simulating random fluctuations. To account for random fluctuations, it is common to replace (9.1) with

$$\dot{v} + \gamma v - \xi(t) = 0 \qquad \text{Langevin equation with Langevin force } \xi(t). \quad (9.3)$$

Here, $\xi(t)$ is a random function with $-\infty < \xi(t) < +\infty$ for $t \geq 0$.[14] It represents a random "Langevin force per unit mass." For a given system the time average of $\xi(t)$ is required to be zero if there is no external force: $\overline{\xi(t)} = \lim_{t_f \to \infty}(\int_0^{t_f} \xi(t)dt/t_f) = 0$.

Key Point 9.19 *A useful mental model is an ensemble (i.e., a collection) of many macroscopically identical systems, each with its own specific (non-random) Langevin force per unit mass $\xi_j(t)$. The index $j = 1, 2, 3 \ldots M$, where M is arbitrarily large, i.e., $M \to \infty$. Such an ensemble is depicted in Fig. 9.15. At each instant of time equal fractions of ensemble members are pushed leftward and rightward, with no preferred direction. This generalises to three dimensions.*

The ensemble average vanishes at each time t, i.e.,

$$\langle \xi(t) \rangle = \lim_{M \to \infty} \left(\frac{\sum_{j=1}^{M} \xi_j}{M} \right) = 0. \quad (9.4)$$

If we include a potential energy function $V(x)$, as in Fig. 9.14(b), this adds an external force $-dV/dx \equiv -V'$, and Eq. (9.3) becomes $\dot{v} + \gamma v + V' - \xi(t) = 0$. It is common to assume that the system is overdamped, keeping the acceleration

---

[14]The Langevin equation is named after French physicist, Paul Langevin (1872-1946).

term negligible, i.e., the term $\dot{v}$ is negligible, and the Langevin equation (9.3) takes the form

$$\dot{x}(t) = \gamma^{-1}[-V'(x) + \xi(t)]. \tag{9.5}$$

Because each ensemble member is subject to its own random force at each instant of time, it will follow its own microscopic path. For a single system, Eq. (9.5) describes a particle's motion. For the full ensemble, at any time t, a fraction $W(x,t)dx$ of ensemble members will have displacements in the interval $(x, x + dx)$. Here $W(x,t) \geq 0$ is interpreted as the probability distribution function (probability per unit length) for the ensemble. Because the particle must be somewhere at each time t, $W(x,t)$ satisfies

$$\int_{-\infty}^{+\infty} W(x,t)dx = 1 \text{ (normalisation condition)}. \tag{9.6}$$

For processes involving Brownian motion and diffusion, the time evolution of the distribution $W(x,t)$ for $t > 0$ is determined by the initial value $W(x,0)$. This type of evolution is called a *Markov process*.

Key Point 9.20 *The distribution function $W(x,t)$ is central in the study of Brownian ratchets. To go further, one must know how that distribution function evolves with time. In addition to the Langevin equation (9.5), two other equations are used in the study of ratchets: (1) a continuity equation similar to that encountered in electromagnetism for the conservation of electric charge,[15] and (2) the so-called Fokker Planck equation.[16]*

In the time-dependent case, where the periodic potential energy can be turned ON or OFF, as for the flashing ratchet, $V(x)$ is replaced by $V(x,t) = g(t)V(x)$ in Eq. (9.5). The function $g(t)$ oscillates between 0 and 1, effecting repeated ON-OFF-ON cycles, resulting in directed motion, $\langle \dot{x}(t) \rangle \neq 0$ for the flashing ratchet. In contrast, for a time-independent *not flashing* ratchet with $g(t) = 1$ and $V(x)$ as in Fig. 9.14, as t becomes large, the distribution $W(x,t)$ becomes independent of the time, and $W(x,t) \to A(T,L)e^{-\beta V(x)}$ and $\langle x(t) \rangle = 0$.

Key Point 9.21 *A Brownian particle in a time-dependent, spatially periodic asymmetric potential energy, $V = V(x,t)$, knowing $W(x,t)$ enables calculation of $\langle \dot{x}(t) \rangle$. A mathematically rich theory of stochastic processes exists. It provides ways to solve the Fokker-Planck partial differential equation.[17] The flashing ratchet has been solved this way, confirming that $\langle \dot{x}(t) \rangle > 0$, i.e., there is directed flow rightward.*

---

[15] $\partial W(x,t)/\partial t = -\partial J(x,t)/\partial x$, where $J(x,t) \equiv \dot{x}(t)W(x,t)$ is the probability current.

[16] For the systems under study, the Fokker-Planck equation is $\partial_t W(x,t) = -\gamma^{-1}\partial_x\{[\partial_x V(x,t) + \xi(t)]W(x,t)\}$.

[17] Risken H., *The Fokker–Planck Equation* 2nd edn (Springer, Berlin, 1989).

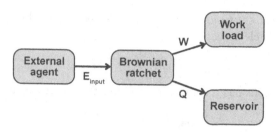

Figure 9.16: Energetics of a Brownian ratchet.

9.4.3 Other Brownian ratchets

An entire field of physics has developed around Brownian ratchets,[18] and a partial list follows.

- **Forced ratchet.** A *forced ratchet*, uses the combination of time-independent potential with a time-varying external driving force $F(t)$ that has a time average of zero. The external force drives the system out of equilibrium.

- **Information ratchet.** This ratchet gathers information on the locations of Brownian particles, enabling it to turn an external potential on and off at times that maximise diffusion in a given direction of motion.

- **Quantum ratchets.** One such ratchet takes advantage of thermal noise at high temperatures, and quantum behaviour at low temperature. At high temperatures, thermal phenomena dominate, and the current is in a preferred direction. However, at sufficiently low temperatures, quantum tunnelling enables a current in the opposite direction.

9.4.4 Brownian ratchets & the 2nd law

Contrary to the heat engines in Ch. 8, Brownian ratchets do not require a temperature difference in order to do work and thus, their efficiency is defined differently. In a constant-pressure, constant temperature environment, these ratchets require an external agent, which supplies internal energy, $|\Delta E_{agent}|$, while doing work $|P\Delta V_{agent}|$ against (an assumed) constant-pressure environment.[19] Assume that the Brownian ratchet ends in its initial state. It undergoes a cyclic process, receiving energy $E_{input} = |\Delta E_{agent}| + |P\Delta V_{agent}|$. During the cyclic process, the ratchet does work $W > 0$ on a load and dumps

---

[18]Cubero, D. & Renzoni, F., *Brownian ratchets: From statistical physics to bio and nano-motors* (Cambridge U.P., New York, 2016).

[19]I am following the analysis in J. M. R. Parrondo & B. J. de Cisneros, "Energetics of Brownian motors: a review," Appl. Phys. **A 75**, 179–191 (2002).

waste energy $Q > 0$ to an (assumed) single reservoir. This is shown in Fig. 9.16, where it is clear that

$$E_{input} = W + Q. \tag{9.7}$$

The external agent has a concomitant Gibbs free energy change,

$$\Delta G_{agent} = -E_{input} - T\Delta S_{agent}. \tag{9.8}$$

Because $\Delta G_{agent} < 0$ when the agent supplies energy, it is helpful to work with the nonnegative energy input to the ratchet,

$$-\Delta G_{agent} = E_{input} + T\Delta S_{agent}. \tag{9.9}$$

I assume that the ratchet operates cyclically and the latter energies are all for one cycle, in which case the total entropy change of the universe is

$$\Delta S_{tot} = \Delta S_{agent} + Q/T \geq 0. \tag{9.10}$$

Things simplify if Eqs. (9.7), (9.8), and (9.10) are solved for W, E_{input}, and Q respectively, and are combined to get $T\Delta S_{tot} = (-\Delta G_{agent}) - W \geq 0$. The efficiency of the ratchet is defined by[20]

$$\eta = \frac{W}{(-\Delta G_{agent})} = 1 - \frac{T\Delta S_{tot}}{(-\Delta G_{agent})}. \tag{9.11}$$

Key Point 9.22 *The work W that a ratchet can do is limited by the Gibbs free energy given up by the external agent, and the efficiency is lower for higher change in total entropy produced during the ratchet's process. The ratchet's operation is consistent with the second law of thermodynamics.*

9.5 ATTEMPTS TO VIOLATE THE 2ND LAW

9.5.1 Perpetual motion machines

The Clausius and Kelvin-Planck statements of the second law of thermodynamics are framed in terms of impossibility, and many have tried to find ways to do the impossible. Further, in statistical mechanics, entropy is inherently statistical, and the second law of thermodynamics does not hold absolutely, which has inspired attempts to systematically violate the law.

Since at least the middle ages, humans have tried to produce *perpetual motion machines*, namely, devices that would run forever without the input of energy via work or fuel. A common form of such a device was an "overbalanced" wheel that was intended to have a preferred direction of rotation. Another was

---

[20]The denominator in Eq. (9.11) accounts for the total energy input needed, e.g., to make the potential flash on and off, for the Brownian ratchet to generate the work W.

Figure 9.17: A perpetual motion machine proposed by Robert Boyle.

Robert Boyle's perpetual vase, illustrated in Fig. 9.17. It cannot work because hydrostatic pressure depends on height and the fluid in the tube cannot rise above that in the funnel.

However, in every case examined, the rationale was not quite right, and no such devices were found that could do positive work on an external load, in violation of the second law of thermodynamics. Nor did they generate energy, in violation of the first law of thermodynamics.

A rather clever *perpetual motion* device was discussed in a recent issue of the American Mathematical Monthly by Tadashi Tokieda. The proposed device is shown in Fig. 9.18. A collection of capsules is attached at equal intervals to a continuous, taut belt that is wrapped over an upper pulley wheel and under the lower pulley wheel. Each capsule consists of a bucket with a sealed, impermeable, elastic membrane with a weight (small solid black circle) attached to its middle. Depending on the orientation of each capsule, its membrane either sags outward (downward), *increasing* the capsule's volume, or sags inward (also downward), which *reduces* the capsule's volume. Notice that for capsules with their membranes sagging at the capsule's top, the compression is greater at greater depths, where the water pressure is higher.

The entire assembly is submerged under water. There are equal weights on each side of the two rotating pulley wheels, but the capsules on the left side have larger volumes than the ones on the right.

Key Point 9.23 *The net buoyant force is larger on the left than on the right. This suggests that the pulley wheels will rotate clockwise perpetually. It is our task to understand why this conclusion is faulty.*

Each capsule has work done on it by gravity, by the water, as the capsule's volume changes, and by the buoyant force. As a capsule moves upward by dz: (1) The work done on a capsule by gravity is $dW_{grav} = -mgdz$. (2) The buoyant force is the weight of the displaced water, $\rho g V$, and the work done by that force is $= dW_{buoyant} = \rho g V dz$. (3) As a capsule's volume changes by dV, the work done on it by the water (in addition to the buoyant force) is $dW_{water} = -P(z)dV$, where $P(z)$ is the hydrostatic pressure at height z, measured from the container bottom. Thus, $dW_{water} > 0$ when $dV < 0$. The

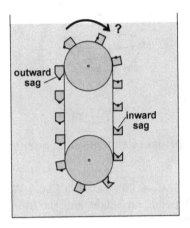

Figure 9.18: Water wheel system that seems to have a larger submerged capsule volume and buoyant force on the left, suggesting perpetual *clockwise* rotation. (Reprinted from T. Tadashi, "A Buoyancy-Driven Perpetual Motion Machine," The American Mathematical Monthly 120, 564–566 (2013) by permission of publisher Taylor & Francis Ltd, http://www.tandfonline.com.)

net work on each capsule is

$$
\begin{aligned}
dW_\text{net} &= dW_\text{gravity} + dW_\text{buoyant} + dW_\text{water} \\
&= (\rho V - m)g\,dz - P(z)dV \\
&= (\rho V - m)g\,dz - d(P(z)V) + V\,dP(z) \\
&= -d((P(z)V) + mgz) \equiv d\phi(z,\theta).
\end{aligned}
\tag{9.12}
$$

In the third line, I've written $-P(z)dV = -d(P(z)V)+V\,dP(z)$, and used the hydrostatic equation $dP(z) + \rho g\,dz = 0$. The fourth line defines the function $\phi(z,\theta)$, which is fully determined by the capsule's height z and its angular orientation θ. These variables imply the capsule volume $V(z,\theta)$. If we follow a capsule through a complete cycle that starts and finishes at height z, the net work done on it is,

$$
\oint dW_\text{net} = \phi(z,\theta)_\text{final} - \phi(z,\theta)_\text{initial} = 0.
\tag{9.13}
$$

Another view that can enhance understanding is to focus attention two capsules located at the same height on each side of the pulley wheel.

1. Applying the second line of Eq. (9.12) to each capsule, with $dz > 0$ on the left and $dz < 0$ on the right, the sum of the two first terms is zero.

2. $P(z)$ is the same for the two capsules because their heights and angular orientations are the same. To lowest order, over the height change dz, dV has opposite signs for the capsules, and adds to zero.

This clever device cannot do any work lifting a load or rotating a shaft because it gains zero energy each cycle. If it did *any* external work, it would deplete its initial energy and would slow down and stop.

Key Point 9.24 *Our initial expectation, based on a higher buoyant force on the left side, was wrong. It ignored the variable volume third term in Eq. (9.12). With that term included, dW_{net} is an exact differential, which guarantees that in one full cycle, $\oint dW_{net} = 0$. No net work is done and the device is not a perpetual motion machine.*

9.5.2   Challenges to the 2nd law

In 2002, Daniel Sheehan of the University of San Diego began what might be called a veritable *movement* to test the second law of thermodynamics in a variety of innovative ways. Sheehan hosted a series of international conferences directed at that task, bringing researchers together to share their ideas and findings, With the late Vladislav Capek, he coauthored a notable volume on second law challenges,[21] and he's written numerous theoretical and experimental articles challenging the second law. The groundbreaking work by Sheehan, Capek, and others has led to much attention given to scrutinising the second law of thermodynamics. In the following sections, I describe some attempts and some proposals challenging the second law. In each case, the system has an asymmetry or gradient and/or a preferred direction. These arise from a specific geometry or external magnetic or gravitational field.

Key Point 9.25 *In some cases, researchers have alleged a violation of the second law. In no case known to me has a violation of the second law been observed experimentally by two or more independent research groups.*

9.5.3   Thermal electrons in a magnetic field

Xinyong Fu and Zitao Fu, of the Shanghai Jiao Tong University, have been working for years on a delicate experiment that challenges the second law of thermodynamics. The experiment entails a custom-made vacuum tube that contains a room-temperature photoelectric metal, Ag-O-Cs. An important property of that metal is that it continually emits electrons at room temperature. A small cooling effect on the metal triggers heating from the constant-temperature environment. The Fu experiment is simple in principle, and can be understood with the help of Fig. 9.19, noting the following points:

1. Electrons in a metal satisfy Fermi-Dirac statistics and the Pauli exclusion principle, which permits only one electron in each energy state

---

[21]Capek, V. & Sheehan, D. P. *Challenges to the Second Law of Thermodynamics: Theory and Experiment* (Springer, Dordrecht, 2005).

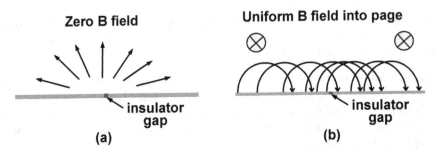

Figure 9.19: Two thermally emitting plates separated by an insulated gap in (a) zero external field (Earth's magnetic field is usually negligible or can be *cancelled* by a subsidiary *magneticfield*); (b) a uniform magnetic field $\vec{B}$ directed into the page. Trajectories are simplified to be two dimensional.

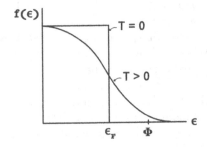

Figure 9.20: Average number of particles at energy ϵ.

(see Sec. 4.1.1, especially Fig. 4.2). The electrons fill the lowest states until the *Fermi sea* is full. The Fermi distribution $f(\epsilon)$, the probability that an electron has energy ϵ, is shown in Fig. 9.20 for the (unattainable) $0\,\mathrm{K}$, and also $T > 0$. For $T = 0$, $f(\epsilon) = 1$ for $\epsilon < \epsilon_F$ and $f(\epsilon) = 0$ for $\epsilon > \epsilon_F$. For $T > 0$, electrons with the highest energies can get into the conduction band, where they can move freely through the metal. If an electron has sufficient energy, it can transcend an energy barrier Φ, called the *work function*, that ordinarily keeps it in the metal. Such electrons leave the metal via *thermionic emission*, which can be substantial at high temperatures. However, if a metal's work function is sufficiently small, thermionic emission can occur, albeit weakly, even at room temperature.

2. In the Fu & Fu experiment, two plates, separated by a mica insulator, are side by side, each emitting electrons at room temperature T with a current density J (in amperes per metre,[2] i.e., $\mathrm{A\,m^{-2}}$), in accordance with the Richardson equation, $J = AT^2 e^{-\Phi/\tau}$. Here, $A \approx 10^6\,\mathrm{A\,m^{-2}}$ and, as before, k is the Boltzmann constant, and $\tau \equiv kT$.

3. The work function Φ is defined as the energy required to remove an electron from the top of the Fermi distribution, the *Fermi energy*, and

bring it to a point outside the metal, a distance from it which is slightly larger than the spacing between atoms in the metal.

4. Figure 9.19(a) shows electrons being emitted in all directions by *both* emitting plates when the external magnetic field is zero. Figure 9.19(b) shows the situation when there is a nonzero, uniform magnetic field, directed into the page. The magnetic field bends the electron trajectories, producing an *asymmetry* that leads to electrons being transferred preferentially from the left to right plate.

5. The asymmetry produced by the magnetic field, $\vec{B}$, results in an electrical potential difference ΔV between the plates. The thermal electron emission is continual, so if the plates are electrically connected through a load, a persistent electric current will flow.

6. With a nonzero magnetic field, a persistent current exists and the emitted electrons remove energy from the plates. Many electrons return to the left or right plate, minimising the net energy loss, which drives the persistent electric current. The lost energy is replaced via heat energy from the constant temperature surroundings.

Key Point 9.26 *The generation of electric work to produce a persistent electric current powered by a single reservoir, violates the Kelvin-Planck form of the second law of thermodynamics. Xinyong Fu and Zitao Fu allege that their experiment demonstrates such a violation.*

For perspective, note that measured work function values for the elements are between about 2 eV and 6 eV, with caesium's being the smallest, $\Phi \approx 2\,\text{eV}$ ($1\,\text{eV} = 1.6 \times 10^{-19}\,\text{J}$). Fu & Fu state that the metal, Ag-O-Cs, has the lowest work function among all the known thermal electron-emitting materials, $\Phi \sim 0.8\,\text{eV}$, and therefore is optimal for maximising thermal electron emission at room temperature. For this reason they custom built vacuum tubes with Ag-O-Cs surfaces having the geometry of Fig. 9.19(b).

Despite the relatively low work function of Ag-O-Cs, thermionic emission at room temperature yields an exceedingly small electric current density. The Richardson equation implies that at the emitted current density J is a monotone increasing function of T, as shown in Fig. 9.21. The temperature range shown corresponds to the measurements made in the Fu experiment, $283.15 - 306.15\,\text{K}$. The range of current densities for the latter temperature range given by the Richardson equation is $0.48 - 6.5 \times 10^{-13}\,\text{A}\,\text{cm}^{-2}$.

The implied rate at which charge is released from area $1\,\text{cm}^2$ is $2 \times 10^{-13}\,\text{A}$ at $T = 293.15\,\text{K}$. These electrons are released with a distribution of velocities. In 1925, Lester Germer, known for the Davisson-Germer experiment on wave-particle duality, showed that the electrons emitted in hot thermionic emission have a distribution that is well approximated by the Maxwell velocity

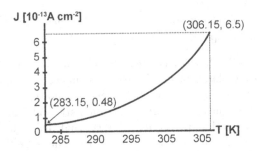

Figure 9.21: Current density, in $A\,cm^{-2}$, vs. T according to the Richardson equation.

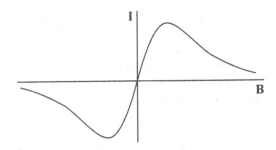

Figure 9.22: Expected behaviour of current I vs magnetic field B.

distribution. It is not known if the same is true for room temperature emission, but the main point is that only a fraction of electrons emitted get to the right plate; i.e., some will have insufficient speed to get to the right plate and others will have large enough speeds to go past the right plate.

When the external magnetic field is zero, electrons emitted from either plate are unlikely to get to the other one. When the magnetic field is small, only electrons with sufficiently low speeds, emitted near the right plate find their way to that plate. For very high magnetic fields, a relatively large fraction of emitted electrons are bent so severely by the field, that they do not get to the right plate. For these reasons, we expect that a graph of the measured current flow in the electric circuit will be highest at some intermediate field strength, as depicted in Fig. 9.22. When the field direction is reversed, $\vec{B} \to -\vec{B}$, the roles of the left and right plates are reversed and the current is also reversed.

Key Point 9.27 *Fu & Fu's experiments involved a custom-made evacuated tube with two plates having Ag-O-Cs surfaces, and the measurement of small currents. The primary result is that with a nonzero magnetic field, persistent currents of 0.39×10^{-13} A, 1.82×10^{-13} A and 16.5×10^{-13} A were measured for temperatures 283.15 K, 293 K and 306.15 K, respectively. Powered by a single reservoir, this appears to violate the second law of thermodynamics.*

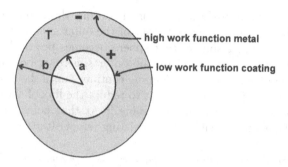

Figure 9.23: Spherical capacitor described in the text.

These results must be considered preliminary for several reasons:

1. For $B = 0$, Fu & Fu measured background currents 0.030×10^{-13} A, 0.040×10^{-13} A, and 0.08×10^{-13} A. These are 10.5%, 1.65%, and 0.467% of the maximum current found at each temperature. The precise source of these is uncertain.

2. The graph of I vs. B has the expected qualitative behaviour in Fig. 9.22, except the measured curve is *not* symmetric, which requires explanation.

3. Potential differences where dissimilar metals are linked can affect results.

4. Fu & Fu's results have not been published in a peer-reviewed journal. A series of articles and a video are available on the Internet.[22]

These concerns do not negate the findings of Fu & Fu, but their results beg for confirmation or rebuttal by one or more independent experimenters. Although the issue is not settled, the Fu & Fu experiment can be a stimulating example for discussion in physics classrooms.

9.5.4 Thermal electrons in a capacitor

In 2010, Germano D'Abramo proposed an experiment to violate the second law of thermodynamics. As with the work of Fu & Fu described in the previous section, his proposal involved room temperature thermionic emission of electrons from a semiconductor. D'Abramo's idea was to construct a spherical capacitor with inner radius a and outer radius $b > a$, as shown in Fig. 9.23. The inner sphere's surface is coated with the same low work function semiconductor, Ag–O–Cs, as in the Fu & Fu experiment. The work function is stated to be $\Phi \lesssim 0.7$ eV, while Fu & Fu cited $\Phi \sim 0.8$ eV. The work function of the outer sphere's inner surface is that of a normal metal.

---

[22]X. Fu & Z. Fu, "Realisation of Maxwell's Hypothesis: A Heat-Electric Conversion in Contradiction to the Kelvin Statement," Preprints 2016, 2016070028 (doi: 10.20944/preprints201607.0028.v6). A video showing the experimental setup is at https://www.youtube.com/watch?v=PyrtC2nQ_UU'.

At room temperature, the electron flux from thermionic emission from the inner spherical surface exceeds that of the outer sphere. As electrons are emitted from the inner sphere, that sphere becomes positively charged, and as those electrons are absorbed by the outer sphere, it becomes negatively charged. This process of *thermo-charging* continues until the electric potential difference between spheres is sufficient to reduce the flux of electrons reaching the outer sphere to an insignificant value.[23] At that point, the capacitor is charged to the extent possible at the existing temperature.

Key Point 9.28 *The spherical capacitor, as imagined, violates the second law of thermodynamics by doing work to separate positive and negative charges using only one reservoir. Connecting the circuit to a load, say, a light emitting diode (LED), would light the LED using energy from a single constant-temperature reservoir, in violation of the Kelvin-Planck statement of the second law of thermodynamics. D'Abramo's work was purely theoretical, and no experiment was done to test his idea.*

An experiment related to the work of Fu & Fu was executed earlier by two Russian physicists, Alexander Perminov and Alexy Nikulov.[24] They used thermionic emission from a surface heated to over $1973\,\text{K}$, with a magnetic field $B = 0.006\,\text{T}$ that generated an asymmetry and electric current. The authors concluded that internal energy can indeed be transformed into power with a single reservoir. In their experiment, the power input exceeded the power output so there was no second law violation. No follow-up work by others confirms or rebuts the principal claims of Perminov & Nikulov.

9.5.5 Theory of air column in gravitational field

A long-standing question that goes back at least to the time of Maxwell and Boltzmann is whether a column of air, enclosed in a well-insulated container in a uniform gravitational field g, has a temperature gradient. An effect of gravity is that the density of the gas decreases with the height from the container's bottom. Two possibilities for the temperature are:

P1. The temperature is independent of the altitude z.

P2. The temperature decreases with the height z from the container's bottom.

Possibility P1 is based on the knowledge if there is a temperature difference between two parts of a gas, a heat process will carry energy from the higher-temperature part to the lower temperature part, thereby reducing the

---

[23]This occurs when the electric potential difference $\Delta V > 1/2mv_{av}^2/e$, where v_{av} is the average speed of the thermally-emitted electrons.

[24]A. Perminov, A. Nikulov & D. P. Sheehan, "Transformation of Thermal Energy into Electric Energy via Thermionic Emission of Electrons from Dielectric Surfaces in Magnetic Fields," in *Second Law of Thermodynamics: Status and Challenges* (AIP Conf. Proc. 1411, 82-100, 2011).

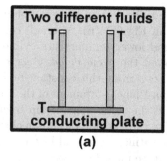 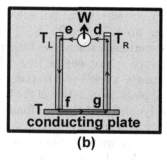

Figure 9.24: (a) Two columns of different fluids in a gravitational field, each with uniform temperature T, as argued by Maxwell and Boltzmann. (b) The same columns, assuming that their temperatures vary with altitude, which would enable running a heat engine, extracting energy from the constant-temperature surroundings, in violation of the second law.

temperature difference. This would lead to temperature equilibration, i.e., a uniform temperature.

Possibility P2 is suggested by the fact that for a dilute (ideal) gas at room temperature, as a molecule rises in a gravitational field, its gravitational potential energy rises and its kinetic energy decreases. If the temperature is proportional to that average kinetic energy, as defined in Eq. (2.13), the temperature would decrease with altitude.

Historically, this issue was debated by some of the great physics researchers. Supporting P2 was Joseph Loschmidt, who was a well-regarded scientist known for his reversibility objection to Boltzmann's statistical mechanics.[25,26] Ludwig Boltzmann and James Clerk Maxwell argued in favour of P1, a uniform temperature, in opposition to Loschmidt, In what follows, I present (a) Maxwell's argument and another argument based on first principles that support position P1, followed by a description of an experiment whose reported results are consistent with position P2, which seems to be a violation of the second law of thermodynamics.

Support for Position P1. Maxwell argued that a gas in a gravitational field would have a uniform temperature. This is shown in Fig 9.24(a), where two cylindrical columns have the same amount of different fluids. The temperatures at their bottoms are forced to be the same by keeping them in contact with a conducting plate at temperature T. Suppose the temperatures

---

[25] A. Bader & L. Parker, "Joseph Loschmidt, physicist and chemist," Phys. Today **54**, 45–50 (2001).

[26] The irreversibility of Boltzmann's work was in conflict with the time-symmetric dynamics of classical mechanics.That objection, together with another by Ernst Zermelo, based upon the Poincaré recurrence theorem, led Boltzmann to adopt the interpretation that his entropy function $S = \ln W$ was a *statistical* function. As such, the second law of thermodynamics was was not an absolute law. Maxwell also realised this and developed his demon example to illustrate the statistical nature of the second law.

at the top surfaces were *not* equal, and the left and right columns have top temperatures T_L and $T_R > T_L$, as in Fig 9.24(b). Maxwell conceived a thought experiment where the higher and lower temperature cylinders were used to power a heat engine that followed the cycle *defgd*. Over time, this would generate work by extracting energy from the constant-temperature conducting plate, in violation of the Kelvin-Planck statement of the second law of thermodynamics. If the two fluids had the same nonuniform temperature variation with altitude, then point e in Fig. (9.24) could be chosen closer to the conducting plate with the same result. Treating the second law as sacrosanct, Maxwell concluded that the gas temperature must be uniform.

A very different argument supporting Position P1 is based on the work of Charles Coombes and Hans Laue.[27] They took $f(z,v)dzdv$ to be the number of molecules in the altitude interval $(z, z+dz)$ with velocities in interval $(v, v+dv)$, and assumed a Maxwell velocity distribution at one height, z_1: $f(z_1, v) = C\exp(-\frac{1}{2}mv^2/\tau)$. Here, the dependence on z_1 is absorbed in the constant C, and $\tau = kT = $ temperpergy. They show that if $\frac{1}{2}mv_1^2 + mgz_1 = \frac{1}{2}mv^2 + mgz$ (kinetic + gravitational potential energy is conserved), then $f(z_1, v_1) = f(z, v)$, which implies

$$f(z, v) = \text{constant} \times \exp(-mgz/\tau)\exp(-\tfrac{1}{2}mv^2/\tau). \qquad (9.14)$$

The distribution scales according to altitude z, so the average kinetic energy $\overline{K}_{\Delta z}$ in an altitude interval Δz and the average number of molecules $\overline{N}_{\Delta z}$ in that interval scale identically, showing the average kinetic energy per molecule is independent of altitude. Recalling the ideal gas definition of temperature in Sec. 1.5, this implies that the gas temperature is the same in all intervals $\Delta z(z)$.

Key Point 9.29 *Gravity causes a density decrease with altitude, and although the total kinetic energy for a group of gas molecules is lower at higher altitudes, there are fewer molecules at higher altitudes. The average kinetic energy per molecule and thus the temperature are independent of altitude. I find this argument compelling.*

Support for Position P2. Consider a classical ideal gas in a gravitational field. Assuming each molecule satisfies energy conservation, $\frac{1}{2}m(v_i^2)_1 + mg(z_i)_1 = \frac{1}{2}m(v_i^2)_2 + mg(z_i)_2$ for positions 1 and 2. Letting $K_i = \frac{1}{2}mv_i^2$ for the ith molecule, this becomes

$$(K_i)_1 - (K_i)_2 = mg[(z_i)_2 - (z_i)_1] < 0 \text{ for } mg[(z_i)_2 > mg(z_i)_1]. \qquad (9.15)$$

Key Point 9.30 *The kinetic energy of each molecule decreases as it rises to higher altitudes. Support for position P2 is based on the notion that this kinetic*

---

[27]C. A. Coombes & H. Laue. "A paradox concerning the temperature distribution of a gas in a gravitational field," Am. J. Phys. **53**, 272–273 (1985).

energy decrease implies that the temperature of the gas must drop with altitude. This is not correct. Our argument in support of Position P1 shows that in an altitude interval $\Delta z(z)$ centred about z, both the average kinetic energy per molecule and the average number of molecules scale as $\exp(-mgz/\tau)$, so the average energy per molecule is actually independent of the altitude z.

Experimental temperature profile. Despite the compelling arguments above that support position P1, a series of experiments were done in recent years by Roderick Graeff, who reports: (a) finding a negative temperature gradient in a vertical column of fluid, and (ii) using such a temperature difference to generate electricity. These results support position P2.[28] I report on this here, not because it proves that the second law is violated (it does not) but rather with the hope that others will repeat the experiments independently. Only with such independent confirmation can an alleged violation of the second law of thermodynamics be taken seriously or ruled out.[29]

The experiments consist of a vertical column of fluid (liquid or gas) with barriers designed to suppress convection. I limit this discussion to the experiments with water, and also air in a Dewar container. These systems were well-insulated from their environments, and temperature gradients between top and bottom were monitored over a period of months. Peltier junctions were connected to the higher and lower temperature regions, and the observed temperature gradient generated an electric voltage. The main results of these experiments were: (1) For air, the temperature gradient was $-0.07\,\mathrm{K\,m^{-1}}$. ((2) For water, the temperature gradient was $-0.05\,\mathrm{K\,m^{-1}}$. (3) Graeff claims production of electric power of about $0.6 \times 10^{-12}\,\mathrm{W}$. (4) Outside the insulated system, where convection was not suppressed, Graeff measured a *positive* temperature gradient.[30]

Key Point 9.31 *Graeff's experiments span years and challenge the second law's sanctity. I am hopeful that one or more laboratories will do similar experiments to settle this age-old debate once and for all.*

9.5.6 Spontaneous pressure differences

In the 1870s, William Crookes, whose skills led him to discover the element thallium using optical spectroscopy in 1861, discovered the radiometer effect.

---

[28] R. W. Graeff, "The Production of Electricity out of a Heat Bath," AIP Conf. Proc. **1411**, 193–220 (2011).

[29] Graeff's several papers, available on the Internet, have evidently not been subjected to peer review. Trained as an engineer, Graeff's papers lack needed scientific rigour and clarity, and contain questionable theoretical analyses.

[30] Higher altitude regions were warmer than lower ones, reminiscent of our experience that "heat rises", a scientifically inaccurate statement of the fact that if a gas is heated from below, it will expand and move upward, being replaced by colder, more dense gas.

Figure 9.25: A radiometer.

The *radiometer* is a device containing typically 2 − 4 vanes that are black on one side and white on the other, and able to rotate with low friction in a sealed glass container with reduced-pressure air. This is depicted in Fig. 9.25. When in the presence of a light source (sunlight or a lamp), the vanes rotate with the white sides leading, as if the black sides were being pushed harder.

At first, it was suspected that the vanes were responding to the pressure of the light hitting the vanes. Recall that a photon gas has nonzero pressure, as given in Table 5.1. Crookes's work was well before the concept of the photon had arisen, but James Clerk Maxwell had shown the existence of light pressure via his theory of electromagnetism. However, if as expected, the black surfaces absorb photons and the white surface reflects them, the radiometer should turn with the black sides leading, as if the white sides were being pushed harder. For example, when a photon with momentum p perpendicular to the surface is reflected, the momentum transfer to the vanes is $2p$. In contrast, if the photon is absorbed, the momentum transfer is only p.

It was not until 1901 that researchers were able to achieve sufficiently low pressures to show that the latter argument did *not* apply. The radiometer actually rotated with the white side leading! The black sides experienced a greater force than the white sides. One explanatory idea was that the black sides are warmer than the white ones, heating the nearby air. This would bring a greater air pressure on the black vanes. However, that idea did not survive.

Crookes and others, including Osborne Reynolds, whose name is attached to the Reynolds number (a measure of turbulent flow in fluids) did a variety of experiments and developed theories to explain the radiometer. Ultimately, it was James Clerk Maxwell, using his legendary analytical skills, who explained why a radiometer rotates with its white vanes leading.[31]

Key Point 9.32 *Maxwell showed that it is not because of a higher pressure against the warmer black surfaces that the Crookes radiometer rotates with its white vanes leading. Rather, he found that such pressure differentials along the bulk of the surface disappear rapidly. The rotation is caused primarily by*

---

[31] An excellent comprehensive exposition of the Crookes radiometer is in A. E. Woodruff, "The Radiometer and How It Does Not Work," The Physics Teacher **6**, 358–363 (1968).

forces exerted by the gas on a narrow strip along the edges of the vanes. That strip gets thinner as the gas pressure rises.[32]

A spinning radiometer *appears* to be a perpetual motion machine. The spinning seems magical. However, the radiometer is not an isolated system. It is continually absorbing radiant energy from an external light source. Turn off that light and the spinning stops. But suppose that a mechanism could be found that produced the needed pressure difference to turn an energetically-isolated radiometer.

In 1998, Daniel Sheehan proposed a scheme that could accomplish this using stationary-state pressure gradients that arise from surface-specific dissociation and recombination of well chosen gases. In 2000, Todd Duncan of Portland State University showed that if Sheehan's plan was viable it would lead to a violation of the second law of thermodynamics. Duncan envisaged a diatomic gas, say hydrogen H_2, in a blackbody cavity at constant temperature. He imagined a turbine with blades of two different materials (B and W), reminiscent of a radiometer's *black* and *white* vanes. Surfaces B and W are assumed release hydrogen molecules adsorbed on them at *different* (desorption) rates. Subsequently, Sheehan likened Duncan's turbine to a radiometer's vanes. Duncan's main point was that under isothermal, steady state conditions, a difference in gas pressure on surfaces B and W could in principle be used to perform external work.

For example, the rotating radiometer (or turbine) could be connected to a pulley system, thereby lifting a weight. With clever design, this process could be perpetual, extracting energy from the system's surroundings, whose outer boundary encloses an isolated system. This extraction would lead to a decrease of the entropy of the isolated system. Two possible conclusions are: (1) If the second law is sacrosanct, the pressure gradients proposed by Sheehan are impossible; or (2) If the pressure gradients exist, there is a violation of Kelvin-Planck form of the second law of thermodynamics.

The thought experiment posed by Duncan is referred to as "Duncan's paradox." In 2014, Sheehan and coworkers reported on an experimental realisation of a variant of Duncan's idea. It involved the dissociation of low pressure hydrogen gas on the metals tungsten and rhenium, both of which have high melting points and are highly heat resistant. That work differed from Duncan's idea in that it entailed reactions that generate a temperature gradient, as depicted in Fig. 9.26, rather than a gradient of pressure.

Sheehan, *et al* use hydrogen gas, which interacts differently at the two metal surfaces. At the lower temperature rhenium surface, the *endothermic* dissociation, $H_2 \rightarrow H + H$ dominates, with energy absorbed to break the bond between two hydrogen atoms, cooling the surface. Near the tungsten surface,

---

[32]At somewhat higher pressures, convective currents move the vanes in the opposite sense, while at atmospheric pressure the edge effects are diminished so much that there is no rotation. At very low pressures, gas pressure effects diminish and a radiometer will rotate with the black side leading because of larger radiation pressure on the white side.

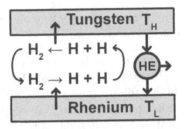

Figure 9.26: Duncan scheme to generate a temperature gradient between two metals, which could in principle be used to power a heat engine (HE).

the *exothermic* reverse process occurs, heating the surface, which creates the desired temperature difference.

Sheehan calls this technique *epicatalysis*, a type of heterogeneous gas-surface catalysis. He believes this technique has the potential to achieve large-scale temperature differences that can be used to drive a heat engine. This would involve two such surfaces being placed near one another to create a stationary temperature difference between them. To dissociate diatomic hydrogen, temperatures above 1700 K were needed. The experiment did reveal a temperature difference between the plates, as desired. Specifically, the measured difference was $\Delta T = 120\,\text{K}$ in the interval, $1700 \leq T \leq 1950\,\text{K}$. This process is characterised by energy flowing *against* the temperature gradient, as shown by upward arrows in Fig. 9.26.

Such techniques are being fine tuned and improved. Sheehan and Glick report on the possibility, based on calculations, of using an epicatalytic diode coupled to a thermoelectric converter (rather than a heat engine) effectively at room temperature. They also envisage scaling the equipment to larger sizes, and speculated that power densities in excess of $10^6\,\text{W}\,\text{m}^{-2}$ of surface area might be achieved.[33]

Key Point 9.33 *The desired temperature difference was achieved using differential hydrogen dissociation on tungsten and rhenium surfaces under high temperature blackbody cavity conditions. Sheehan, et al wrote that they do not know any credible way to reconcile these results with standard interpretations of the second law of thermodynamics. Despite predictions of large-scale epicatalytic thermal devices in the future, as with the other challenges to the second law, independent validation is needed.*[34]

---

[33]A good description of the current state of affairs, with citations to past articles, is in D. P. Sheehan & T. M. Welsh, "Epicatalytic thermal diode: Harvesting ambient thermal energy," Sustainable Energy Technologies and Assessments **31**, 355–368 (2019).

[34]Even without a second law violation, such devices might find useful applications.

CHAPTER 10

Reflections & Extensions

CONTENTS

10.1 REFLECTIONS

Mechanics is inadequate for thermodynamics processes because it does not account for the internal energy E stored within macroscopic objects. That internal energy is insufficient to describe physical behaviour without its partner, entropy $S(E)$. An appreciation of the strong linkage between internal energy and entropy can bring a deeper understanding of thermodynamics and of our world.

The history of thermodynamics put *heat* in the forefront as an entity, rather than a process. Once it became understood that internal energy could be altered by both heat and work processes,[1] and the first law of thermodynamics $\Delta E = Q + W$ was accepted, the errors of the past should have faded away. Energy is the *stuff* that gets transferred by the processes heat and work. However, remnants of past confusion still exist. This includes treating heat as a stored entity (which it is not) and using the inappropriate metaphor *disorder* for entropy, which is often misleading. Throughout this book I have given arguments intended to help reduce future errors.

All thermodynamics processes entail redistributions, i.e., *spreading* of energy. An example is the mixing of two distinguishable gases, where each species spreads its energy to a larger spatial volume. The result is an *equitable*, unchanging energy distribution of an equilibrium state. The concept of *equity* is not commonly used. I have discussed it qualitatively with examples, and quantitatively in Key Point 3.9, using the Boltzmann entropy equation, $S = k \ln \Omega$. The latter expression can help us understand what the entropy in

---

[1] E can also be altered also by mass transfer, as discussed in Sec. 1.3.

an equilibrium state represents. For processes, entropy changes are the result of increased spatial energy spreading, which leads to a more equitable energy distribution. Once in an equilibrium state at a given temperature, entropy is a measure of *temporal* spreading over a system's accessible microstates.

The important concept of reservoirs, and subtle differences between reversible and irreversible processes, especially for heating and cooling cycles, are clarified. Also covered are the important concepts of free energy and available energy, including what happens when reservoirs are finite. Issues like these often nag at teachers and students. Similarly, the concept of *entropic force*, which is unfamiliar to many, is clarified through a number of examples.

I have described various models of gases and solids, with emphasis on their energy and entropy functions. Analysis of the ubiquitous photon gas shows that not all equations of state are the familiar ideal gas ones. The photon gas is used to highlight a variety of process types. Relevant examples involving photons are covered, from blackbody radiation to the common incandescent lamp to the cosmic background and black hole radiation phenomena.

The relatively new and mathematically sophisticated Jarzynski equality, Eq. (4.76), is explained with examples. That equality entails an average over a large number of identical experiments, a procedure that enables analysis of nonequilibrium systems with a high speed computer.

I have emphasised that entropy is not merely a concept. All matter in equilibrium has a numerical entropy value that can be measured and/or calculated, and relevant data are tabulated in compilations such as the *CRC Handbook of Chemistry and Physics*. Added well after the first and second laws of thermodynamics, the third law of thermodynamics pins down zero temperature entropy for pure crystalline solids. Some materials, like glasses, become "stuck" in frozen-in states as they are cooled, which results is a nonzero residual entropy as the environment's temperature approaches 0 K.

Entropy has the strange units $J K^{-1}$, but notably, those units come from the factor k in $S = k \ln W$. Thus entropy *could* be defined to be dimensionless, which would give temperature the units of energy. Historically, entropy's units are an artefact of temperature scales devised independently of the energy concept. Had temperature been defined as an energy like $kT \equiv \tau$, which I call tempergy, entropy would have been dimensionless. I have emphasised that temperature is *not* always proportional to the energy per particle, as it is for a classical ideal gas, and showed that tempergy is almost always the energy that it takes to increase the dimensionless entropy of a macroscopic system by one. At extremely low temperatures, systems typically behave otherwise.

Over the years, cycles that do external work or perform heating or cooling tasks have received much attention, with a focus on energy efficiencies. I have examined both reversible and irreversible cycles. The latter led to an introduction of the "second law efficiency," which accounts for any increase in the entropy of the universe. Technological progress that utilises the interplay of energy and entropy led to powerful methods that cool to temperatures near absolute zero. In turn, that led to the discovery of remarkable properties of helium-4 at temperatures under 4.2 K and helium-3 at a few minikelvins. Remarkably, nuclear adiabatic demagnetisation can bring a system to the microkelvin region. The interplay of energy and entropy is central in these endeavours.

Maxwell, and later Boltzmann, realised that small scale fluctuations exist and that the principle of entropy increase is not absolute. Maxwell's demon, which was conceived to show that the second law is *statistical*, has received considerable attention over the last century and a half, and have been directed largely toward attempts to violate the second law, although that was *not* Maxwell's intention. The smallness of the fluctuations that Maxwell's demon employs is what makes statistical mechanics a viable predictor of equilibrium thermodynamics for macroscopic systems. However, fluctuation phenomena on the nanoscale have made possible biological processes and human-made devices that require nonequilibrium states for their very operation. Despite their dependence on thermal fluctuations, these Brownian ratchets operate safely within the bounds of the second law of thermodynamics.

There has been a substantial attempt by researchers to find violations of the second law. Some of the interesting experiments and ideas along those lines are described herein. Although some of the attempts seem promising, to date, no violation has been demonstrated conclusively by two or more independent researchers. Nonetheless, such attempts can serve as valuable teaching tools as well as stimuli for further research.

10.2 EXTENSIONS

10.2.1 Lieb-Yngvason formulation of thermodynamics

As shown here, thermodynamics is typically presented by discussing heat processes and Carnot cycles to get from energy to entropy. But the subject has also been developed using a postulation-based approach by a number of people. The first time I taught thermodynamics, I followed Herbert Callen's book, based on the postulational approach by his mentor, Laszlo Tisza. That approach is relatively clean mathematically, but left me seeking a physical basis for the postulates, which led me ultimately to the concept of energy spreading. I learned later that chemist Frank Lambert independently had proposed the similar idea of "energy dispersal." An axiomatic formulation by Greek mathematician Constantin Carathéodory in 1909 was based on the postulate that in the neighbourhood of any given state, some states are not accessible by an adiabatic process. That approach was abstract and too mathematically sophisticated for many physicists.

In 1999, Elliott Lieb and Jakob Yngvason (L-Y) extended some of Carathéodory's ideas, and derived the principles of thermodynamics in a mathematically precise set of articles that deserves to be called a *tour de force.* The L-Y theory uses well-defined axioms and mathematically rigorous proofs. Despite its mathematical level, the L-Y formulation is based on the highly physical quantities of energy, volume and entropy. I encourage curious readers to examine the L-Y article in Physics Today and the follow up letters to the editor and reply from Lieb and Yngvason.[2]

Of primary interest here, energy, work, and work coordinates such as volume are central to the analysis, and the concept of heat is not used at all! Also unusual is that temperature is defined only *after* the existence of entropy is established. All of this is notable, even if the L-Y mathematical analysis is not fully digestible for many teachers and students. The concept of "preorder" for adiabatic processes is used throughout. Physically, this relates to Carathéodory's insight that some states cannot be reached adiabatically from a given state. If state Y is adiabatically accessible from state X, then L-Y say X *precedes* Y. They establish that for any two states X and Y, either X precedes Y, Y precedes X, or both. Interestingly that relation was needed in Sec. 2.2 of this book to define heat Q in terms of work W.

The goal of L-Y was in large measure to discover where the entropy function and its non-decreasing property come from. They found that entropy arises from relations between the equilibrium states of macroscopic systems. The L-Y work is restricted to equilibrium states, and follow up articles pointed out the difficulty of extending such analyses to non-equilibrium states.

10.2.2 Quantum mechanics and the second law

Along with attempts to challenge the sacrosanct nature of the second law, are investigations of whether the second law of thermodynamics is fundamentally quantum mechanical. The many successes of quantum theory make this prospect alluring.

Perhaps irreversibility is directly related to the act of making a measurement on a system. One view is that the measurement process causes a collapse of the system's wave function, which leads from a situation of quantum uncertainty to certainty. Measurement breaks the time reversal symmetry of the Schrödinger equation, introducing a time asymmetry that is reminiscent of irreversible processes in thermodynamics.

In his book *Time and Chance*,[3] David Albert provides an interesting view that avoids the *measurement problem* of quantum theory. In contrast, Albert avoids the need for measurement by appealing to a relatively new, untested theory known as Ghirardi-Rimini-Weber (GRW) *spontaneous collapse* theory. GRW theory posits that individual particles undergo *spontaneous* wave function collapse. Spontaneous collapse is postulated to occur only every 100

---

[2]E. H. Lieb & J. Yngvason, "A Fresh Look at Entropy and the Second Law of Thermodynamics," Phys. Today **53**, 32–32 (2000); The letters are in D. Siminovitch *et al.*, "Entropy Revisited, Gorilla and All," Physics Today **53**, 11–15 (2000).

[3]Albert, D., *Time and Chance* (Harvard University Press, 2003).

million years for single-particle observations, and GRW theory is compatible with experiments that show no spontaneous wave-function collapse.

The fact that there must still be a collapse when individual particles are measured is explained by the belief that there is quantum entanglement between the measured particle and the *macroscopic* measuring apparatus. If that apparatus has say, 10^{15} particles (there are likely more), each of which undergoes collapse every 10^8 years $\approx 10^{15}$ seconds, then the collapse of one of the measuring apparatus atoms is expected once during each second of the measurement. A recent article shows such behaviour under some circumstances.[4]

---

[4]D. Bedingham & P. Pearle, "Continuous-spontaneous-localization scalar-field relativistic collapse model," Physical Review Research **1**, 033040–10pp (2019).

CHAPTER 11

Appendices: Mathematical Identities

CONTENTS

11.1 DERIVATIVES & GIBBS-DUHEM EQUATION

Internal energy, Helmholtz free energy, Gibbs free energy and enthalpy each have *natural* variables for which they obey useful partial derivative relations. Entropy does too, but gives the same information as the internal energy.

$$dE(S,V,N) = TdS - PdV + \mu dN$$
$$dA(T,V,N) = -PdV - SdT + \mu dN$$
$$dG(T,P,N) = -SdT + VdP + \mu dN \tag{11.1}$$
$$dH(S,P,N) = TdS + VdP + \mu dN$$

These four energy equations imply the following useful expressions:

$$dE \implies (\partial E/\partial S)_{V,N} = T, \ (\partial E/\partial V)_{S,N} = -p, \ (\partial E/\partial N)_{S,V} = \mu$$
$$dA \implies (\partial A/\partial V)_{T,N} = -P, \ (\partial A/\partial T)_{V,N} = -S, \ (\partial A/\partial N)_{T,V} = \mu$$
$$dG \implies (\partial G/\partial T)_{P,N} = -S, \ (\partial G/\partial P)_{T,N} = V, \ (\partial G/\partial N)_{T,P} = \mu \tag{11.2}$$
$$dH \implies (\partial H/\partial S)_{P,N} = T, \ (\partial H/\partial P)_{S,N} = V, \ (\partial H/\partial N)_{S,P} = \mu$$

If we confine our attention to a single phase, the four energy functions E, S, A, G and their derivatives will be continuous. Under these conditions, the mixed second derivatives must be equal. For example,[1] $\partial(\partial E/\partial S)/\partial V = \partial(\partial E/\partial V)/\partial S$ implies the *Maxwell equation,*

$$(\partial T/\partial V)_S = -(\partial P/\partial S)_V \tag{11.3}$$

---

[1] I've omitted the subscripts for constant V, S to reduce notational clutter.

Three other most common Maxwell relations are:

$$(\partial T/\partial P)_S = -(\partial V/\partial S)_P$$
$$(\partial S/\partial V)_T = -(\partial P/\partial T)_V \qquad (11.4)$$
$$(\partial S/\partial P)_T = -(\partial V/\partial T)_P$$

The equations above relate to changes in the *extensive* thermodynamic variables E, A, G, and H for a single phase. An important equation that relates changes in *intensive* variables is the Gibbs-Duhem equation,[2]

$$N d\mu + S dT + V dP = 0. \qquad (11.5)$$

This can be seen using the third lines in both Eqs. (11.1) and (11.2), where the Gibbs function $G(T, P, N)$ is a function of the intensive variables temperature and pressure and the extensive number of particles N. Thus G is a function $f(T, P)$ times N, but because of (11.2), $f(T, P) = \mu(T, P)$. Given $G = \mu N$, $dG = N d\mu + \mu dN$. Combining this with the third of Eqs. (11.1) results in the Gibbs-Duhem equation, 11.5.

---

[2]Pierre Duhem was a thermodynamicist, hydrodynamicist, and historian of science.

Subject Index

Author Index

Adler, E., 100
Agrawal, D. C., 169
Ahlborn, B., 227
Aizenman, M., 189
Albert, D., 296
Aldrich, B., 208
Allen, J. F., 243

Bacon, F., 6
Baierlein, R., 61, 110, 210
Bardeen, J., 246
Bartels, R. A., 176, 177
Bedingham, D., 297
Bekenstein, J., 91, 175
Bennett, C. H., 259, 263, 265
Bernard, W. H., 201
Bernstein, J., 6
Berry, R. S., 228
Bertrand, F. L., 210
Black, J., 7
Bohr, N., 243
Bohren, C., 208
Boltzmann, L., 61, 68
Boonchui, S., 111
Boyle, R., 279
Brillouin, L., 265
Brissaud, J. B., 210
Brown, R., 5
Bustamante, C., 151

Cailletet, L., 238
Callen, H. B., 240, 295
Capek, V., 281
Carathéodory, C., 295
Carnot, S., 8, 45, 54

Chang, H., 155
Chipot, Ch., 151
Clapeyron, B. P. E., 246
Clausius, R., 42, 50, 54, 210, 214
Clement, N., 8
Colding, L., 8, 13
Coombes, C., 288
Cooper, L., 246
Crookes, W., 289
Crooks, G. E., 148
Cubero, D., 277
Curzon, F. L., 227

D'Abramo, G., 285
Darrow, K., 208
de Cisneros, B. J., 277
Debye, P., 124
Denbigh, K., 210
Dewar, J., 241, 242
Dickerson, R. H., 224
Dingle, H., 208
Dirac, P. A. M., 98
Duhem, P., 299
Dumont, S., 151
Duncan, T., 291

Earman, J., 264
Easson, D. Q., 93
Ehrenfest, P., 243
Einstein, A., xvi, 6, 120, 185
Eshuis, P, 272
Español, P., 270

Falk, H., 100
Fermi, E., 98, 214
Feynman, R. P., 269, 270

Printed in the United States
by Baker & Taylor Publisher Services

Printed in the United States
by Baker & Taylor Publisher Services